Lecture Notes in Computer Science   1044

Edited by G. Goos, J. Hartmanis and J. van Leeuwen

Advisory Board: W. Brauer   D. Gries   J. Stoer

T0254688

Springer
Berlin
Heidelberg
New York
Barcelona
Budapest
Hong Kong
London
Milan
Paris
Santa Clara
Singapore
Tokyo

Bernhard Plattner (Ed.)

Broadband Communications

Networks, Services, Applications, Future Directions

1996 International Zurich Seminar
on Digital Communications, IZS'96
Zurich, Switzerland, February 21-23, 1996
Proceedings

Springer

Series Editors

Gerhard Goos, Karlsruhe University, Germany

Juris Hartmanis, Cornell University, NY, USA

Jan van Leeuwen, Utrecht University, The Netherlands

Volume Editor

Bernhard Plattner
Institut für Technische Informatik und Kommunikationsnetze
Eidgenössische Technische Hochschule Zürich
ETH Zentrum, CH-8092 Zürich, Switzerland

Cataloging-in-Publication data applied for

Die Deutsche Bibliothek - CIP-Einheitsaufnahme

Broadband communications : networks, services, applications,
future directions ; proceedings / 1996 International Zurich
Seminar on Digital Communications, Zurich, Switzerland,
February 21 - 23, 1996. Bernhard Plattner (ed.). - Berlin ;
Heidelberg ; New York ; Barcelona ; Budapest ; Hong Kong ;
London ; Paris ; Santa Clara ; Singapore ; Tokyo : Springer,
1996
  (Lecture notes in computer science ; Vol. 1044)
  ISBN 3-540-60895-8
NE: Plattner, Bernhard [Hrsg.]; International Zurich Seminar on
  Digital Communications <14, 1996>; GT

CR Subject Classification (1991): C.2-3, E.4, H.1, H.4.3, B.4.1

ISBN 3-540-60895-8 Springer-Verlag Berlin Heidelberg New York

© Springer-Verlag Berlin Heidelberg 1996
Printed in Germany

Typesetting: Camera-ready by author
SPIN 10512601   06/3142 – 5 4 3 2 1 0   Printed on acid-free paper

Foreword

In the second half of the 1990s broadband networks will evolve to become an indispensable resource in national economies and our society as a whole. It is obvious at the time of this writing that ATM will be deployed as a solution for broadband networks in wide and local area environments; as a matter of fact various broadband network testbeds have been set up throughout the world, and in Europe a wide-area ATM pilot network encompassing eighteen countries has been put to service and is about to be replaced by commercial services in some countries. It can be expected that before long real applications will be running on top of various broadband networks and experience with their use will become available.

It is for the reasons outlined above that the theme of IZS '96 was selected to be "Broadband Communications", emphasizing the three areas of *networks*, *experience with services*, and *application technology*.

Selecting such a broad theme implies that the conference will provide a snapshot of the state of the art in communication technology. The response to the Call for Papers indeed met the expectations of the Programme Committee: Close to 70 manuscripts dealing with many aspects of broadband communications were submitted. The manuscripts were reviewed rigorously by the members of the PC or other knowledgeable reviewers, such that each manuscript received four independent opinions, which were instrumental in the decision process conducted by the PC. Finally, the PC chose 26 papers from the set of submitted manuscripts. Many good contributions had to be rejected, but we anticipate that most of these will be published elsewhere.

By explicitly including *services* and *applications* as subtopics in the Call for Papers the Programme Committee hoped to attract reports of systems in which the emerging broadband communication infrastructure was actually used productively. However, a brief look a the conference programme and the table of contents of the proceedings shows that interesting and well-written reports about applications of broadband networks are not readily found. There is just one session which discusses applications, which actually includes two papers invited by the PC.

We are indebted to many individuals who contributed to the IZS '96: Specifically, we thank the members of the Organizing Committee for their dedication and hard work prior and during the conference, the members of the Programme Committee for their invaluable advice in selecting the best papers and creating the programme, and the authors of manuscripts and speakers for publishing and presenting their research results.

We also acknowledge the contributions of the sponsors, which helped alleviate the problems that are always associated with a restricted budget, and the support of the patrons, which made this event possible.

January 1996   Bernhard Plattner, President and Programme Chair
                                         Beat Keller, Chairmann

1996 International Zurich Seminar On Digital Communications

Broadband Communications, Networks, Services, Applications, Future Directions

ETH Zürich, Switzerland, February 19-23, 1996

Patronage: IEEE Switzerland Chapter on Digital Communication Systems and IEEE The Institute of Electrical and Electronics Engineers

President and Programme Chair:

Bernhard Plattner, ETH Zürich, CH

Technical Programme Committee:

Bernhard Plattner, ETH Zürich, CH, Programme Chair
Thomas Walter, ETH Zürich, CH, Deputy Programme Chair

Ernst Biersack, Eurecom, France
Laurie Cuthbert, Queen Mary and Westfield College, U.K.
André Danthine, Université de Liège, Belgium
Konrad Froitzheim, Universität Ulm, Germany
Jean-Pierre Hubaux, EPF Lausanne, CH
David Hutchison, Lancaster University, U.K.
Hansjörg Kley, Siemens-Albis AG, CH
Luciano Lenzini, University of Pisa, Italy
Hannes Lubich, SWITCH Geschäftsstelle, CH
Riccardo Melen, Politecnico di Milano, Italy
Guru Parulkar, Washington University, USA
Thomas Plagemann, University of Oslo (UNIK), Norway
Martin Potts, Ascom Tech AG, CH
B. D. Pradhan, Centre for Development of Telematics, India
G. Ramamurthy, NEC USA, Inc., USA
Harry Rudin, IBM Research, CH
Jonathan Smith, University of Pennsylvania, USA
Walter Steinlin, Telecom PTT, CH
Thierry Van Landegem, Alcatel Bell Telephone, Belgium
Martina Zitterbart, Universität Braunschweig, Germany

Organizing Committee:

Beat Keller, Ascom Tech, Chair

Raymond Bandle, University of Zürich, Local Arrangements
Sergio Bellucci, Technopark Zürich, Treasurer
Hannes Lubich, SWITCH, Publicity
Annette Schicker, Hinwil, Registration
Caterina Sposato, ETH Zürich, Secretariat
K.H. von Grote, Ascom Tech, Tutorial

The conference was generously supported by:

Ascom Ltd., Berne
Swiss Federal Institute of Technology (ETH), Zürich
Management and Technology Institute Ltd. (MTI), Zürich
Schweizerischer Elektrotechnischer Verein (SEV), Zürich
Union Bank of Switzerland
University of Zurich

Reviewers

Serious, accurate, and detailed reviews are essential for the success of any conference. It is a great pleasure to thank the reviewers listed below and the members of the Programme Committee for their precious contribution to this important task.

Aepli, T.	Garcia, F.
Ball, F.	Gehrhard, V.
Bauer, D.	Goebel, V.
Bhatnagar, J.	Grieder, R.
Biersack, E.	Hubaux, J.-P.
Bocci, M.	Humblet, P.
Bonaventure, O.	Hutchison, D.
Braun, F.	Iliadis, I.
Braun, T.	Kandrical, M.
Bregni, S.	Kley, H.
Broennimann, R.	Klingler, C.
Buddhiko, M.	Koerner, E.
Campbell, A. T.	Kuepfer, H.
Cosmas, J.	Kure, O.
Coulson, G.	Lemppenau, W.
Crochat, O.	Lenzini, L.
Cuthbert, L.	Leuthold, P. E.
Danthine. A.	Li, B.
Davies, N.	Lubich, H.
Dugelay, J.-L.	Luise, M.
Duverney, P.	Lunn, A.
Edmaier, B.	Manthorpe, S.
Engbersen, T.	Martins, J.
Demierre, E.	Mathy, L.
Fantacci, R.	Mauthe, A.
Ferro, E.	Melen, R.
Froitzheim, K.	Miah, B.
Gähwiler, W.	Nonnenmacher, J.
Gara, S.	Palazzo, S.
Garcia Adanez, X.	Scott, A.

Papadopoulos, C.

Parulkar, G.

Petit, G.

Petri, S.

Pitts, J.

Plagemann, T.

Plattner, B.

Potts, M.

Pradhan, B. D.

Ramamurthy, G.

Raman, G.

Robert, S.

Rudin, H.

Ruprecht, J.

Sawwaf, R.

Scharf, E.

Schoenwaelder, J.

Schormans, J.

Schott, W.

Sigg, K.

Simpson, S.

Smith, J.

Steffen, A.

Steinlin, W.

Sun, Q.

van As Harmen, R.

Van Landegem, T.

Van Mieghem, P.

Vermeulen, C.

Voeroes, P.

Waldner, R.

Walter, T.

Winstanley, S. B.

Wittmann, R.

Zitterbart, M.

Contents

Broadband Network Architecture

Designing for Quality of Service Guarantees

Protocol Support for Multimedia/Multipoint Services

Traffic Modeling and Performance Evaluation

Fairness in Resource Allocation

Applications

Server Functions in ATM

Satellite and Wireless Networks

Broadband Access and Switching

Acceptance and Congestion Control

A Path Selection Method in ATM Using Pre-Computation

Olivier Crochat[1], Jean-Yves Le Boudec[1], Tony Przygienda[2]

Abstract - We propose a method for path selection using pre-computation in an ATM network. The method is based on an algorithm that pre-computes a list of routes for each node in the network. We then study the lack of diversity in the paths chosen by our method, and the resulting risk of call blocking. Improvements to the method are proposed, and the decrease in call blocking is evaluated by simulation. We show that, with a load-balancing assumption, the method is able to provide paths equivalent to those of the On-Demand method.

Keywords - On-demand routing, pre-computed routing, path selection, path finding, resilience, blocking

1 Introduction

We consider the choice of a path when establishing a call in a communication network supporting *guaranteed bit rates*. Such connection-oriented services will be widely used to transport video transmission and other real-time applications. ATM, but also IP with some extensions (ST.II or RSVP) offer connection-oriented services [Top90, Zha93].

To respond to a connection request, a path satisfying the connection's constraints in bandwidth, delay, hops number has to be found. We call *routing* the act of finding such a path (also called a route in this article). The routing function can be seen as shown in figure 1.

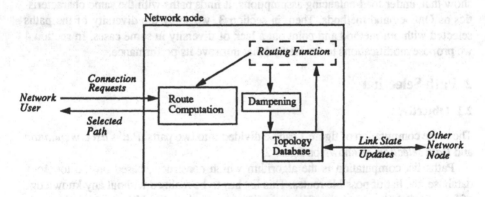

Fig. 1. Conceptual view of Routing Function

---

1. Laboratoire de Réseaux de Communication, DI - EPFL, CH - 1015 Lausanne, http://lrcwww.epfl.ch/
2. IBM Zurich Research Laboratory, CH - 8803 Rueschlikon,

Routing could be in the network nodes or in a central server. Its location does not affect the way the routing is done with our method. We assume that, for connection-oriented services, the routing is done at connection request time. This leads to the need for a global image of the network where the routing occurs. This image (which contains a description of all the links, their starting and ending points, their capacity) is stored in the topology database. The database maintains a correct view of the network through Link State Updates, which are sent by all nodes to give information about the status of their attached links. In order not to have too many of these updates, the *Dampening* function decides whether the changes in the database are important enough to require an update. A more documented discussion on the routing and its different parts can be found in [Leb94, Cyp71, OSI89].

A major choice in routing is whether it is done *On-Demand* or on a *Pre-Computed* basis. When On-Demand is used, the route computation is done at the time of the request, leading to a delay due to the calculation time. In contrast, when Pre-computation is used, connection setup time is faster [Gau95], but the lack of knowledge on the precise characteristics of the connection to be routed is a challenge. In section 2.3, we will see that our method is able to address that aspect without any a-priori classification of connection characteristics. Our method applies in particular to the selection of bi-directional, but asymmetric paths (namely, paths with different bandwidth requirements in the forward and return directions, but that require both directions to use the same links), as it is the case with ATM. Only peak bandwidth allocation is described in our examples, though the method applies to more general cases.

Pre-computation is attractive when the pre-computed routes can be used more than once without recalculating them. Such is the case when connection requests are frequent, but relatively small (in terms of required bandwidth), leading to high loads on real-time resources, but infrequent link state updates (thanks to the Dampening function).

In section 2 we define a path selection method that uses path pre-computation. We show that, under load-balancing assumptions, it finds paths with the same characteristics as On-Demand methods. Then, in section 3, we study the diversity of the paths selected with our method and point out a lack of diversity in some cases. In section 4 we propose modifications to the method that improve its performance.

2 Path Selection

2.1 Objectives

The route computation of figure 1 can be divided into two parts, *Paths list computation* and *Path selection*, as shown on figure 2.

Paths list computation is the algorithm which constructs, based on the topology database, the list of possible routes. This list has to be realized without any knowledge of future calls' characteristics. The paths list computation algorithm used in this article is described in [Prz95]. It analyses the network image and produces, for each destination, a list of optimal paths. This list is not dependent on any a-priori classification of the requests, but it guarantees that no better route can be found than those in the list, regardless of connection requirements at setup time.

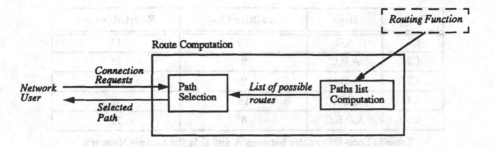

Fig. 2. Conceptual view of Route Computation Function

Path selection will then, on a per communication basis, choose a path among the ones proposed, depending on the particular requirements of each request.

2.2 Non-Dominated Routes

Before defining the method to select a route from a list, we have to see what kind of route is given by the paths list computation algorithm. We first explain the concept of *non-dominated* routes, and how it is possible to characterize and compare routes without having any knowledge of the future requests' characteristics.

Consider the sample network illustrated in figure 3. We assume that there exists

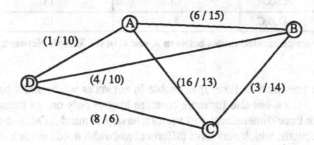

Fig. 3. Sample Network (additive cost / restrictive cost)

only one link between each node pair. Each link has a *cost vector*, containing two costs in this example. The first one is additive; the first cost of a path will be the addition of those of the links it goes through. It could be the number of VP hops a connection has to do in a network where nodes are ATM Virtual Channel Switches and links are Virtual Path (VP) Connections. The second one is restrictive; the second cost of a path going through a list of links is the minimum of their second cost. It may represent the available bandwidth of a link at a certain point in time, and can vary a lot.

All the possible loop-free routes between A and C and their costs are shown in Table 1.

A route is defined to be *strictly better* than another one if it offers an equal or a higher restrictive cost on its worst link, has a lesser or equal additive cost and one of them at least is not equal. In our example, routes C2 and C5 are strictly better than all

Route	Hops	Additive Cost	Restrictive Cost
C1	A,C	16	13
C2	A,B,C	9	14
C3	A,D,C	9	6
C4	A,B,D,C	18	6
C5	A,D,B,C	8	10

Table 1: Loop-free routes between A and C in the Sample Network

the other ones. However they cannot be compared since C5 has a smaller additive cost, while C2 is able to route larger calls, from a restrictive point of view. So, intuition would tell us to use C5 for calls having bandwidth requirements smaller than 10, and C2 for calls with requirements up to 14.

Those routes are said to be *non-dominated* (or extremal in [Prz95]), which means that there exists no other route that is strictly better. Generally, the structure of such routes is a lattice [Car79] and the set of non-dominated routes the *lower bound* of the lattice. Table 2 lists the non-dominated routes of the sample network shown in figure 3.

Route	Hops	Additive Cost	Restrictive Cost
C2	A,B,C	15	14
C5	A,D,B,C	8	10

Table 2: Non-dominated routes between A and C in the Sample Network

The algorithm presented in [Prz95] is usable in networks where links have two-dimensional cost vectors, but also for more costs, as long as only one of them is additive. We use it with three-dimensional cost vectors, having in mind ATM bi-directional asymmetric connections, which may have different bandwidth needs in each direction, but are required to follow the same path. The additive cost is the delay or number of hops, and the other two are the available bandwidth in each of the directions. With two restrictive costs, the method supports only peak-bandwidth asymmetric, bi-directional allocation. With more dimensions it is able to support allocation mechanisms that go beyond peak rate allocation.

This algorithm is used for the paths list computation. This is because it allows pre-computation in polynomial time of the list of all the non-dominated routes from a source to the other nodes in a network.

2.3 Path Selection with Load-Balancing

Based on the list given by the paths list computation algorithm, the path selection module chooses the best route for each different connection request. This is done in three steps:

1 - The first choice made by the policy is to eliminate all the paths that do not have enough bandwidth (in both directions) to fulfil the requirements of the request.

2- Then, among all the possible ones, the cheapest (in terms of additive cost) is chosen. This 'shortest path' is the one that will cause the least damage to future calls [Key90].

3 - In case of more than one path with minimal additive cost, the path preserving load-balancing is chosen.

This term, *load-balancing*, deserves an explanation. The aim is to try not to fill a certain path entirely, leaving the others empty. So, when more than one path is left after the two first steps, the path chosen is the one which has, after adding the pending connection, the largest capacity left for other connections. This is done by selecting the one which has the smallest of its restrictive costs larger than the one of every other possible routes. An example is shown in Table 3. Route X3 is chosen, since it has the most capacity left for future requests on its most loaded link.

	X=(25 / 9 / 9) is to be routed		
Route	Proposed routes (of minimal additive cost)	After adding the request	Minimum of Restrictive Costs
X1	(19 / 30 / 10)	(19 / 21 / 1)	1
X2	(19 / 15 / 25)	(19 / 6 / 16)	6
X3	(19 / 20 / 20)	(19 / 11 / 11)	11
X4	(19 / 10 / 30)	(19 / 1 / 21)	1
(x / y / z) = (additive cost / first restrictive cost / second restrictive cost)			

Table 3: Example of the load-balancing method

Formally, the path selection module is defined as follows.

Path Selection Module

```
Let R be the connection request and P the path chosen at a certain iteration.
Let C(x), Cap1(x) and Cap2(x) be the additive, first and second restrictive
costs of x, respectively

P={};
C(P)=Max;                    /*Additive Cost */
Cap1(P)=Cap2(P)=0;           /* First & Second Restrictive Cost */
FOR each route I of the list DO
     IF C(I) < C(R) THEN
          IF C(I) < C(P) THEN
               P=I;
          ELSE IF C(I)==C(P) THEN
               IF Min[Cap1(I)-Cap1(R), Cap2(I)-Cap2(R)] >
                    Min[Cap1(P)-Cap1(R), Cap2(P)-Cap2(R)] THEN
                    P=I;
               ENDIF;
          ENDIF;
     ENDIF;
ENDFOR;
```

We want to compare our path selection method with the Load-Balancing On-Demand method. On-Demand has the advantage of potentially finding paths that better suit the individual connection request. As we show below, this is not true in cases where a load-balancing option is taken. Let the request be (C_{req} / Cap_{req}) and the solutions found by On-Demand (C_{OD} / Cap_{OD}) and by Pre-Computed routing (C_{pre} / Cap_{pre}).

The method applied by the Load-Balancing On-Demand to find a route is the following:

Find (e.g. using Dijkstra) the shortest path
with the constraint of having $Cap_{OD} > Cap_{req}$.
If more than one path with minimal cost is found,
use the load-balancing policy described above.

The method applied by pre-computation is the following:

Using the path selection policy defined above, choose one path
from the list of non-dominated routes given by the paths list computation.

Proposition

The route proposed by our path selection method (shortest path with load-balancing on the list of non-dominated routes) is equivalent to the one which would be proposed by the Load-Balancing On-Demand Calculation.

Proof

The proof, done here for a two-dimensional cost vector, can be easily extended to more dimensions. Table 4 shows the possibilities of differences between the costs of the On-Demand and Pre-Computed solutions.

C_{OD} ? C_{pre}	Cap_{OD} ? Cap_{pre}	Is it possible?
<	<	No, (C_{OD} / Cap_{OD}) would have been chosen by pre-comp. too
<	=	No, (C_{OD} / Cap_{OD}) would be in non-dominated routes list
<	>	No, (C_{OD} / Cap_{OD}) would be in non-dominated routes list
=	<	No, to be discussed[a]
=	=	Possible, to be discussed[b]
=	>	No, (C_{OD} / Cap_{OD}) would be in non-dominated routes list
>	<	No, O.-D. would not have chosen a path that is not the shortest
>	=	No, O.-D. would not have chosen a path that is not the shortest
>	>	No, O.-D. would not have chosen a path that is not the shortest

Table 4: Loop-free routes between A and C in the Sample Network

So the only possibility is C_{OD} = C_{pre} / Cap_{OD} = Cap_{pre} showing that pre-computa-

a. If there is more than one path with the same cost and different capacities, then the load-balancing policy always chooses the largest capacity, so, as both use the same load-balancing method, it is not possible to have different capacities.
b. It is the same path, or an equivalent one. If two or more paths with the same cost vector are present in the graph, only one of them is stored in the non-dominated routes list.

tion with load-balancing gives equivalent results to On-Demand with load-balancing.

Load-balancing is not always optimal. In particular it may block large calls due to the fragmentation of large paths. However, we focus here mainly on cases where traffic consists of a lot of small connections. Large connections require a different treatment anyway, because they trigger the Dampening mechanism, and force a topology update and a new paths list computation.

3 Evaluation of Diversity for our Path Selection Method

3.1 Node-to-Node Diversity

The point we want to focus on is the *diversity* of the non-dominated routes. Two routes are said to have a *good diversity* if they have no link in common and a bad one if they share one or more. An example is shown on figure 4.

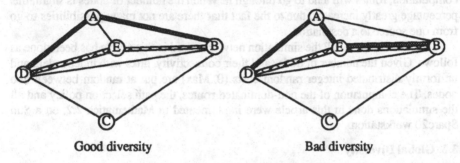

Good diversity   Bad diversity

Fig. 4. Diversity between two routes

This is very important for two reasons. The first one is the resilience to component failures, for which it is good to have only a small amount of paths affected by a link/node failure (failure of link E-B in figure 4 will only cause one route to be out-of-order to the 'good diversity' case). The second is avoiding overload of only parts of a network. If many paths all want to go through a small number of links, these will soon become overloaded, and pending connections will be blocked until a new topology update.

The non-dominated routes have been tested to see if their diversity was good, or if they were all going through the same links. The results are presented in table 5. It shows the percentage of non-dominated routes going through the most used link (the link through which most of the routes go through) as a function of the number of nodes in the network, and their connectivity. The diversity comparison has been performed for non-dominated paths going from one node (the source) to one destination. The result is the mean over all the destinations.

Number of nodes in the network	Connectivity	Average number of optimal paths between two nodes	Percentage of paths having a bad diversity
15	0.2	2.7	81.3
15	0.3	4.5	80.0
20	0.2	5.2	60.7
20	0.3	7.2	63.1
25	0.2	7.3	70.0
25	0.3	18.3	53.1
30	0.2	11.6	53.0
30	0.3	11.9	50.5

Table 5: Percentage of paths going through the most used link

We can see that between 50% and 80% of the non-dominated routes all go through a certain link. This percentage can be explained by the fact that if a link is very "good" (high restrictive costs and very small additive one), it is normal that in the shortest path computation, routes will tend to go through it. When the number of nodes is small, this percentage greatly increases, due to the fact that there are not many possibilities to go from one source to a destination.

The construction of all the simulation networks used in this article has been done as follows. Given the number of nodes and their connectivity, links with independent and uniformly distributed integer random costs (0..Max) are put at random between two nodes. The computation of the non-dominated routes, the path selection policy and all the simulations done in this article were implemented in Mathematica 2.2, on a Sun Sparc20 workstation.

3.2 Global Diversity

The results obtained in 3.1 reveal the lack of diversity of the non-dominated routes calculated for a single node of the network. But in a real situation, each node is the source of only a small number of connections, and the interesting point is the diversity between connections originated in different nodes of the network. Note that, even in this case, each node has only local knowledge of what is going on, which means that, besides the topology database, it has no knowledge at all of the connections starting in the other nodes. Full information passing is essentially equivalent to a centralized solution of the problem [Key90] and is not the aim of this article, which is to find a good path selection method that uses only local knowledge (topology database), on a node per node basis.

To test the diversity on a global basis we did the following simulations.

We define a connection request to be a list containing the source node, the end node, the first restrictive component and the second one $(S,D,C1,C2)$. Each node of the network is then the source for two requests. The destination, D, is chosen randomly. C1 and C2 are fixed to 1/20 of Max, so as to have small capacity requests in line with the assumption made in section 1. This traffic is then added to the network, one connection after the other, and the number of rejected requests is stored. It is done in this way:

FOR each node in the network **DO**

　　　Compute the list of its non-dominated routes; /* using paths list computation */
　　　Find the best route from the source to the destination,
　　　　　using the path selection policy;
　　　Remove the capacity used by this request from the links followed by it;
　　　IF capacity of one of the link < 0 **THEN**
　　　　　restore the capacity;
　　　　　reject the request;
　　　　　increment number of rejected routes;
　　ENDIF;
　ENDFOR

Table 6 shows the percentage of rejected connections as a function of the number of nodes, the connectivity, and the number of requests.

Number of nodes in the network	Connectivity	Number of requests	Rejected requests (in %)
15	0.2	30	35.0±6.3
15	0.3	30	39.4±12.2
20	0.2	40	38.3±6.5
20	0.3	40	24.0±12.6
25	0.2	50	32.4±4.1
25	0.3	50	31.6±5.7
30	0.2	60	38.8±16.0
30	0.3	60	25[a]

Table 6: Rejected requests due to blocking, with 90% confidence intervals

a. One run only, no confidence interval computed

The results are quite bad, with around a third of requests blocked. This proves that the diversity must also be poor at the network level. An overload of the most used links occurs rapidly, which leads to the blocking of many connections, and which would trigger a Link State Update. This would bring on a recomputation of all the routes, which has to be avoided if possible, because pre-computation routing is helpful only if the paths list computations are not too frequent.

It is interesting to note that the rest of the links are most surely not used a lot, and a solution could be to use them more, by having alternate paths. The next chapter will compare two methods that we propose to overcome these blocking problems.

4 Improvement of Path Diversity

4.1 Proposed Methods

The policy to choose a route given in 2.3 is not modified, because it always gives the shortest path and, when used in a lightly loaded network, it gives a route having no problem of blocking.

Below we propose two methods for improving resilience to blocking. The idea of both proposed methods is to add, for each non-dominated route given by the paths list computation, an *alternate* route, used after the rejection of the first try through the shortest possible non-dominated route [Ash90, Key90]. This 'second chance' solution is, as for the non-dominated routes, pre-computed.

Simple method

This solution consists of having another of the non-dominated routes as 'second choice', using the path selection policy shown in 2.3 to choose it, but after removing the 'first choice' route from the list. The implementation of this method is very simple and has very small computation needs, but the problems of diversity lack between non-dominated routes from the same source (shown in 3.1) could lead to a high probability of going through the same overloaded link.

Complex method

The second solution consists in computing, for each non-dominated route, a shortest path using Dijkstra on the additive cost, but on a modified topology database. This one is obtained by removing from the actual topology all the links with at least one restrictive cost smaller than the ones of the route we are trying to find an alternative to, and all the links being part of the 'first choice' route. The alternate route computed in this way is, if it exists, at least as good as the original one, and has no link in common with it.

Algorithm for the complex method

Let A be the alternate path of the route I in the list
Let $C(x)$, $Cap1(x)$ and $Cap2(x)$ be the additive, first and second restrictive costs of x, respectively
Let T be the original topology database, and T_{tmp} the modified one.

FOR each route I of the list DO
$\quad$ $T_{tmp} = T$ - all links with $Cap1 < Cap1(I)$ or $Cap2 < Cap2(I)$;
$\quad$ $T_{tmp} = T_{tmp}$ - all links used by I;
$\quad$ A = shortest path (using Dijkstra on T_{tmp}) to the same destination as I;
ENDFOR;

IF A exists THEN it is the alternate path with lowest C ENDIF;

Pre-computation is, of course, used for those alternate paths, but the complexity of this method, $O(N^4)$ (where N is equal to the number of nodes in the network) for three-dimensions cost vectors could be a problem. No optimization work has been done on it and improvements should be possible.

The simulation results for the two methods are shown in 4.2, along with a comparison between them.

4.2 Efficiency

In Table 7 we compare the results obtained by the three proposed methods:

The first one, with no protection against blocking, as shown in 2.3.

The second one, simple, with alternate route chosen among the other non-dominated routes.

And the third one, complex, with a completely disjoint alternate path for each non-dominated route (the last two methods being described in 4.1).

Number of nodes in the network	Connectivity	Number of requests	First method[a][b]	Simple method[a]	Complex method[a]
15	0.2	30	35.0±6.3	23.3±9.0	29.4±9.5
15	0.3	30	39.4±12.2	21.7±8.6	22.8±7.7
20	0.2	40	38.3±6.5	22.9±7.6	22.5±7.5
20	0.3	40	24.0±12.6	12.5±9.0	7.0±4.1
25	0.2	50	32.4±4.1	17.2±6.9	13.2±5.9
25	0.3	50	31.6±5.7	16.8±6.4	6.8±5.3
30	0.2	60	38.8±16.0	20.0±37.0	10.4±4.0
30	0.3	60	25[c]	13.3[c]	5[c]

Table 7: Rejected requests due to blocking, with 90% confidence intervals

a. Percentage of rejected requests (with confidence interval in %)
b. Same results as shown in table 5
c. One run only, no confidence interval computed

The results obtained by the first method are quite bad, going up to 40% of requests rejected. The second and third ones show an improvement compared to the first one. It is worth noticing the following three points:

- The percentage of rejection decreases when the connectivity increases. If there are more links, there are also more paths, and the probability of overloading only certain links decreases.
- When the connectivity is high, there are a lot of non-optimal paths between each pair of nodes, and the complex method gives better results than the simple one.
- When the number of nodes is small, the complex method is not better than the simple one. This is because completely disjoint paths are often impossible in such small networks, and so, no alternate paths are found.

It is the difference between the results obtained by the second and the third method when the network is large that confirms the assumption that the overloaded links must be the ones through which most of the non-dominated routes go. Indeed, the fact of forbidding their use to find alternate paths (third method) leads to an evident improvement in the results, unfortunately to the detriment of the computation complexity.

Another way of considering this is to see when a Link State Update becomes necessary so to have a blocking probability lower than a certain threshold. The third method then becomes much more competitive, because its low blocking rate allows longer periods between Link State Update, requests being able to be satisfied even if the network is more loaded than with the other two methods. This could compensate its complexity, because it spaces the dates between paths list computation.

5 Conclusion

We have shown that it is possible to route communications in an ATM network using route pre-computation and that the results are the same as those that would be obtained by On-Demand computation if a load-balancing option is chosen for the path selection. Furthermore tests have revealed problems of request failures because of the lack of diversity of the routes given by paths computation. Modifications to the path selection policy proposed in chapter 3 have been tested, and an improvement to the blocking problem has been shown.

Further research focuses on finding the lower theoretical bound to the blocking of connections.

References

[Ash90]   Design and Control of Network with Dynamic Nonhierarchical Routing, *G.R. Ash*, IEEE Communications Magazine, Oct. 1990

[Car79]   Graphs and Networks, *B. Carré*, Clarendon Press - Oxford, 1979

[Cyp71]   Communications Architecture for Distributed Systems, *R.J. Cypser*, ISBN 0-201-14458-1, 1971

[Gau95]   Scalability Enhancements for Connection-Oriented Networks, *E. Gauthier, J.-Y. Le Boudec*, TR 95/127, DI-EPFL, 1995

[Key90]   Distributed Dynamic Routing Schemes, *P.B. Key, G.A. Cope*, IEEE Communications Magazine, Oct. 1990

[Leb94]   Routing Metric for Connections with Reserved Bandwidth, *J.-Y. Le Boudec, T. Przygienda*, EFOC-N, 1994

[OSI89]   ISO/IEC TR 9575, Information Technology - Telecommunications and Information Exchange Between Systems - OSI Routeing Framework, *ISO*, 1989

[Prz95]   A Route Pre-Computation Algorithm for Integrated Services Networks, *J.-Y. Le Boudec, T. Przygienda*, TR 95/113, DI-EPFL, 1995

[Top90]   Experimental Internet Stream Protocol, Version 2, (ST.II), *C. Topolcic*, IETF, 1990

[Zha93]   RSVP: A New Resource ReSerVation Protocol, *L. Zhang, S. Deering, D. Estrin, S. Shenker, D. Zapalla*, IEEE Network, September 1993

Design and evaluation of distributed link and path restoration algorithms for ATM meshed networks

Kris Struyve, Piet Demeester
University of Gent - IMEC
Department of Information Technology (INTEC)
Sint-Pietersnieuwstraat 41, 9000 Gent, Belgium
Tel: +32.9.264.33.16, Fax: +32.9.264.35.93
Email: kris.struyve@intec.rug.ac.be

Leo Nederlof, Luc Van Hauwermeiren
Alcatel Corporate Research Centre, Antwerp, Belgium

Abstract

Two distributed restoration algorithms are presented in this paper to restore disrupted traffic in meshed broadband networks due to link or node failures. Both algorithms are based on a multi-sender approach. The first algorithm uses *link restoration*, whereas the second algorithm uses *path restoration*. The performances of these algorithms are analyzed on two test networks. It is shown that they achieve fairly high restoration ratios within seconds. The algorithms are compared with other distributed restoration algorithms published in literature and with graph theoretical algorithms.

Keywords

ATM, distributed, self-healing, link, path, network recovery

1. Introduction

The deployment of high-bandwidth fiber-optic transmission systems and reconfigurable high-speed digital cross-connects, on one hand, and the increasing demand of customers, on the other hand, urge the provisioning of survivability techniques in future broadband telecommunication networks[1]. A number of distributed restoration algorithms have been proposed to protect traffic in STM[2,3,4,5,6] and ATM[7,8,9] networks against network failures. Centralized restoration requires reliable communication means between the network operation center and the cross-connects. Furthermore, distributed restoration achieves much higher restoration speeds than centralized restoration.

This paper presents two dynamic distributed restoration algorithms for meshed broadband networks, i.e. LINKRES and PATHRES. The approach taken in LINKRES is based on an initial distributed *link restoration* algorithm[10]. The most significant difference is the extension of the confirm phase to optimize the restoration of node failures. PATHRES is derived from LINKRES, but uses *path restoration*. The rest of the paper is organized as follows: section 2 discusses link versus path restoration. Sections 3 and 4 respectively describe the restoration algorithms LINKRES and PATHRES. Section 5 analyses the performance of the algorithms on two test networks and compares these results with results of other distributed restoration algorithms published in literature. Finally, conclusions are presented in section 6.

2. Link versus path restoration

The first algorithm, LINKRES, uses a *link restoration* approach, i.e. it reroutes the initial routes (IR) of the affected connections to alternative routes (AR) between the nodes adjacent to the failure (Figures 1 and 2). LINKRES is able to recover from single and multiple link failures as well as node failures. Most distributed link restoration algorithms presented in literature are inadequate for multiple link failures and node failures. Komine's algorithm[5] handles node failures, but assumes unidirectional trails. However, in order to comply with ITU recommendations[11], ATM trails are to be bidirectional and both directions should follow the same route. Hence, the designation of sender and chooser roles to the nodes adjacent to the failure, based on upstream and downstream positioning, is not free of ambiguity. If, on the other hand, the designation is based on an arbitration convention such as node identifier (ID), deadlock occurs if the node with the lowest ID fails. Indeed, the neighbour nodes, having higher IDs, cannot distinguish a link failure from a node failure. They play the role of choosers and await in vain request messages coming from the failed sender node.

The second algorithm, PATHRES, on the other hand, uses a *path restoration* approach, i.e. it reroutes the initial routes (IR) of the affected connections to alternative routes (AR) between their end nodes (Figure 3). As any other path restoration algorithm, PATHRES makes more efficient use of spare resources in the network and is able to recover from single as well as multiple link and node failures. A major drawback of dynamic distributed path restoration algorithms is that for every failed connection a restoration process is invoked. So, many more restoration messages are generated which slows down the overall restoration speed. However, PATHRES is to be used in combination with rapid back-up VPC restoration techniques[8,9]. Hence, during the restoration phase most of the disrupted VP connections (VPC) are recovered via their preassigned back-up VPCs while the other VP connections, which do not have preassigned back-up VPCs, are recovered using PATHRES. Next, during the post-restoration phase PATHRES is used to reconstruct back-up VPCs. In order to further reduce the number of messages, failed connections between the same pair of end nodes may be aggregated into VP groups (VPG).

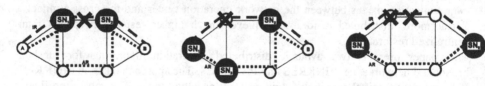

Fig. 1 LINKRES due to a link failure.

Fig. 2 LINKRES due to a node failure.

Fig. 3 PATHRES due to a link and/or a node failure.

Both LINKRES and PATHRES assume bidirectional VP connections. Since all restoration messages express bandwidth information in pairs, the algorithms can handle symmetrical as well as asymmetrical connections. Furthermore, whereas conventional SENDER - CHOOSER algorithms flood request messages from one single sender node all the way to the chooser node, LINKRES and PATHRES flood request messages from multiple sender nodes towards each other. Hence, these algorithms can set the hop count limit to almost half of the conventional algorithms which reduces the size of the restoration area. The algorithms are designed for ATM networks, but with some minor modifications they can also be applied in STM networks.

3. Link restoration algorithm LINKRES

LINKRES utilizes four phases (Figure 4) to restore affected traffic between the nodes adjacent to the failure, i.e. a *request phase*; a *confirm and decision phase*; a *connect phase* and a *release phase*. In the following paragraphs each phase of the algorithm is discussed in turn.

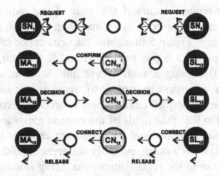

Fig. 4 The four phases of LINKRES.

3.1 Request phase

The nodes adjacent to the network failure, called *sender nodes*, detect the failure and invoke the request phase. In case of a link failure (Figure 1), the two end nodes of the link act as sender nodes, while in case of a node failure (Figure 2), the neighbour nodes of the failing node act as sender nodes. During the request phase, nodes search for candidate alternative routes by selectively flooding *request messages*. Each sender node selectively floods request messages on all of its outgoing links excluding the link on which it detected the alarm. *Tandem nodes* selectively rebroadcast updated copies of received request messages. Hence, logically, N trees TR_n (n=1,...,N) are constructed branch-per-branch originating from the N sender nodes SN_n. Sender as well as tandem nodes store all received and forwarded request messages. Furthermore, nodes explicitly reserve spare capacity requested in these messages. The *requested capacity* is set to the minimum of the required capacity and the unreserved spare capacity of the link on which the request message is being flooded. Since a sender node cannot distinguish a link failure from a node failure, the required capacity equals the affected capacity between the sender node SN_i and the node at the other end of the link on which the sender node detected the alarm. In case of a node failure, only part of the required capacity can be restored due to affected connections terminating at the failed node. A *hop count* field is included in a request message to limit the extent of the restoration area. Since different request messages originating from the same sender node SN_i may be flooded on a link, the image of TR_i is more a mesh than a tree. *Signatures* are used to distinguish between different messages on the same link belonging to the same tree. The signature of a request message is unique with respect to a particular link, similar to the VPI identifier of a VP connection.

Recovery of link failures requires that each node has at least knowledge of the IDs of its neighbour nodes as well as of the bandwidth of all the VPs on all of its incident links. Recovery of node failures demands in addition that each node knows the *second node* of its VPs. Based on this data, the sender node SN_i initializes the *list of other sender nodes* with the IDs of all other sender nodes which are one or two hops away and between which at least one connection is lost. Tandem nodes add to the *route* of every outgoing request message their ID as well as the requested capacity and the signature of the matching incoming request message. The route of a request message prevents looping and enables location of spare capacity contention due to so called *over-requests*. An over-request occurs if the sum of the requested capacity of all forwarded request messages in a node is greater than the requested capacity of the matching received message. Figure 5 illustrates the occurrence of an over-request due to selective reflooding on links B (i.e. req_10/5) and C (i.e. req_10/10) of a request message (i.e. req_10/10) arriving at node N via link A. Note, all messages express bandwidth information in pairs in order to recover symmetrical as well as asymmetrical bidirectional disrupted connections. Moreover, extra entries (i.e. OVR: N, 10/10) are appended to the route fields of the request messages flooded on links B and C indicating the node identifier and the requested capacity of the corresponding incoming request message. As a result, possibly different candidate alternative routes may be found, in response to the request messages being flooded on links B and C, with a total aggregate capacity (i.e. capacity 20/15) exceeding the requested capacity (i.e. capacity 10/10) on link A. In order to resolve such overlapping (in terms of capacity) candidate alternative routes, a confirm and decision phase following the request phase is required. If, on the other hand, as is shown in Figure 6, the total outgoing flooded requested capacity (i.e. capacity 7/7) is less than the incoming requested capacity (i.e. capacity 10/10) due to lack of unreserved spare resource, then this capacity difference (i.e. capacity 3/3) is stored in a *hold buffer*. Later on, if reserved spare capacity is released, this message is reflooded.

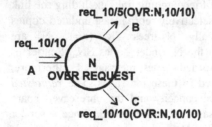
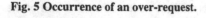
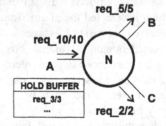

Fig. 5 Occurrence of an over-request.	Fig. 6 Use of holdbuffer.

While expanding, branches of different trees, e.g. TR_i and TR_j, meet at *collision nodes* CN_{ij}^k (k=1,...,K). A collision identifies a candidate alternative route CAR_{ij}^k and triggers the next phase of the algorithm, i.e. the confirm and decision phase. A tandem node determines whether two request messages collide based on the contents of the list of other sender nodes of the messages: two trees TR_i and TR_j meet if, and only if, the sender node ID SN_i is listed in the list of other sender nodes of the other request message and vice versa. In the course of the restoration process, the contents of these lists are updated in the tandem and sender nodes to reflect, for example, that no more affected connections between the sender nodes SN_i and SN_j need to be restored.

3.2 Confirm and decision phase

The two originating sender nodes SN_i and SN_j of the interfering trees TR_i and TR_j are designated *master node* MA_{ij} (e.g. SN_i) and *slave node* SL_{ij} (e.g. SN_j) based on an arbitration convention (e.g. node ID). The collision node $CN_{ij}{}^k$ assigns an unique key to the candidate alternative route $CAR_{ij}{}^k$ and forwards a *confirm message* towards the master node MA_{ij}. The confirm message is a combination of the two colliding request messages and informs the master node concerning the other end node (i.e. SL_{ij}), the capacity and the route of $CAR_{ij}{}^k$. Due to over-requests which have previously been connected, the actual available capacity of the candidate alternative route equals at most the capacity indicated in the confirm message. In order to resolve capacity contention, each sender node stores all the over-requests registered in received confirm (and decision) messages in a *contention table*, and keeps track of how much of an over-request has been connected. Hence, the master node MA_{ij} is able to detect and resolve all possible capacity overlap in the *master part* of the candidate route. However, the master node MA_{ij} only detects capacity overlap in the *slave part* due to previously connected alternative routes between MA_{ij} and SL_{ij}. Indeed, MA_{ij} is unaware of contention in the slave part due to over-requests which are utilized by alternative routes between SL_{ij} ($=SN_j$) and another sender node SN_m ($m \neq i$). It is clear that this situation does not occur in the case of single link failures.

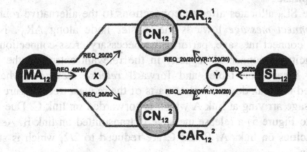

Fig. 7 Contention detection in case of a link failure.

These contention detection problems are illustrated in Figures 7 and 8. In Figure 7 an over-request occurs at node Y (i.e. incoming requested capacity 20/20, total outgoing requested capacity 40/40). Suppose the master node MA_{12} first fully connects (i.e. capacity 20/20) candidate alternative route $CAR_{12}{}^1$ and updates its contention table accordingly. Next, MA_{12} detects the previously connected over-request at node Y in the slave part and does not connect $CAR_{12}{}^2$. In Figure 8, on the other hand, an over-request occurs at node Z (i.e. incoming requested capacity 40/40, total outgoing requested capacity 80/80). Similar to Figure 7, MA_{12} first fully connects $CAR_{12}{}^1$ (i.e. capacity 20/20). In the meantime MA_{32} fully connects $CAR_{32}{}^1$ (i.e. capacity 20/20). Next, MA_{12} is unaware of $CAR_{32}{}^1$ and wrongly connects $CAR_{12}{}^2$ (i.e. capacity 20/20). To prevent such an error, the master node MA_{ij} transmits its partial decision by means of a *decision message* to the slave node SL_{ij} via the candidate alternative route $CAR_{ij}{}^k$.

Upon receiving the decision message, the slave node, similar to the master node, updates its contention table and finally calculates the available capacity of the route. As such, the CAR_{ij}^k becomes permanent and the connect phase is instigated.

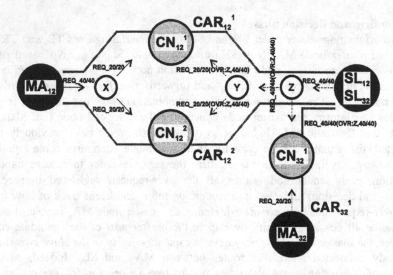

Fig. 8 Contention detection problem in case of a node failure.

3.3 Connect phase

The slave node SL_{ij} allocates affected connections to the alternative route AR_{ij}^k and transmits a *connect message* towards the master node along AR_{ij}^k. Nodes, upon processing the connect message, perform the necessary cross-connections, mark the requested capacity as connected capacity in the relevant entries in the request and spare capacity reservation tables and forward *release messages* wherever over-requests occured to break down obsolete parts of the request trees. Figure 9 illustrates a connect message arriving at link A which is forwarded on link C. Due to the over-request (refer to Figure 5) a release message is transmitted on link B. As a result the requested capacities on links A, B and C are reduced to 2/2, which is still an over-request situation.

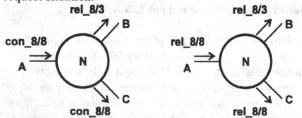 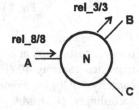

Fig. 9 (refers to Fig. 5) Connect in an over-request situation.

Fig. 10 (refers to Fig. 5) Release in an over-request situation.

Fig. 11 (refers to Fig. 6) Release in a non-over-request situation

3.4 Release phase

Connect messages cause tandem as well as sender nodes to forward, if appropriate, release messages to break down obsolete parts of a tree. Releasing a branch of a tree TR_i, i.e. releasing reserved spare capacity on a link, stimulates the elaboration of another tree TR_j and thus the continuation of the search process.

The signature of a release message identifies the matching request branch. Whenever a node receives a release message, it decreases the requested capacity of the incoming branch and forwards new release messages on outgoing branches to release the abundant requested capacity compared to the requested capacity of the incoming branch (Figures 10 and 11). The node immediately re-utilizes released reserved capacity to flood other request messages which are memorized in the hold buffer.

4. Path restoration algorithm PATHRES

Though PATHRES is similar in approach to LINKRES, there are some major differences which simplify the algorithm (Figures 3 and 12). Suppose that each VPC or VPG, from now on shortly denoted as restoration unit (RU), has an unique identifier known by both of its end nodes. Next, consider a particular restoration unit RU_{ij}. Upon detecting alarm signals, the end nodes of RU_{ij} act as sender nodes (i.e. SN_i and SN_j) and selectively flood request messages. Hence, two request trees TR_{ij} and TR_{ji} are constructed during the request phase originating respectively at SN_i and SN_j. Whenever these complementary trees collide at an intermediate node CN_{ij}^{k} (k=1,...K) a candidate alternative route CAR_{ij}^{k} for RU_{ij} is identified. Note, possibly multiple request trees TR_{ij} (j=1,...,i-1,i+1,...,N) originate at a single sender node SN_i. Each tree TR_{ij} corresponds to a particular restoration unit RU_{ij} which terminates at SN_j. This is unlike LINKRES where only one request tree TR_i originates per sender node SN_i. Moreover, this tree TR_i collides with trees TR_j originating from other different sender nodes SN_j (j=1,...,i-1,i+1,...N) dependent on the *list of other sender nodes*. Since in PATHRES the tree TR_{ij} only collides with its complementary tree TR_{ji}, request messages must contain the unique identifier of RU_{ij} instead of a list of other sender nodes. Furthermore, the selective flooding of request messages in PATHRES is even more restricted compared to LINKRES as a restoration unit cannot be split up onto different alternative routes. Indeed, apart from hop count limitation and loop prevention, only a single request message of a tree TR_{ij} is ever flooded on a link, even if reserved spare capacity is released. Hence, request messages no longer contain *signatures* to distinguish between different branches of the same tree on a link. As a result the image of a request tree TR_{ij} matches an actual tree, whereas in LINKRES the image of a request tree TR_i is more alike a mesh. In addition, a request message is flooded on a link if, and only if, the unreserved spare capacity of the link equals at least the capacity of RU_{ij}. Over-requests are still possible.

To retain a single alternative route from the multiple collisions CAR_{ij}^{k} (k=1,...,K) between TR_{ij} and TR_{ji}, each collision node transmits a confirm message towards the master node MA_{ij}. The alternative route is selected based on a first-come-first-served basis, i.e. the first confirm message arriving at MA_{ij} determines the CAR_{ij}^{k} which is used to reroute RU_{ij}. A tandem node propagates a confirm message towards the master node if, and only if, previously it has not received any confirm message related to other candidate alternative routes between MA_{ij} and SL_{ij}. If so, the node releases also the corresponding outgoing branches, with exception of the branches which are part of

CAR_{ij}^{k}. The first confirm message arriving at the master node is made permanent by forwarding a connect message towards the slave node. Upon processing the connect message, nodes perform the cross-connections. While obsolete parts of the (master) tree TR_{ij} are released during the confirm phase, the obsolete parts of the (slave) tree TR_{ji} are released whenever a node in the slave part receives a connect message. Upon receiving the connect message at the slave node, the rerouting of the affected RU_{ij} is completed.

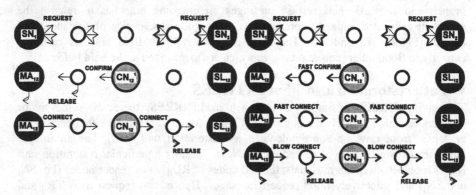

Fig. 12 The three phases of (slow) PATHRES.

Fig. 13 The four phases of (fast) PATHRES.

Focusing on ATM networks in particular, high-speed restoration might be achieved by performing the cross-connections during the request phase, similar to the distributed Flood-Connecting restoration algorithm[8]. Hence, if a collision occurs, the collision node cross-connects the master part of the candidate alternative route to the slave part and directly forwards through the switch fabrics, without processing at tandem nodes, a confirm message towards the master node (Figure 13). The first confirm message arriving at the master node designates the matching CAR_{ij}^{k} as permanent. Next, the master node transmits a fast connect notification to the slave node which completes the actual rerouting of RU_{ij}. In parallel, the master node transmits a (slow) connect message to the slave node intended to release obsolete branches of the master TR_{ij} and slave TR_{ji} trees.

5. Simulation results

In this section we will analyze LINKRES and PATHRES on two different test networks using an object-oriented discrete-event simulator, ATMSIM, which we have developed in C++. The spare capacity provisioning of both networks, i.e. the New Jersey LATA network (11 nodes, 23 links) and the CATS network (32 nodes, 54 links, 340 VPCs), permit full restorability of any single link failure. Each node in the network is equiped with a controller to process restoration messages. A controller is modeled as a single-server-FIFO-queuing system. We adopted following assumptions to be inline with the RREACT paper[4]: all messages are 64 bytes long, all messages have equal priority, messages are serviced by a controller in the order they were received, it requires 10ms to service any incoming message and an additional 10 ms to generate each outgoing message, the propagation delay on any link is set to 0.5 ms in the New Jersey LATA network and to 1 ms in the CATS network, the signal transmission rate is set to 64 kbit/s, bandwidth is expressed in *equivalent units*, the bandwidth of any VPC equals one unit, all VPCs are symmetric and fault detection times are neglected.

In the following simulations we have also solved the restoration problems using graph theoretical algorithms, i.e. Busacker-Gowen and Dijkstra (not taking time aspects into consideration nor any hop limit restriction). This allows for a resource utilization and restoration ratio comparison between dynamic distributed algorithms and centralized algorithms. It turns out that distributed restoration algorithms are suboptimal, but on the other hand the restoration speed of distributed algorithms is shown to be on the order of seconds, whereas the restoration speed of centralized algorithms is expected to be on the order of minutes or more[12].

5.1 Evaluations on the New Jersey LATA network

The RREACT paper presents simulation results of link failures in the New Jersey LATA network using different link restoration algorithms, i.e. FITNESS[6], Two-Prong[3] and RREACT[4]. Table 1 provides a comparison of the restoration performances of these three algorithms and of LINKRES for a particular example link failure. These results show that LINKRES attains full restoration and is as cost efficient as RREACT and Two-Prong. Though the restoration speed is lower compared to Two-Prong, LINKRES outperforms FITNESS by far. LINKRES achieves full restoration of any single link failure if the hop limit is set to 5. An average restoration ratio of 98% is achieved within the typical 2 second restoration requirement, though full restoration of the worst case link failure requires 5 seconds. LINKRES assigns affected VPCs to alternative routes on a first-come-first-served basis and hence, as argued by Anderson et al[13], is likely to be suboptimal. The optimal solution for the single-commodity minimum-cost restoration problem, which we have calculated with the Busacker-Gowen algorithm, confirms this statement, i.e. LINKRES uses a total of 3% more spare resources than the optimal algorithm. In summary, the computer evaluations on the New Jersey LATA network show that the performance of LINKRES is comparable to RREACT and Two-Prong while offering the extra asset of restoring node failures. This is demonstrated on the CATS network.

Restoration algorithm	link failure N05-N08 (81 working channels lost)		
	Restoration ratio (%)	Spare resource usage	Restoration time (ms)
RREACT	100	204	402
FITNESS	100	289	1756
Two-Prong	100	210	273
LINKRES	100	204	647

Table 1 Comparison of four dynamic distributed link restoration algorithms for a particular link failure.

5.2 Evaluations on the CATS network

We will first consider single link failures and next node failures. For the evaluation of (slow) PATHRES, we grouped all VP connections between the same pair of end nodes in one VPG and used these VPGs as restoration units. Moreover, we assumed that before the PATHRES algorithm is invoked all affected VPCs are torn down. So, more spare capacity is available for restoration.

Single link failures

Table 2 as well as Figure 14 show that, for the given default assumptions, LINKRES (hop limit set to 3) restores any single link failure within 1.4 s, whereas PATHRES (hop limit set to 6) attains full average restoration within 7.1 s. PATHRES is clearly much slower than LINKRES. The worse time performance of PATHRES is due to the higher number of restoration processes invoked (i.e. summed over all link failures: 54 for LINKRES versus 836 for PATHRES) and consequently the larger amount of restoration messages generated. Furthermore, in case of PATHRES the restoration information has to propagate between the end nodes of the VPGs, whilst in case of LINKRES the information propagates between the nodes adjacent to the failure.

	Link restoration		Path restoration	
	LINKRES (HL = 3)	BUSACKER-GOWEN	PATHRES (HL = 6)	DIJKSTRA
Average restoration ratio (%)	100	100	100	100
Restoration time (s)	1.40	-	7.10	-
Total spare resource usage	6719	6679	6017	5111

Table 2 LINKRES versus PATHRES for single link failures.

Table 2 also indicates the optimal solution obtained with the Busacker-Gowen algorithm for link restoration. For path restoration we sequentially searched for all affected VPGs the shortest (in terms of hops) alternative path using the Dijkstra algorithm, each time updating the spare capacity of the links in the network. We cannot apply Busacker-Gowen as this algorithm possibly splits up the affected flow into multiple alternative flows. However, this is in contradiction with the assumption that an affected VPG cannot be split up and is restored as one unit. It should be stressed that the iterative Dijkstra restoration solution[14] does not provide the optimal minimum-cost solution, i.e. the outcome depends on the order in which the VPGs are restored. Path restoration is more cost efficient than link restoration, i.e. path restoration uses less spare capacity to reroute all the affected connections. Moreover, LINKRES and PATHRES are shown to be less cost efficient compared to respectively Busacker-Gowen and Dijkstra.

Average restoration ratio (%)	Restoration time t(s) (SER = 10 ms, GEN = 10 ms)	Restoration time t(s) (SER = 10 ms, GEN = 0.1 ms)	Restoration time t(s) (SER = 100 ms, GEN = 0.1 ms)
85	0.60	0.35	2.15
90	0.70	0.40	2.50
95	0.85	0.55	3.05
100	1.40	0.85	5.05

Table 3 Influence of differing service and generation delays on LINKRES performance.

Table 3 illustrates the influence of the service and generation delays of restoration messages in the node controllers on the restoration speed. Neglecting the delays associated with the generation of outgoing new restoration messages, reduces the restoration time with almost 35%. Detailed studies of the processing of restoration messages in an ATM cross-connect are ongoing. Preliminary simulation results indicate that the service delays measure approximately 100 ms and that the generation delays are indeed negligible. Hence, full average restoration would take 5 s instead of 1.4 s.

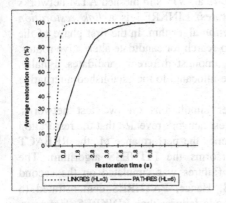

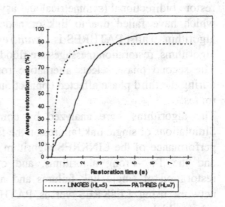

Fig. 14 LINKRES versus PATHRES for single link failures.

Fig. 15 LINKRES versus PATHRES for single node failures.

Node failures

Evaluation results of node failures using LINKRES and PATHRES are summarized in Table 4. Note that for the calculation of the metrics only restorable connections (i.e. connections not terminating at the failing node) are taken into account. PATHRES restores node failures more adequately than LINKRES, i.e. LINKRES restores less than 90%, whereas PATHRES restores 97% of the affected connections. Again we note that link restoration is much faster than path restoration: LINKRES reaches an average restoration ratio of 85% after 1.8 s whilst PATHRES needs 3.5 s (Figure 15).

	Link restoration		Path restoration	
	LINKRES (HL = 5)	BUSACKER-GOWEN	PATHRES (HL=7)	DIJKSTRA
Average restoration ratio (%)	88.67	97.79	97.53	99.05
Restoration time (s)	3.75	-	8.20	-
Total spare resource usage	3728	4443	4026	3812

Table 4 LINKRES versus PATHRES for single node failures.

Whereas link restoration of a single link failure is a single-commodity flow rerouting problem (just one pair of nodes adjacent to the link failure), link restoration of a node failure is a multi-commodity flow rerouting problem as there are multiple pairs of nodes adjacent to the node failure between which affected flow needs to be restored. Hence, Busacker-Gowen does not provide us with the optimal minimum-cost solution. Moreover, we applied Busacker-Gowen iteratively for the different flows. Neither Busacker-Gowen (link restoration) nor Dijkstra (path restoration) succeed full restoration of all node failures. Whereas Busacker-Gowen performs remarkably better than LINKRES, the difference between PATHRES and Dijkstra is far less.

6. Summary and conclusion

Two distributed dynamic restoration algorithms were presented. Both algorithms restore bidirectional (symmetrical and asymmetrical) VPCs in meshed ATM networks which have failed due to link or node failures. LINKRES is a *link restoration* algorithm, whilst PATHRES is a *path restoration* algorithm. In the first phase of the algorithms, restoration messages are flooded to search for candidate alternative routes. The second phase selects alternative routes amongst different candidates. Finally, during the third phase affected connections are allocated to the established alternative routes.

The algorithms were analyzed by computer simulations on two test networks. Simulations of single link failures on the first test network revealed that the restoration performance of the LINKRES algorithm matches the performance of the RREACT and the Two-Prong algorithms, and outperforms the FITNESS algorithm. The restoration of single link failures and node failures was simulated on the second network using LINKRES versus PATHRES. Hence, PATHRES demonstrated to achieve higher average restoration ratios for node failures than LINKRES. However, LINKRES proved to be much faster. Concerning future work, the obtained results encourage us to enhance the restoration of link and node failures using LINKRES, and to evaluate the performance of PATHRES in co-operation with backup-VPC restoration techniques.

Acknowledgement

Part of this research has been supported by the RACE II project IMMUNE[1] ("End-to-end Survivable Broadband Networks").

References

[1] Leo Nederlof, Kris Struyve, Chris O'Shea, Howard Misser, Yonggang Du, Braulio Tamayo, "End-to-end Survivable Broadband Networks", IEEE Communications Magazine, September 1995

[2] W.D. Grover, "The Self-Healing Network: a fast distributed restoration technique for networks using digital cross-connect machines", proceedings of Globecom '87, pp. 28.2.1-28.2.6, 1987

[3] Chow, J. Bicknell, S. McCaughy, "A fast distributed network restoration algorithm", International Phoenix Conference on Computer and Communications, March 22-26, 1993, Tempe, Az.

[4] Chow, S. McCaughey, S. Syed, "RREACT: A distributed network restoration protocol for rapid restoration of active communication trunks", proceedings of 2nd IEEE Network Management and Control Workshop, Westchester, 1993

[5] H.Komine, T. Chujo, T. Ogura, K. Miyazaki, T. Soejima, "A distributed restoration algorithm for multiple link and node failures of transport netorks", proceedings of Globecom '90, pp. 403.4.1-5, December 1990

[6] C.H. Yang, S. Hasegawa, "FITNESS: Failure Immunization Technology for Network Service Survivability", proceedings of Globecom '88, pp 47.3.1-47.3.6, November 1988

[7] H.Fujii, N. Yoshikai, "Restoration Message Transfer mechanism and Restoration Characteristics of the Double-Search Self-Healing ATM network", IEEE JSAC, vol. 12, no.1, pp. 149-158, January 1994

[8] R. Kawamura, K. Sato, I. Tokizawa, "Self-healing ATM network techniques utilizing VPs", 5th International Network Planning Symposium, Kobe, Japan, May 1992

[9] P.A. Veitch, D.G. Smith, I. Hawker, "A distributed protocol for fast and robust VP restoration", 12th IEE UK Teletraffic Symposium, Windsor, March 1995

[10] Leo Nederlof, Hans Vanderstraeten, Patrick Vankwikelberge, "A new distributed restoration algorithm to protect ATM meshed networks against link and node failures", proceedings of ISS '95, April 1995

[11] ITU-T Recommendation M.3100, "Generic network information model", Geneva 1992

[12] T-H. WU, "Fiber Network Service Survivability", Artech House, May 1992

[13] J. Anderson, B. Doshi, S. Dravida, P. Harshavardhana, "Fast Restoration of ATM Networks", IEEE-JSAC, vol 12, no 1, January 1994

[14] D. Vercauteren, P. Demeester, J. Luystermans, E. Houtrelle, "Availability analysis of multi-layer networks", 3rd International Conference on Telecommunication Systems, March 16-19, 1995

Scalability Enhancements for Connection-Oriented Networks

E. Gauthier

gauthier@di.epfl.ch

J.-Y. Le Boudec

leboudec@di.epfl.ch

LRC

EPFL CH-1015 Lausanne

Abstract. We consider the issue of increasing the number of connections that connection oriented networks, such as ATM, can handle. We describe one step that aims at reducing connection awareness inside the network. To that end, connections between the same pair of access nodes are grouped together and made indistinguishable inside the network. The concept of dynamic virtual path trunks is introduced as a support mechanism, and it is shown how virtual path links can be setup and maintained without additional round-trip delays.

Keywords: connection awareness, multiplexed signaling, bandwidth renegotiation.

1 Connection Limits for Network Nodes

Support for real time traffic in communication networks requires that a connection oriented philosophy be employed. Some network technologies are connection oriented right from their definition: this is the case for ATM [1, 2], or Narrowband ISDN [3]. Some others initially based on a connection-less principle, like bridged LANs or Internet [4], are introducing resource reservation methods that end up using a concept of connection (or analogous concepts such as "soft state" for RSVP [5, 6]). As a result, and as multimedia communication becomes a mass market, the networking industry will be confronted with the requirement to build and operate connection oriented networks of increasing sizes.

Designing and operating connection oriented networks is not simple, because of the complexity required to support connections. Every connection established through the network is associated at every node with label swapping tables, with capacity reserved in queues, and with connection control blocks (or equivalent denominations) used by the signaling or control protocol; connection establishment, tear-down, or simply keeping the connection alive requires some processing [7].

Last but not least, every connection is visible to network management, and thus comes with a horde of various data structures to represent the connection's static and dynamic attributes, together with measurement results for statistical purposes. Due to these elements, network nodes have limits on their connection handling capabilities. Such limits depend strongly on the node design, and generally depend on its configuration (such as amount of memory installed, or number of ports if some of the processing is distributed on the port cards). The limits can be expressed in a number of ways, representing limitations in both memory and processing capacity. For example, a local area ATM switch may be limited to support at most 10,000 connections in steady state, and at most 150 connection establishment attempts per second. (In general, exact limitations are expressed in a more complex way because the problem is multi-dimensional.) Of course, node capacity is not only limited by connections, but also by other factors such as link capacity and number of ports.

We believe that the success of connection oriented networks and especially the introduction of ATM to the end users will require that these limits be pushed as far as possible. Ideally, we would like to be able to design network nodes that are not limited by the connection handling aspects; to that end, we are developing the *scalability enhancements for connection-oriented networks* (SCONE) project. The goal of SCONE is to develop solutions that enable the use of many connections per end-user. We consider the four following directions for developing a SCONE solution:

1. **Basic performance improvement:** with faster processor, faster, larger and cheaper memory, the performance of existing implementations can be improved.

2. **Hardware assistance:** some functions traditionally implemented in software can gain performance if casted into silicon; this would typically concern basic protocol processing functions such as timer management.

3. **Function distribution:** distributing functions among several processors and associated memory is a well known, promising method for increasing performance. Some network architectures lend themselves well to function distribution because they define functional blocks with loose coupling. For example, the ATM-forum P-NNI signaling method can be implemented with one control instance per port. Similarly, the proprietary control system of [2] defines functions for connection signaling, for bandwidth allocation that can be easily implemented as parallel modules.

4. **Architecture solutions:** by architecture solutions we mean protocols and methods that speed up the network operation. Architecture solutions can improve the performance of connection handling by reducing connection awareness wherever this is not necessary.

All these four directions are not, of course, exclusive and their effect at increasing the number of connections a network can support can be cumulated. In this paper, we consider one architectural solution (Section 2). We do not claim to be exhaustive and there are many other aspects that need consideration, such as, for example, the use of precomputation rather than on demand computation for path selection [8, 9], or Quality of Service management issues arising from multiplexing [10].

2 Reducing Connection Awareness

In this paper, we present a method that reduces connection awareness at transit nodes in an ATM network. The method is based on the following ideas:

- Connections that have same source and destination nodes are grouped together and made indistinguishable at transit points;
- Virtual Path concepts are used for grouping connections.

The method is described for point-to-point connections. Application to multicast connections is under study. In this paper, we apply our ideas to a private network environment, and adopt the framework of the ATM Forum P-NNI working group, which assumes that signaling between nodes inside a private network uses a simple extension of the UNI signaling. However, we believe that our method can equally be applied to public networks.

2.1 Virtual Path Trunks

The method is based on the concept of dynamic Virtual Path Trunks (VPTs). We use the term " trunk " to denote a link between adjacent nodes inside the network (as opposed to access links). VPTs are virtual path connections (VPCs) setup inside the network, by the network itself, for the sole purpose of reducing connection awareness inside the network. Once established, VPTs are used as trunks, namely, the network nodes connected by VPTs act as though they were physically adjacent. VPTs are not visible at the UNI. Figure 1 shows an example of VPT used by an end-to-end connection, and illustrates related concepts. Transit nodes that are not the end points of VPTs act as Virtual Path switches. We assume, as is usual in a private environment, that ATM switching nodes can act both as transit and access nodes, and can perform both virtual path and virtual channel switching [11]. We also assume that every VPT carries a signaling link (using for example VCI = 5). The signaling link needs to be established between the two entities at the end of the VPT.

In principle, VCCs can exist only inside a VPC, since the ATM cell header requires the VPI field to be populated. However, in many cases in private networks, especially at the UNI, there is no real VPC supporting the VCC, in that the VPC exists only on one single physical link. In such cases we speak of *implicit* VPCs. Implicit VPCs exist only in that a VPI value is allocated; however, they need not be visible a separate network management entities, there is no need to establish, maintain or release them, and they have no associated resources. Figure 2 shows implicit VPCs.

VPTs could in principle be concatenated and more complex scenarios could be envisioned than shown on Figure 2 (with VPTs between transit nodes for example). In this paper, we consider only cases where VPTs are used between access nodes. In particular, VPTs apply to scenarios where end users establish Virtual Channel Connections; the concept cannot be used to carry end-to-end Virtual Path Connections.

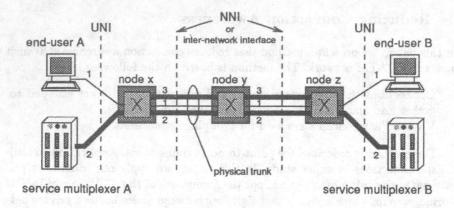

Fig. 1. Example of VCCs, VPCs and VPTs: Virtual Channel Connection 1 is setup between systems at UNIs A and B; it is made of 4 VC links (all numbered 1). Similarly Virtual Path Connection 2 is setup between systems at UNI A and B, and is made of 4 VP links. There also exists a Virtual Path Connection used as a trunk (VPT 3) between node x and z. The VPT is used internally and is not visible at UNIs.

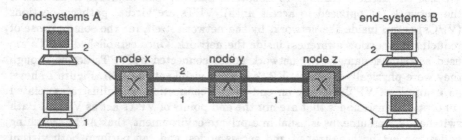

Fig. 2. Virtual Channel Connection 1 is setup hop-by-hop, using only implicit VPCs; VCC 1 is made of 4 VC links. Virtual Channel Connection 2 uses an already established Virtual Path Trunk; VCC 2 is made of 3 VC links.

2.2 Dynamic Virtual Path Trunks

Assume, as an example, that end-system A, served by node x, is establishing a virtual channel connection to end-system B. Assume also that the routing and topology function has determined that B is served by node y and that the connection should be attempted through node y (the routing and topology function is outside the scope of this paper [12]). If no VPT exists between node x and y, then a hop-by-hop setup flow like in Figure 3 would be used (with implicit VPCs on all VC links). There, node y is active at setting up, maintaining and releasing the connection. In contrast, if a VPT already exists between node x and y, and has sufficient capacity to accommodate the new VCC, then the setup flow involves only access nodes x and y (Figure 4). In that case, node y does not see the new connection.

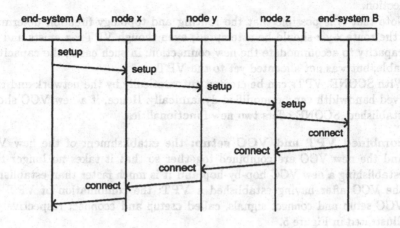

Fig. 3. VCC establishment hop-by-hop

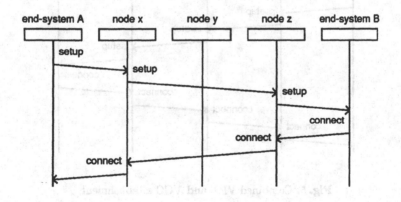

Fig. 4. VCC establishment over an existing VPT between x and z

To make the presentation simpler, signals call-proceeding and connect-ack are not shown on the figures.

In current architectures, VPTs must be created manually, in an *ad hoc* manner, and their capacity is permanently reserved. If there are no VPTs with enough bandwidth to support a new VCC, either a new VPT should be created manually or the new VCC should be established hop-by-hop.

With SCONE, we would like to take advantage of VPTs in order to decrease connection awareness at the transit point. The challenge is (1) to avoid adding overhead round-trips at VCC creation and (2) be able to use an existing VPT that does not have enough bandwidth reserved to accommodate the new connection.

Note that it is possible that the routing and topology function determines that the route x-y-z should be attempted, even though VPT x-z exists and has the capacity to accommodate the new connection: in such cases, the capacity is available, but was not allocated yet to the VPT.

With SCONE, VPTs can be created automatically by the network and their reserved bandwidth can be modified dynamically. Hence, if a new VCC should be established, SCONE offers two new functionalities:

– **combined VPT and VCC setup:** the establishment of the new VPT and the new VCC are combined together so that it takes no longer than establishing a new VCC hop-by-hop and it is much faster than establishing the VCC after having established a VPT; the combination of VPT and VCC setup and connect signals, called csetup and cconnect respectively, is illustrated in Figure 5.

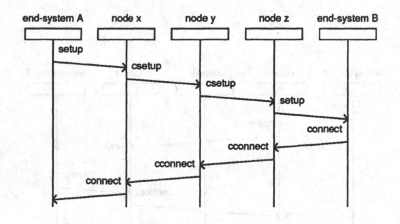

Fig. 5. Combined VPT and VCC establishment

– **simultaneous VCC setup and VPT bandwidth increase:** if an existing VPT has not enough bandwidth, this function establishes a new VCC as fast as if the VPT did not have enough bandwidth as shown in Figure 6; two new signals are introduced for this purpose: reservreq which increases the reserved VPT bandwidth in each node along the VPT and reservack which acknowledges successful reservation of all nodes as specified in the fast reservation protocol (FRP) [13]; the dotted region indicates that signals setup and reservreq are both sent and received simultaneously.[1]

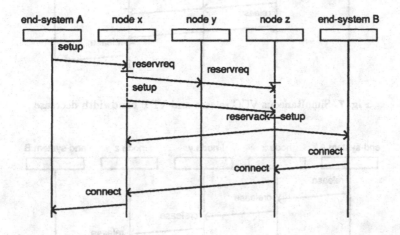

Fig. 6. Simultaneous VPT new bandwidth reservation and VCC establishment : signals within the same dotted region are simultaneous.

To simplify the presentation, signals validation-request and reservation-complete are not shown in the figures.

From a connection handling point of view, it is worth noticing that modifying the bandwidth of one VPT is considerably simpler than setting up a new connection. In particular, it can be assumed in some implementations that FRP used for that purpose is handled by the switch hardware [13].

Similarly, if a VCC should be released, SCONE offers two new functionalities :

- **simultaneous VCC release and VPT bandwidth decrease:** as shown in Figure 7, this new function takes no longer than a simple VCC release; signal release requests that resources reserved by a VCC be returned to its VPT while signal releasreq decreases the VPT bandwidth by the amount reserved by the VCC;
- **combined VPT and last VCC release:** the release of the VPT and of its last VCC are combined together so that it takes no longer than releasing a VPT and it is much faster than releasing the VPT after having released its last VCC; the combination of VCC and VPT release signals, called crelease, is shown in Figure 8.

Again, to make the presentation simpler, signals release-complete and releasreq-complete are not shown on the picture.

---

[1] Note that in Figure 6 the setup is forwarded to end-system B once both reservreq and setup are received at node z.

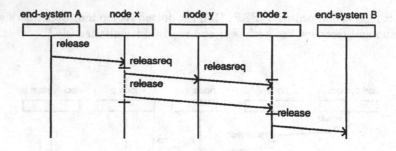

Fig. 7. Simultaneous VCC release and VPT bandwidth decrease

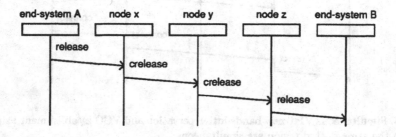

Fig. 8. Combined VPT and last VCC release

In essence, the method proposed here consists of replacing bundles of individual VCCs that follow the same route by single VPTs with variable capacity. We assume the functional model illustrated in Figure 9 for connection handling. The signaling function is called DCE on the network side of the UNI, and DXE at the P-NNI. Resources requested by setup flow are owned by Resource Managers (RM); routes are computed by the "Topology and Route Selection" (TRS). The SCONE method has a lot of potential for routing large number of VCCs in a transparent way. One particularly promising application to re-routing scenarios (for example after failure) is for future study. The new signaling functionalities are described in detail in the following section.

3 SCONE Architecture for ATM

We now introduce the different functional blocks involved in setting up and releasing VPTs and VCCs. We will only study in detail the DXE function since this is the only function affected by the addition of dynamic virtual path trunks to current architectures. A DXE supports all signaling carried by a particular trunk and is composed of the following two signaling agents:

  – a *trunk agent* (TA) that supports the signaling with the next adjacent node ;

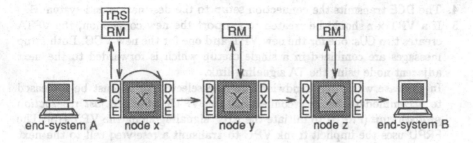

Fig. 9. Functional model for connection handling

– a *virtual path trunk agent* (VPTA) that supports the signaling with the other
extremity of each VPT starting at that trunk.

3.1 Connection Request Example

Figure 10 shows the request of a point to point VCC, where each circled number
corresponds to one of the following steps.

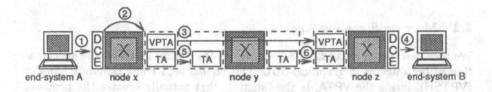

Fig. 10. Connection Request Flows: DXE functions are dashed.

1. A connection setup is issued by end-system A across the UNI to its network
   access node x; the signal is received at x by the DCE which supports ATM
   signaling with A.
2. The DCE function forwards the connection request to the DXE function which
   supports all signaling down the computed route x-y-z.
3. If VPT x-z exists and is selected to accommodate the new connection, the
   VPTA creates a *connection unit* (CU) to represent the P-NNI signaling state
   of the new connection. The VPTA uses the VPT signaling link to transmit
   a setup to the peer VPTA at the other extremity of the VPT. When the
   setup is received by the peer VPTA, a peer CU is created and receives the
   connection request.
   The VPTA is also responsible for allocating VPT bandwidth to the new
   connection.

4. The DCE transmits the connection setup to the destination end-system B.

5. If a VPT x-z should be created to support the new connection, the VPTA creates two CUs: one for the new VPT and one for the new VCC. Both setup messages are combined in a single csetup which is forwarded to the next adjacent node using the TA signaling link.

   In the case where the bandwidth for the selected VPT must be increased to accommodate the new connection, the VPTA creates a *fast reservation protocol unit* (FRPU) to initiate the FRP signaling along the VPT [13]. The FRPU uses the implicit trunk VPC to transmit a reservreq cell to the next adjacent node.

6. The processing at transit nodes is much simpler. A reservreq signal is received by the *fast reservation protocol signaling handler* (FRPSH) entity which is responsible for reserving the new VPT bandwidth. The FRPSH then forwards the reservreq to the next adjacent node using the implicit trunk VPC.

   A csetup signal is just forwarded to the next adjacent node using the TA signaling link. When the csetup is received by the destination DXE at the other extremity of the VPT, the VPTA creates two peer CUs: one for the new VPT and one for the new VCC.

In the case where a VCC is forced to be established hop-by-hop, the setup propagates through the TA signaling links and a new CU is created in the TA of both extreme nodes x and z.

3.2 More on Signaling Agents

We now describe in detail the TA and the VPTA. Figure 11 shows the specification of both agents in graphical-SDL. The *virtual path trunk signaling handler* (VPTSH), inside the VPTA, is the function that actually creates the protocol stack responsible for a VPT signaling link; this protocol stack is composed of:

- a *connection coordinator* (CC) that creates the CUs of the connections carried by a particular VPT ;
- a *service specific coordination function* (SSCF) which serves as an interface to the underlying SSCOP;
- a *service specific connection oriented protocol* (SSCOP) that provides, among other services, assured data transfer and keep-alive functions with the peer entity.

The DCE/DXE channel is used inside the node for signaling between a DCE and a DXE or between two DXEs. The VPTA uses the T channel to transmit and receive signals, such as csetup, through the TA signaling link. The NNI channel is used for signaling between nodes along the computed route.

Initially, the TA function contains one instance of all processes since we assume the trunk signaling link to be already established. However the VPTA function contains initially only the VPTSH since no pre-setup VPT signaling link is assumed.

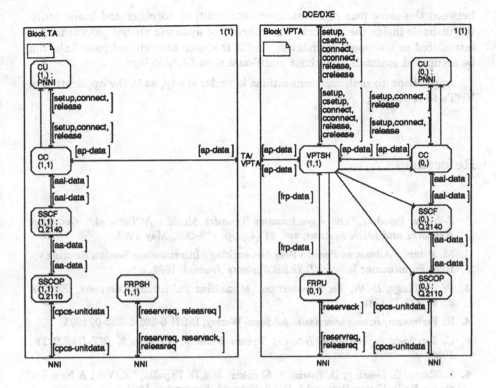

Fig. 11. Trunk and Virtual Path Trunk Agents: signals are listed between brackets, plain arrows indicate signal directions, process names are followed by their initial and maximum number of instances and sometimes by the process type, dashed arrows indicate process creation (a dashed arrow is directed from the parent process to the child process), and channel identifiers are indicated outside the block frame.

To make the figure simpler, signals call-proceeding, connect-ack, alerting, status, status-enquiry, notify, validation-request, reservation-complete, reservation-denied, release-complete and releasreq-complete are not shown. Moreover, signal lists ap-data, frp-data, aal-data, aa-data and cpcs-unitdata are not further detailed, also for the sake of simplicity.

SCONE specifications can be obtained in program-SDL as well as for SDT3.0 via anonymous ftp at `lrcftp.epfl.ch` in the directory `/pub/scone/`.

4 Conclusion

We consider the issue of increasing the number of connections that connection oriented networks, such as ATM, can handle. We describe one step that aims at reducing connection awareness inside the network. To that end, connections between the same pair of access nodes are grouped together and made indistinguishable inside the network. The concept of dynamic virtual path trunks is introduced as a support mechanism, and it is shown how virtual path links can be setup and maintained without additional round-trip delays.

Application to multicast connections is under study, as is the application of VPTs to re-routing.

References

1. J.-Y. Le Boudec, "The Asynchronous Transfer Mode : A Tutorial," *Computer Networks and ISDN Systems*, vol. 24 (4), pp. 279–309, May 1992.

2. M. Peters, "Advanced Peer-to-Peer Networking : Intermediate Session Routing vs. High Performance Routing," *IEEE Systems Journal*, 1993.

3. W. Stallings, *ISDN, An Introduction*. Macmillan Publishing Company, ISBN 0-02-415471-7, 1989.

4. R. Perlmann, *Interconnections*. Addison-Wesley, ISBN 0-201-56332-0, 1993.

5. C. Topolcic, *Experimental Internet Stream Protocol, Version 2, (ST-II)*. IETF, 1990.

6. L. Zhang, S. Deering, D. Estrin, S. Shenker, and D. Zapalla, "RSVP : A New Resource ReSerVation Protocol," *IEEE Network*, September 1993.

7. L. Gün and R. Guérin, "Bandwidth Management and Congestion Control framework of the Broadband Network architecture," *Computer Networks and ISDN Systems*, vol. 26 (1), pp. 61–78, 1993.

8. J.-Y. Le Boudec and T. Przygienda, "A Route Pre-computation Algorithm for Integrated Services Networks," Technical Report 95/113, DI-EPFL, CH-1015 Lausanne, Switzerland, February 1995.

9. O. Crochat, J.-Y. Le Boudec, and T. Przygienda, "Path Selection in ATM using Route Pre-Computation," Technical Report 95/128, DI-EPFL, CH-1015 Lausanne, Switzerland, May 1995.

10. D. Tennenhouse, "Layered Multiplexing Considered Harmful," in *Protocols for High-Speed Networks* (H. Rudin and R. Williamson, eds.), pp. 143–148, IFIP, North-Holland, 1989.

11. J.-Y. Le Boudec, E. Port, and L. Truong, "Flight of the FALCON," *IEEE Comm Mag*, pp. 50–56, February 1993.

12. J.-Y. Le Boudec and T. Przygienda, "Routing Metric for Connections with Reserved Bandwidth," *EFOC-N*, 1994.

13. P. Boyer and D. Tranchier, "A Reservation Principle with Applications to the ATM Traffic Control," *Computer Networks and ISDN Systems*, vol. 24, pp. 321–334, 1992.

Fair Queueing Algorithms for Packet Scheduling in BISDN

S. Jamaloddin Golestani

Bellcore
445 South Street
Morristown, NJ 07960, USA
jamal@thumper.bellcore.com

Abstract

This paper discusses several algorithms related to fair queueing and studies the application of fair queueing to the provision of quality of service (QOS) in broadband multi-media networks. We develop an intuitive understanding of several recent packet scheduling algorithms related to fair queueing, and show that one of them, the self-clocked fair queueing (SCFQ) algorithm, is both technically correct and practically implementable. The SCFQ algorithm provides a good analytic bound on delay, even in a multi-hop network; provides a bound on the discrepancy in service among sources (i.e., fairness); and provides isolation between sources to guarantee each source access to its allocated resources. The only computationally complex part of the algorithm is a sorting function, which can be implemented in hardware. We conclude that fair queueing provides a feasible and reliable way of satisfying QOS requirements for diverse traffic sources and applications, while maintaining a high level of network utilization. These properties make fair queueing a useful component of ATM network technology.

Keywords: Fair Queueing, Packet Scheduling, ATM, BISDN, Quality of Service.

1 Introduction

This paper studies the application of fair queueing to broadband multi-media packet networks, such as the ATM.

The packet scheduling discipline employed at the switching points of a broadband network serves as an essential control element in regulating the flow of traffic and facilitating the provision of quality of service (QOS) appropriate to different applications. The majority of current packet switched network technologies use the first-in first-out (FIFO) queueing. FIFO queueing is the simplest and most natural form of packet scheduling[1]. However, for broadband networks offering

---

[1] We use the term *packet* generically as an atomic unit of data. In an ATM network the scheduling algorithm would be applied at the ATM cell level, while in the Internet it may be applied at the IP datagram level as well.

truly integrated services, more sophisticated solutions to packet scheduling and traffic control are necessary. Priority queueing, round robin scheduling[9], virtual clock algorithm[14], fair queueing [4, 11, 7], stop-and-go queueing [6], and delay EDD [5] are among the packet scheduling disciplines that have been proposed to address different problems in providing the quality of services required for multi-media traffic in high speed communication networks.

Our purpose in this paper is to examine the application of fair queueing to packet scheduling and provision of service quality in broadband networks. In Section 2 we review and compare existing fair queueing algorithms and identify the scheme which is most suitable for broadband implementation. Section 3 discusses how fair queueing facilitates the provision of QOS in a broadband multi-media network, without sacrificing network utilization. Section 4 summarizes the conclusions of this paper.

2 Fair Queueing Algorithms

The fair queueing service discipline was originally developed in an attempt to maintain fairness in the amount of service provided by a server to the competing users. Unlike FIFO queueing where a session can increase its share of service by presenting more demand and keeping a larger number of packets in the queue, the underlying idea in fair queueing is to serve each session in proportion to its service share, independent of the load presented. The service share of sessions are negotiated and assigned to them in advance. *Round robin* service discipline [9], which is an early form of fair queueing, assumes an equal service share for all the sessions and an equal length for all the packets. It therefore provides service to the sessions in a round robin fashion, picking one packet for service from each session with backlogged traffic, and proceeding to the next session.

When the length of packets are not the same and/or the service shares assigned to the sessions are not equal, the definition of the fair queueing[2] and the right order of providing service to the sessions become a more subtle matter. Over the past few years, several studies have been devoted to the formulation and development of fair queueing algorithms suitable for this more general scenario. However, most of the schemes proposed either fail to provide fairness in a true sense or involve a high degree of computational complexity. In this section, we review and compare some of the schemes proposed in connection with fair queueing, and then focus on a more recent proposal [7] which combines fairness with generality of operating conditions and ease of implementation.

2.1 Fluid-Flow Idealization

One approach to formulating fair queueing for the case of unequal session shares is the *packet-by-packet fair queueing* (PFQ) scheme proposed by Demers et al. [4],

---

[2] Sometimes, the term fair queueing is used to refer to the special case where session shares are equal and *weighted fair queueing* is used for the general case where session shares are not equal. In this paper, we use fair queueing to refer to both cases.

and later used by Parekh and Gallager [11]. In this approach, the notion of fairness is first applied to an idealized fluid-flow network model, and then the outcome is used to specify fair queueing for the actual packet-based traffic scenario. With a fluid flow model of traffic, the service may be offered to sessions in arbitrarily small increments. Equivalently, it may be assumed that multiple sessions can receive service in parallel. As the result, it is possible to divide the service among the sessions, at all times, exactly in proportion to the specified service shares. We refer to this idealized form of service discipline as *fluid-flow fair queueing* (FFQ). This scheme is referred to in [11, 12] as *generalized processor sharing* (GPS).

To describe the FFQ scheme more clearly, consider the set of sessions using a given transmission link. Denote r_k as the share of service assigned to session k, and C as the transmission rate of the link. Let $W_k(t_1, t_2)$ stand for the amount of traffic served for session k during (t_1, t_2). We refer to a session as backlogged if it has packets either receiving service or waiting for service in the queue. As said earlier, in the fluid-flow traffic scenario, it is possible to make sure that, for all backlogged sessions, the received service normalized to the assigned share remain equal. In other words, given an arbitrary interval (t_1, t_2), and for any pair of sessions k and j that are backlogged during this entire interval, we may require that the following holds:

$$\frac{W_k(t_1, t_2)}{W_j(t_1, t_2)} = \frac{r_k}{r_j}. \tag{1}$$

Or, equivalently,

$$\frac{W_k(t_1, t_2)}{r_k} = \text{Constant}, \tag{2}$$

for all sessions k which are backlogged during (t_1, t_2). Eqs. (1) or (2) constitute the definition of fairness for the case of fluid flow model [11].

To illustrate the FFQ scheme, consider an example of three sessions a, b, and c with assigned service shares $r_a = r_b = 1$, and $r_c = 2$, sharing a link of rate $C = 4$. Let all packets have the length $L = 4$. Starting with an idle system, the sessions a, b, and c become active at times $t = 0$, $t = 1.5$, and $t = 4$, respectively, as shown in Fig. 1a. During the interval from $t = 0$ to $t = 1.5$, the transmission capacity is exclusively used by session a, and its packet is served in time $L/C = 1$ sec. During the interval from $t = 1.5$ to $t = 4$, the capacity is equally shared between sessions a and b. Then, in the interval after $t = 4$, the link is shared among all three sessions.

Let us now consider the service received by session a, which remains backlogged during all three intervals. This session is served, first at the link's full transmission rate C, then at the rate $C/2$, and finally at the rate $C/4$. Fig. 1b illustrates the rate of service provided to session a after time $t = 0$. Because the queueing scheme is fair, any other backlogged session with an equal session share would have been served at the same rate during the above intervals. Since $r_a = 1$, Fig. 1b also represents the rate of service in the system per unit share of service or, alternatively put, the normalized rate of service to a backlogged session.

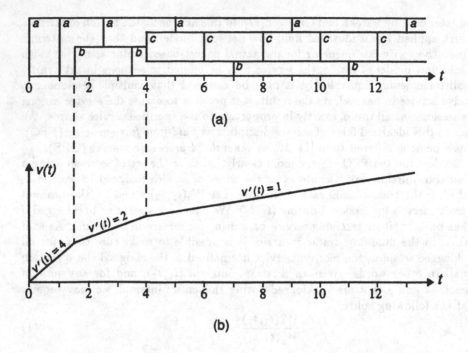

Figure 1: a) Example of a fluid-flow fair queueing system showing how service is provided to the backlogged sesions in parallel and in proportion to the corresponding service shares. b) The virtual time function $v(t)$ representing the rate of service, per unit share, for any backlogged session.

In general, for any FFQ system, a curve similar to Fig. 1b may be constructed. This curve, which is denoted by $v(t)$, is called *virtual time* of the system [11]. $v(t)$ is a precise indication of the progress of service per unit share for a backlogged user in the FFQ system. It can be shown that $v(t)$ is a piecewise linear function with its slope at any point of time t inversely proportional to the sum of the service shares of the sessions which are backlogged at t.

2.2 Packet-by-Packet Fair Queueing (PFQ)

Obviously, fluid-flow fair queueing cannot be applied to the actual packet-based traffic scenarios, where only one session can receive service at a time, and where an entire packet must be served before another one is picked up for service. In this case, Eqs. (1) and (2) cannot be satisfied exactly and for all periods (t_1, t_2). However, it is possible to keep the normalized services $W_k(t_1, t_2)/r_k$ received by different backlogged sessions k close to each other. Depending on exactly how one tries to accomplish this objective, different fair queueing algorithms emerge. The *packet-by-packet fair queueing* (PFQ) algorithm, proposed by Demers86 et al. [4], and Parekh and Gallager [11], is defined in the following manner. First a hypothetical FFQ system with the same server speed and an identical pattern of traffic arrivals is considered and the order that packets *finish service* in accordance with the fair queueing rule of Eq. (2) is determined. In the actual packet-based network, the same order is used to offer service to the packets in the queue. This definition is identical to Parekh's *packet-by-packet generalized processor sharing* (PGPS) [11].

The most compact realization known for the PFQ scheme is an algorithm developed in [11], which is based on calculating a service tag for each arriving packet, and serving the queued packets in increasing order of the associated service tags. For a given session k, the service tag of different packets are recursively computed as

$$F_k^i = \frac{L_k^i}{r_k} + \max\left(F_k^{i-1}, v(a_k^i)\right),\qquad (3)$$

where

P_k^i = the ith packet of session k,
L_k^i = the length of packet P_k^i,
a_k^i = the arrival time of packet P_k^i,
$v(t)$ = the virtual time function of the corresponding FFQ system,
$F_k^0 = 0$, for all k.

In order to compute the packet service tags, we need to know $v(t)$ for the corresponding FFQ system. Since $v(t)$ is a piecewise linear function, the complexity of its computation depends on the frequency of its breakpoints. Although the breakpoints of $v(t)$ may be infrequent on average, it is possible for a large number of breakpoints to occur during a single packet transmission time, or time slot [7]. This behavior will prevent an implementation of PFQ from operating reliably at broadband transmission speeds.

2.3 Virtual Clock (VC)

Zhang's *virtual clock* (VC) algorithm [14] constitutes an important, albeit incomplete step towards the formulation of fair queueing based on the notion of virtual time. This scheme also serves packets in increasing order of embedded service tags. The algorithm for computing service tags has a few variations but the main idea is captured by the following version:

$$F_k^i = \frac{L_k^i}{r_k} + \max\left(F_k^{i-1}, a_k^i\right).\qquad (4)$$

Although Eqs. (3) and (4) are very similar, the VC algorithm does not share the fairness property of the PFQ algorithm. This is due to a small, albeit important, difference between the two algorithms, i.e., the use of the virtual time $v(a_k^i)$ in the PFQ algorithm in lieu of the real time a_k^i used in the VC algorithm. The negative impact of this replacement on the fairness property of the algorithm will be intuitively clarified in Appendix A, where we discuss a simple example in which the VC algorithm fails to achieve fairness.

2.4 Self-Clocked Fair Queueing (SCFQ)

The virtual clock scheme fails to provide fairness among the users and the PFQ scheme leads to considerable computational complexity, especially at high transmission speeds, as explained earlier. The desired properties of fairness and ease of implementation are uniquely combined in the self-clocked fair queueing (SCFQ) scheme proposed recently[3] [7]. The source of complexity in the PFQ scheme is that it defines fairness in reference to the events in the hypothetical FFQ system, which creates the need for simulating events in that system and computing the corresponding virtual time $v(t)$. This problem is alleviated in the SCFQ scheme by adopting a self-contained approach to the definition of fairness. Like the PFQ scheme, the SCFQ scheme is also based on the notion of system's virtual time, viewed as the indicator of progress of work in the system, except that the measure of virtual time here is found in the actual queueing system itself, rather than being derived from a hypothetical system.

For the SCFQ scheme the service tag is computed as

$$F_k^i = \frac{L_k^i}{r_k} + \max\left(F_k^{i-1}, \hat{v}(a_k^i)\right), \qquad (5)$$

where the new virtual time $\hat{v}(t)$ is defined as

$\hat{v}(t) =$ the service tag of the packet in service at time t.

The same definitions and conditions used for Eq. 3 apply here. Whenever the server becomes idle and no packets are queued, the algorithm may be re-initialized by setting $\hat{v}(t) = 0$, and resetting the last tag for each session, $F_k^i = 0$ for all k.

The above algorithm is similar to the algorithm needed to implement the PFQ scheme, except for the important difference that $v(a_k^i)$ in (3) has been replaced by $\hat{v}(a_k^i)$ in (5). The block diagram of Fig. 2 illustrates the relationship between the service tag assigned to the arriving packets, the virtual time $\hat{v}(t)$ of the system, and the service tag of the packet in transmission. Unlike the extensive computations needed to evaluate $v(t)$ in Eq. (3), in the case of the SCFQ algorithm the virtual time $\hat{v}(t)$ is simply extracted from the packet situated at the head of the queue. It follows that the only major complexity in the implementation of the SCFQ algorithm is the maintenance of a sorted list of packets, based on the corresponding tags. A software search may be unreasonably complex at broadband speeds (e.g. 2.73 μsec for an ATM cell at 155.5 Mb/s). A VLSI chip has been prototyped which can perform this function at an order of magnitude faster than required for ATM [1].

We now consider the issue of fairness and see whether the SCFQ algorithm conforms to our expectations. It has been shown [7] that the services provided to

---

[3] We subsequently learned that an algorithm similar to the SCFQ was used in a simulation experiment [10] but the algorithm itself was never explained in the literature. We have determined that the version of the algorithm used in this simulation study is not fair due a fine but important difference with the SCFQ algorithm described in this paper.

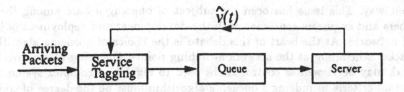

Figure 2: Schematic illustration of the SCFQ algorithm. The system's virtual time at t, which is needed to compute the service tag of a packet arriving at t, is simply extracted from the packet receiving service at t.

different sessions by the above algorithm always satisfy the following relationship:

$$\left| \frac{W_k(t_1, t_2)}{r_k} - \frac{W_j(t_1, t_2)}{r_j} \right| < \frac{L_k^{max}}{r_k} + \frac{L_j^{max}}{r_j}. \tag{6}$$

Here, k and j are a pair of sessions which are backlogged during the interval (t_1, t_2), and L_k^{max} and L_j^{max} are the maximum possible length of a packet from sessions k and j, respectively. Eq. (6) is the statement of fairness for the SCFQ algorithm. It confirms that the normalized services provided to different sessions over *any* interval of time will remain within a fixed maximum discrepancy from each other, as long as the compared sessions remain backlogged during that interval. As the duration of the interval (t_1, t_2) increases, the impact of this discrepancy on the average service rates vanishes. To take an example we consider an ATM network and compare two backlogged sessions k and j with the service shares $r_k = r_j = 1$. It follows from (6) that

$$\left| W_k(t_1, t_2) - W_j(t_1, t_2) \right| < 2L, \tag{7}$$

where L is the size of an ATM cell. It is conceivable that this bound could be reduced to one cell through a more constrained algorithm, however it cannot be reduced below L. Consider (t_1, t_2) to be the period of service to a cell from session k. Then, during this interval, session k receives service L and no other session gets any service, by the non-preemptive nature of packet switching. It is shown [7] that, in general, the SCFQ bound of Eq. (6) is within a factor 2 of the best possible bound for any queueing scheme.

While a formal proof of the fairness properties of the SCFQ algorithm can be found in [7], in Appendix A we use an example to develop intuitive understanding about how fairness is achieved by this algorithm.

3   Packet Scheduling and Quality of Service

In this section, we discuss how the provision of quality of service (QOS) to the multi-media traffic can be facilitated by fair queueing. One of the most challenging issues in the design of broadband packet networks, such as the ATM, is the provision of the QOS required by the multi-media applications, in a reliable and efficient way. This issue has been the subject of ongoing debate among the researchers and engineers concerned with the development and deployment of high speed networks. At the heart of this debate is the choice of queueing algorithms for packet scheduling at the network switching points. Queueing algorithms in a network may be viewed as control knobs used to regulate a complex system. An important criteria in judging a queueing algorithm must be the degree of control it provides over the course of events and the QOS in the network. Another criteria is the simplicity of its implementation.

The most natural form of packet scheduling used in the majority of present packet networks is first-in first-out (FIFO) queueing. While very simple to implement, FIFO queueing suffers from severe limitations in the amount of control that it provides over the provision of service in the network. These limitations stem from the basic fact that, in FIFO queueing, service is not provided to the competing sessions in accordance with any pre-specified set of service parameters. Any session which loads more packets into a FIFO queue is guaranteed to receive more service. This type of queue behavior encourages heavier traffic from the users, which contributes to the build up of congestion.

The degree of control over the QOS in a FIFO queueing packet network may be considerably improved by enforcing stringent packet admission and call set up policies. Packet admission policies, such as *leaky bucket* [13], usually place a cap on the long-term average rate, as well as the short-term burstiness of the traffic. By subjecting each session's traffic to such limitations, the burstiness of the total traffic at the edge of the network is effectively controlled, helping to achieve a target quality of service in terms of delay, loss, and etc. However, this approach is impaired by two limitations. First, enforcement of stringent admission policies decreases the room for statistical multiplexing and lowers the network utilization. Second, limiting the traffic burstiness at the edge of a FIFO queueing network is not always enough to predict the burstiness of the traffic as it proceeds inside the network. Packet clustering and the formation of longer bursts is a well-known behavior of networks equipped with FIFO queueing [3, 6].

A second approach to improving controllability of the QOS in a FIFO queueing network is to depart from the pure FIFO discipline and to incorporate priorities into the queueing mechanism. This solution helps to attain more predictable QOS for the high priority traffic, at the cost of less reliability in serving the traffic with low priority. While adding priorities improves the control over the QOS yielded by FIFO queueing, it may not always be sufficient to address the stringent requirements of heterogeneous multi-media traffic, in a sufficiently reliable and efficient way.

3.1 Provision of Quality of Service by Fair Queueing

In so far as the provision of QOS and control over the network is concerned, fair queueing is in sharp contrast to FIFO queueing. By serving the traffic from different sessions in proportion to the assigned service shares, fair queueing creates a mechanism for resource sharing while maintaining a minimum throughput for each backlogged user. More generally, the session service shares, which may be assigned and modified by the network management algorithm, supply a vast set of control knobs with which the course of events in a fair queueing network can be shaped.

The above intuitive properties of fair queueing have recently been translated into solid mathematical bounds on the maximum end-to-end delays in the network. It is shown that fair queueing, when employed in conjunction with an appropriate packet admission policy such as the leaky bucket, can guarantee maximum session backlogs and end-to-end delays in the network. This important result, originally established for the FFQ and PFQ schemes [11, 12], remains valid under the SCFQ scheme [8]. The bound on the session backlogs at the network nodes can be converted into a bound on packet loss probabilities, once the cell discarding policy is specified. In the case of FIFO queueing networks, it has not been possible to derive similar guarantees on the loss and delay performance.

Another important property of fair queueing networks is the following: session service shares may be assigned in such a way that the loss and delay guarantees for a given session remain valid independent of the admission policies enforced on other sessions [12, 8]. With such an arrangement, the worst-case delay and loss performance of a session would only depend on the arrival of traffic from the session itself. It turns out that the choice of packet admission policy for a given session is less critical in the case of fair queueing networks. The network can be engineered so that the main impact of the admitted traffic from a session (or class of sessions) is on the performance of that session (or class of sessions) itself. This property may be exploited to improve the statistical multiplexing gain, by allowing more bursty arrivals from sessions with less stringent service quality requirements.

We conclude that fair queueing offers an attractive packet scheduling mechanism for resource sharing among users with incompatible traffic characteristics and QOS demands. It provides a reliable way of satisfying QOS requirements in a multi-media environment while maintaining a high level of network utilization.

3.2 Fair Queueing Applied to Different Classes of Sessions

So far in this paper, we have looked at fair queueing as a way of achieving fairness among individual *sessions* which share a given resource. In a broadband network, the number of sessions sharing a particular communication link may be in the hundreds, or even higher. While applying fair queueing among individual sessions provides the highest number of control parameters to regulate traffic, for most cases this degree of control is not necessary. An alternative and more pragmatic way of applying fair queueing is to first classify sessions into *classes* each consisting of sessions with compatible traffic characteristics and service requirements. Fair queueing can then be used to achieve fairness among the services offered to these session classes, based on service shares collectively assigned to each class. Within each class, a simpler discipline such as the FIFO may be applied to divide the available service among individual sessions, since the characteristics and QOS requirements of sessions within a class are compatible. This approach has been proposed by Clark, Shenker, and Zhang [2]. It greatly simplifies the implementation of fair queueing, since there will be far fewer session classes than there are sessions in a network.

4 Conclusion

We have discussed the basic operation of four packet scheduling algorithms related to fair queueing, FFQ, PFQ, VC, and SCFQ. One of these algorithms, SCFQ, combines fairness with feasibility of implementation at broadband speeds. We have compared fair queueing with FIFO queueing and concluded that fair queueing greatly facilitates the reliable provision of QOS to multi-media traffic while maintaining a higher level of network utilization. While, the implementation of SCFQ algorithm is feasible at broadband speeds, the associated cost is higher compared to FIFO queueing. However, this additional cost should be weighed against the potential benefit of higher network utilization under fair queueing. We hope that this paper, by providing a clear perspective on different aspects of fair queueing, contributes to the objective choice of packet scheduling algorithms for broadband multi-media networks.

Appendix A   An Illustrative Example

In this appendix we develop a qualitative understanding as to how the fairness is enforced by means of computing service tags, in accordance with (5). It is in fact surprising that, in spite of the drastic difference between Eqs. (3) and (5) regarding the interpretation of virtual time and the replacement of $v(t)$ by $\hat{v}(t)$, the SCFQ scheme retains fairness.

Some insight into the issue may be gained by first considering a simpler queueing scheme, herein referred to as the *naive* algorithm, in which the packet service tags are computed as,

$$F_k^i = \frac{1}{r_k}L_k^i + F_k^{i-1}, \qquad i = 1, 2, \cdots, \tag{8}$$

with $F_k^0 = 0$. Consider now a fixed session k and the sequence of packets P_k^1, P_k^2, $P_k^3, \ldots$, arriving on this session. The service tag given to the first packet is equal to L_k^1/r_k, the tag assigned to the second packet is equal to $L_k^1/r_k + L_k^2/r_k$, and so on. In general, each time a new packet arrives, the service tag is incremented by the normalized length of that packet. It follows that in this scheme, each time a packet P_k^i finishes service, its service tag F_k^i becomes equal to the total normalized service provided to session k, up to that time. Therefore, by always offering service to the packet with the lowest service tag in the queue, the scheme tries to equate the normalized services of all the sessions. This attempt is made regardless of how long each session has been backlogged or absent.

The above arrangement in effect leads to the accumulation of service credit by the absent sessions. To see this point more clearly, we consider an example. Let there be three sessions a, b, and c, with the session shares $r_a = 1$, $r_b = 2$, and $r_c = 4$, sharing a transmission link with the rate $C = 8$. For computational simplicity, let the length of packets of any session be always $L = 8$. Accordingly, each packet takes 1 second to be transmitted. Notice that the term L_k^i/r_k in (8), which represents the service tag increments, is equal to 8, 4, and 2, for packets of sessions a, b, and c, respectively. Assume that, at time $t = 0$, sessions a and b become backlogged with the arrival of 4 packets from a and 8 packets from b. We further assume that session c does not become backlogged until $t = 6$. According to (8), packets of session a will receive the service tags 8, 16, 24, and 32. Similarly, the service tags $4, 8, \ldots,$ and 32 will be assigned to the packets

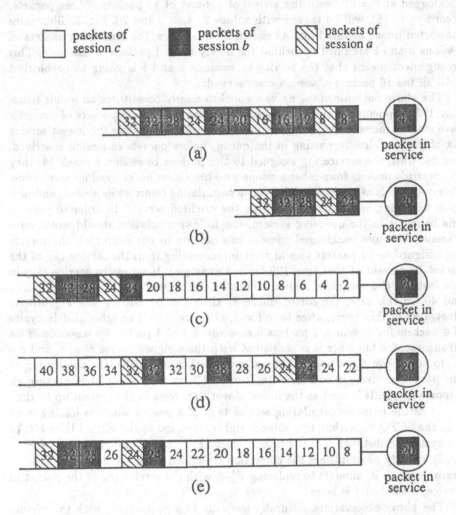

Figure 3: Comparison of packet scheduling based on the naive, SCFQ, and virtual clock algorithms for the example considered in the text. Illustrated in the figures are the queue at different times, sorted in terms of the packet service tags. a) Queue at $t = 0$, for either of the 3 algorithms. b) Queue at $t = 6$, for either of the 3 algorithms. c) Queue just afetr $t = 6$, for the naive algorithm. d) Queue just afetr $t = 6$, for the SCFQ algorithm. e) Queue just afetr $t = 6$, for the virtual clock algorithm.

arriving from session b. In Fig. 3a, we have illustrated the packets in the queue at time $t = 0$, sorted in increasing order of their service tags. This figure shows the order in which these packets will be served, as long as there is no new packet arrivals. We see that, so far, the provision of service to the sessions in accordance with the naive algorithm (8) is fair since 1 packet from session a and 2 packets from session b will be served, alternately.

At time $t = 6$, after 6 packets have been served and the 7'th packet has just been picked up for service, the queue will be like Fig. 3b. Now let session c become backlogged at $t = 6^+$, with the arrival of a burst of 10 packets. These packets, according to (8), will be tagged with values 2, 4, ..., and 20. Fig. 3c illustrates the sorted queue at this time. As shown in this figure, the remaining packets of sessions a and b are lined up behind the newly arrived packets of session c. This arrangement means that the service to sessions a and b is going to be blocked until all the 10 packets of session c are served.

The above outcome of the naive algorithm clearly constitutes an unfair situation. It is happening because the service tags assigned to the packets of session c have started incrementing from the initial value $F_c^0 = 0$, while the lowest service tag of packets already waiting in the queue, before packets of session c arrived, was 24. Until the service tag assigned to the packets of session c reach 24, they will override packets from other sessions who have been backlogged for some time. This outcome is as if session c has been accumulating credit while absent, and now upon becoming backlogged is receiving the credited service. In order to prevent this behavior in the queueing system, the following solution should work: once a session becomes backlogged anew (like session c in our example), the service tag assigned to its packets should start incrementing from the service tag of the packet being sent at that time (20 for our example). If we assign service tags in this fashion, the packets of session c in our example will receive tags 22, 24, ..., and 40. In this case, the sorted queue at time $t = 6^+$ will be like Fig. 3d. As illustrated in this figure, after time $t = 6$, the service will be scheduled in cycles of 4 packets for session c, 2 packets for session b, and 1 packet for session a. This arrangement is fair since it is consistent with the assigned shares of a, b, and c.

In the above correction to the naive algorithm of Eq. (8), the service tag of the packet receiving transmission at time t (20 in our example) is playing an important role. It is used as the index of work progress in the system up to time t, and as the basis for calculating service tags of a session which is joining in at t. In the SCFQ algorithm, this value is rightly denoted as the virtual time $\hat{v}(t)$ of the system. Substitution of the term $max\left(F_k^{i-1}, \hat{v}(a_k^i)\right)$ in (5), in lieu of F_k^{i-1} in (8), exactly accomplishes the correction described in the context of the above example, since it amounts to replacing F_k^{i-1} with the service tag of the packet in service, if the latter is larger.

The above observations naturally leads us to a comparison with the virtual clock algorithm. Zhang [14] after noting the unfair outcome of computing service tags in accordance with the naive algorithm of Eq. (8), suggests that the term F_k^{i-1} in (8) should be replaced with $max\left(F_k^{i-1}, a_k^i\right)$, to circumvent the problem of credit accumulation by the bursty users. The result is the service tag computation algorithm (4). Compared with the SCFQ scheme, the virtual clock algorithm uses the real time a_k^i in (4), in lieu of the virtual time $\hat{v}(a_k^i)$ in (5), in order to account for the time a session has been absent prior to becoming backlogged at a_k^i. The best way to see why this approach does not work is to carry on with our previous example and consider the performance delivered by the virtual clock algorithm. The application of virtual clock algorithm to our example does not cause any difference in the service tag of packets arrived before $t = 6$. When the first packet of session c arrives at time $t = 6^+$, we have $a_c^1 = 6$. It follows from (4) that packets of session c will receive tags 8, 10, ..., and 26. The sorted queue at $t = 6^+$ will now be like Fig. 3e. According to this new arrangement of the queue, 9 consecutive packets of session c will have to be served before the service of sessions a and b is resumed. By comparing Fig. 3c and Fig. 3e, we see that virtual clock algorithm does not provide sufficient improvement over the naive algorithm. Specifically, after $t = 7$, it fails to provide service to sessions a and b at the allocated rate. The reason is that unlike the virtual time $\hat{v}(a_k^i)$, the real time a_k^i is not an authentic measure of the progress of work in the queueing system up to time a_k^i.

References

1. H. J. Chao. A novel architecture for queue management in the ATM network. *IEEE Journal on Selected Areas in Communications*, 9(7):1110–1118, September 1991.
2. D. D. Clark, S. Shenker, and L. Zhang. Supporting real-time applications in an integrated services packet network: Architecture and mechanism. In *ACM SIGComm Symp.*, pages 14–26, 1992.
3. R. L. Cruz. A calculus for network delay, part I: Network elements in isolation. *IEEE Transactions on Information Theory*, 37(1):114–131, January 1991.
4. A. Demers, S. Keshav, and S. Shenkar. Analysis and simulation of a fair queueing algorithm. In *Proc. SIGCOMM'89*, pages 1–12, Austin, Texas, September 1989.
5. D. Ferrari and D. Verma. A scheme for real-time channel establishment in wide-area networks. *IEEE Journal on Selected Areas in Communications*, 8(3):368–379, 1990.
6. S. J. Golestani. Congestion free communication in high speed packet networks. *IEEE Transactions on Communications*, 32(12):1802–1812, December 1991.
7. S. J. Golestani. A self-clocked fair queueing scheme for broadband applications. In *IEEE INFOCOM'94*, pages 636–646, 1994.
8. S. J. Golestani. Network delay analysis of a class of fair queueing algorithms. *IEEE Journal on Selected Areas in Communications*, 13(6):1057–1070, August 1995.
9. E. L. Hahne. *Round Robin Scheduling for Fair Flow Control*. PhD thesis, Department of Electrical Engineering and Computer Science, MIT, December 1986.
10. A. T. Heybey and J. R. Davin. A simulation study of fair queueing and policy enforcement. *ACM Comp. Comm. Rev.*, 20(5), October 1990.
11. A. K. Parekh and R. G. Gallager. A generalized processor sharing approach to flow control in integrated services networks: The multiple node case. In *Proc. IEEE INFOCOM'93*, pages 521–530, 1993.
12. A. K. Parekh and R. G. Gallager. A generalized processor sharing approach to flow control in integrated services networks: The multiple node case. *ACM/IEEE Transaction on Networking*, 2(2):137–150, April 1994.
13. J. S. Turner. New directions in communications or which way to the information age. *IEEE Communications Magazine*, 24(10):8–15, October 1986.
14. L. Zhang. Virtual clock: A new traffic control algorithm for packet switching. *ACM Transactions on Computer Systems*, 9(2):101–124, May 1991.

Burstiness Bounds Based Multiplexing Schemes for VBR Video Connections in the B-ISDN

Maher Hamdi* and James W. Roberts**

* Télécom Bretagne, Department of Networks and Multimedia Services

BP 78 35512 Cesson Sévigné, France

Tel: (+33) 99 12 70 23 Fax: (+33) 99 12 70 30 Email: Maher.Hamdi@enst-bretagne.fr

**France Télécom (CNET)

38 Avenue du Général Leclerc, 92131 Issy les Moulineaux, France

Tel: (+33) 1 45 29 57 01 Fax: (+33) 1 45 29 60 69 Email: James.Roberts@issy.cnet.fr

Abstract

We consider variable bit rate video connections in the B-ISDN where the quality of service in terms of data loss and network delay must be guaranteed using appropriate congestion control mechanisms. We follow a preventive traffic control strategy based on the notion of traffic contract and connection accecptance control. We suggest that video connections be characterized by a set of burstiness constraints that define the traffic contract. We show how video coders can ensure that their output conforms to the traffic contract thus avoiding data loss at the user-network interface. Based on these burstiness constraints, we propose simple and easily implementable ressource allocation schemes that allow the QoS parameters to be guaranteed.
Keywords: B-ISDN, Traffic Control, Mpeg, Burstiness Function, GCRA, QoS Guarantee, VBR.

1 Introduction

The Broadband ISDN is expected to provide an efficient solution to the transport of Variable Bit Rate (VBR) video traffic. Statistical multiplexing of video traffic, however, has proved to be a rather difficult problem in terms of resource management and traffic control. While it is widely accepted that a statistical multiplexing gain can be achieved from the superposition of independant video sources [9], [10], appropriate traffic control procedures remain to be designed to ensure the Quality of Service (QoS) required by video applications.

We assume traffic control is restricted to preventive actions which aim to guarantee QoS standards by admission control and policing [14]. New connections are set up only if their declared traffic, added to existing traffic, would not lead to congestion. The source traffic must be policed to ensure that the traffic declaration is adhered to. This is the principle adopted for the B-ISDN (with the exception of the newly defined ABR service class).

Network performance depends on traffic characteristics. Many models have been proposed to characterize VBR video traffic [10], [15], [18]. Most of these models are based on specific video trace and cannot be easily policed by the network (see Heeke [9] for a policeable model based on a Markov chain).

In this paper we propose the use of a non-stochastic parameters as traffic descriptor for VBR video. The burstiness function defines a set of bounds on video traffic burstiness. These bounds are both enforceable at the coder and controllable by the network.

The paper is organized as follows: In section 2 we discuss the ability to control ATM multiplexer performance and introduce the burstiness function of a video traffic. In section 3 we describe the Mpeg standard and its rate control algorithms. In section 4 we show how Mpeg coders can enforce their output to fit in the traffic contract and how conformity can be controlled by the network. Section 5 presents multiplexing schemes based on the traffic contract parameters and allowing QoS guarantees.

2 Multiplexer Performance and QoS

ATM network performance cannot be studied in isolation from its impact on video communications. For video connections, data loss rate and network delay are the main QoS parameters.

The effect on end to end image quality of packet loss is not yet well defined. The effect of cell loss is not only dependent on the average cell loss rate but also on the distribution of the cell losses over time. Periods of high cell loss due to network congestion can have a serious detrimental impact on image quality.

Delay requirements clearly vary depending on the application. For interactive video communication applications a maximum end to end delay of some 250 ms is appropriate while a much longer delay would be tolerable for a user simply watching a video sequence. This delay must be known beforehands to be able to dimension the decoder playback buffer. It is for these reasons that a number of video applications (especially interactive video) require a guaranteed service that ensures the cell loss rate and multiplexer delays to be within some negotiated values.

There clearly arises a need for compromise between network efficiency and image quality. In the following we consider the relation between traffic characteristics and network multiplexer performance and discuss the possibilities for traffic control to guarantee QoS standards.

Studies on the performance of ATM multiplexers handling variable bit rate traffic show that there are broadly two types of congestion leading to cell delays (e.g., [19]).

The first type of congestion occurs when the combined rate of all multiplexed sources is less than multiplexer output rate. Delays then occur due to the coincident arrival of cells from different sources. These delays are of short duration, generally less than the time to transmit a few tens of cells (e.g. less than 1 ms). This type of congestion is referred to as *cell scale congestion*.

The second type of congestion occurs when the combined rate exceeds the output rate, for a period greater than the duration of a video frame, say. The delays here are typically much longer than those occurring in cell scale congestion. This type of congestion is known as *burst scale congestion*.

Multiplexer performance can be controlled by ensuring that the probability of a rate overload leading to burst scale congestion is negligible. The only delay to be taken into account is then the small delay due to cell scale congestion. If the buffer is long enough only to absorb cell scale congestion, periods of burst scale congestion lead to cell loss. The cell loss ratio is then given by the following fluid approximation. Consider a set of VBR sources and let their combined bit rate at time t be Λ_t (we assume the notion of bit rate is clear from the source type, e.g., a video source may have a constant bit rate throughout a given frame). The cell loss ratio is then approximated as:

$$CLR = \frac{E\{(\Lambda_t - C)^+\}}{E\{\Lambda_t\}} \tag{1}$$

where C is the output link capacity and $E\{\cdot\}$ denotes the expectation function. If the bit rate distribution of individual sources is known, it is possible to guarantee the cell loss ratio by performing connection admission control.

Burst scale congestion can be controlled by imposing constraints on the input traffic. In particular if a source i whose instantaneous bit rate is denoted by $R_i(t)$ entering a multiplexer verifies the burstiness constraint:

$$\forall s \leq t, \int_s^t (R_i(t) - c_i)dt \leq b_i \tag{2}$$

and the multiplexer output is greater than $\sum_i c_i$, then the queue length is bounded by $\sum_i b_i$ [2]. Delay is bounded and data loss can be avoided by providing a buffer greater than $\sum_i b_i$.

More generally, given a connection characterized by its instantaneous bit rate $R(t)$, a function $b(c)$ is called a *burstiness function* of the underlying connection if it satisfies:

$$\forall s \leq t, \int_s^t (R(t) - c)dt \leq b(c)$$

This function defines a set of burstiness constraints and has already been used in [2]. Note that if the traffic defined by $R(t)$ is offered to a server of fixed service rate c, data loss is avoided if the buffer length is greater than or equal to $b(c)$.

We suggest that the traffic contract for video connections be defined by the burstiness function $b(c)$, described at the end of the previous section. The importance of this function has already been mentioned in number of papers [2] [17] [16]. In particular, it has been used to compare the burstiness of two video connections [17]. We believe it is well suited to defining a traffic contract because it is closely related to the widely used leaky bucket mechanism.

Because traffic passing through LB(c,b) satisfies the burstiness constraint of expression 2, a traffic having $b(c)$ as burstiness function necessarily conforms to the set of leaky bucket controllers defined by LB($c,b(c)$). The advantage of this relationship between burstiness function and leaky bucket is that we can make use of the already standardized Generic Cell Rate Algorithm[14].

In the next section, we describe the essential features of the Mpeg standard and its main rate control algorithms.

3   Mpeg Coding and Rate Control

Mpeg is the ISO/IEC standard for video compression and has been designed to satisfy a large variety of video applications. It is becoming very popular and is currently used in a number of video communication services including video on demand and WWW browsers. Mpeg-1 is suited for mass video storage and retrieval systems and is adapted to rates up to 1.5 Mbits/s. More recently Mpeg-2 was chosen as a broadcast TV quality standard.

The Mpeg video syntax defines the group of pictures structure (GoP) containing three types of frames I, P and B.

The coding algorithm is complex and based on a division of each picture into blocks, groups of blocks and macroblocks. Each macroblock is coded with respect to the choice of a quantization parameter Q which determines spatial resolution. Bit rate and image quality decrease with increasing Q.

The Mpeg standard offers two coding options: CBR coding allowing the generated signal to be transmitted at constant rate with bounded delay and VBR coding where output rate variations are only constrained by the peak rate.

Full details can be found in the standards [11] [12] [13]; for a more readable presentation, see[3].

3.1   CBR Coding

Codecs for video transmission have traditionally aimed to produce a constant bit rate stream suitable for transport over circuit switched telecommunication networks.

The Mpeg closed loop algorithm is essentially based on the quantization parameter Q determining the resolution of the currently coded macroblock. A fixed quantity of bits is allocated to each GoP and apportioned progressively to successive pictures and, within pictures, to successive macroblocks. Bit rate variability persists even at the GoP scale since the number of bits used may be different to the a priori assignment. The difference is taken into account in fixing the bit allocation of the next macroblock or GoP. Details of the algorithm are given in [11].

To re-synchronize the received signal it must pass by a playback buffer in the decoder. The length of the buffer (to avoid loss) and the initially imposed delay (to avoid buffer starvation) must be determined from known characteristics of the coder output and required quality of service. It is a requirement of the encoder that the bit stream it produces will not cause the decoder buffer to either overflow or underflow.

The drawback of CBR coding is that the same bit rate is generated independently of the scene contents thereby resulting in variable quality.

3.2 Open Loop Coding

VBR video can be generated using open loop coding where the same quantization parameter is used for all macroblocks naturally resulting in variable bit rate output. The rate depends on image complexity and activity and image quality is said to be constant since the quality reduction is assumed to be the same for all scenes. The output of open loop coders is unconstrained and the traffic behaviour is only guided by the arbitrarily evolving scenes contents.

When observing open loop video coder output, we can distinguish variability occuring over different time scales [4] [1]. For such video sequences it has been noted that the distribution of output rate varies substantially for different minutes long segments of the same sequence [9] [4] as well as for sequences of different types [8]. Indeed, long video sequences seem to systematically exhibit long range dependence whose significant detrimental impact on performance is beginning to be well understood [1], [4]. In particular, when multiplexing such sources, congestion can hardly be controlled.

We would therefore argue for a compromise between network efficiency and traffic burstiness and suggest the use of an appropriate rate control that can provide variable bit rate and at the same time 'network friendly' video traffic.

4 Constrained VBR Video

In this section, we discuss how video coders can be controlled to make their output fit a precise burstiness function.

For illustration purposes, we used an open loop Mpeg coder to compress a 10,000 frames long video sequence which was captured in 384x288 format and represents a music video clip showing high image complexity and active scenes. The peak rate was fixed at $p = 5Mbits/s$ and the obtained mean rate was $r = 1.85Mbits/s$. The bustiness function of a traffic characterized by its instantaneous bit rate $R(t)$ is given by: $b(c) = \max_{s<t} \int_s^t (R(t) - c)\, dt$ where $r \leq c \leq p$.

4.1 Burstiness Reduction

To be able to control the burstiness function, we suggest using a closed loop control in the coder to ensure that $b(r)$ never exceeds a given value M. This means that the output traffic must conform to a leaky bucket of leak rate r and bucket size M. The rate control can be realized by acting on the quantizer Q

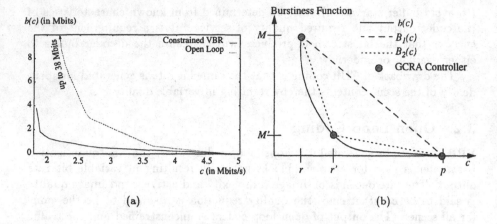

(a)

(b)

Figure 1: (a) Burstiness function of Open Loop and Constrained VBR coding, (b) Burstiness Function Control at the UNI

proportionally to the bucket fulness as suggested in [21]. A detailed shaping algorithm is presented in [6] and [7]. Obviously, we assume the coder is able to control its peak rate using a smoothing output buffer. The advantage of this coding is that it fixes the two extreme points of $b(c)$ giving $b(r) = M$ and $b(p) = 0$. In addition, traffic variability is controlled by the parameter M: the lower the value of M, the less bursty is the traffic. Indeed, the particular case of $M = 0$ produces CBR traffic of rate r. In Figure (1.a) we plot the $b(c)$ function of the sequence mentioned below when coded both in open loop and contrained VBR mode with $M = 3.75Mbits$.

4.2 Visual Quality

It is clear that constraining the output bit rate will have an effect on the visual quality of the coded signal. However we believe that full variability of open loop coding is not necessary to maintain visual video quality. Subjective quality from the user point of view depends mainly on its capacity to perceive the information displayed on the monitor. In particular, it has been shown that the ability of human eyes to detect image degradations on a television screen decreases with increasing image spatial frequency and motion [20]. It is thus possible for the network to take advantage from reducing the scale of traffic variability at the cost of slight quality modification. The desired visual quality is then tuned according to parameter M whose value can be chosen depending on user requirements.

4.3 Policing Function

Ideally, the traffic contract would be defined by the function $b(c)$. The standardized GCRA algorithm allows the network to control traffic conformance to a leaky bucket. It is therefore possible to control conformance to the traffic contract by using a set of GCRA algorithms with appropriate parameters: Sustainable Bit Rate= c_i ($r \leq c_i \leq p$) and Burst Tolerance $b(c_i)$ acting simultaneously on the video connection. In practice, we will be constrained by the UNI hardware complexity which may not provide for as many GCRA controllers as wanted. There is a compromise between the UNI complexity and the precision of the traffic control procedure.

In the current state of standardization, traffic control for VBR connections allows the network to control the two extreme points of the burstiness function corresponding to (r, M) and $(p, 0)$. Since the burstiness function is always a non increasing and convex function [2], it is always bounded by the line segment that joins its two extreme points. The network then considers the worst case burstiness function $B_1(\cdot)$ whose equation is (see Figure (1.b)):
$B_1(c) = M(p-c)/(p-r)$.

If an additional GCRA controller can be used, the $b(\cdot)$ function could be controlled in a more precise manner. A tighter upper bound of $b(\cdot)$ would be given by the conjunction of two line segments and defined by the function $B_2(\cdot)$ as shown in Figure (1.b). The additional GCRA controller ensures the conformance to a leaky bucket of parameters r' and M' where $r \leq r' \leq p$ and $M' = b(r')$. In this case traffic contract is defined by $B_2(\cdot)$ whose equation is:
$B_2(c) = (M - b(r'))(r' - c)/(r' - r) + M'$
if $r \leq c \leq r'$ and
$B_2(c) = M'(p-c)/(p-r')$ if $r' \leq c \leq p$
The rate control algorithm inside the coder has to ensure the conformance of the output traffic to the new constraint i.e. to a leaky bucket of parameters r' and M'. Again these parameters are set by the user according to the desired quality.

5 Multiplexing Schemes

Based on the knowledge of burstiness function of video connections, we propose some resource management schemes that allow the network to guarantee the negociated QoS by performing Connection Acceptance Control. As stated in section 2, multiplexer performance control depends on the delay tolerated by the application.

5.1 Bufferless Multiplexing

If video applications are highly interactive, their traffic cannot suffer burst scale congestion and multiplexers should allow for cell scale congestion only.

Assume a multiplexer of rate C handles a set of N statistically identical video sources characterized by their peak rate p, mean rate r and burstiness function $b(\cdot)$. The function $B(r) = Nb(r)$ is a burstiness function of the aggregated traffic. Interesting bounds on the loss process can be derived from $B(\cdot)$. Indeed, the effect of cell loss on the visual quality is not only dependent on the average

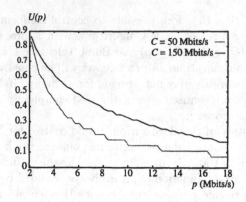

Figure 2: Bufferless multiplexing: link utilization

loss rate but also on the way data is lost. Data loss occurs in bursts. Clearly the burst size L is bounded by $B(C)$. The duration l of a cell loss burst is also bounded as [5]: $l \leq \min_{r<C} \frac{B(r)}{C-r}$.

Based on peak and mean cell rates, multiplexer performance can be evaluated by making the following worst case traffic assumption: traffic is emitted in peak rate bursts and the probability a source is active is equal to the ratio of mean rate to peak rate. The probability distribution of the combined input rate is then binomial and the cell loss ratio derived from (1) is given by:

$$CLR(N,C,r,p) = \frac{\sum_{n=\lceil \frac{C}{p} \rceil}^{N} (np - C) \binom{N}{n} \alpha^n (1-\alpha)^{(N-n)}}{Nr} \qquad (3)$$

where $\alpha = \frac{r}{p}$. For fixed values of C, r and ϵ (the data loss ratio), the allowed number of multiplexed sources, denoted by $N(p)$, depends only on source peak rate p and is given by:

$$N(p) = \sup_{n \geq 0} \{n, CLR(n, C, r, p) < \epsilon\} \qquad (4)$$

Let $U(p)$ denote the multiplexer link occupancy defined as

$$U(p) = \frac{N(p)r}{C} \qquad (5)$$

For $\epsilon = 10^{-6}$, Figure 2 shows the $U(p)$ function for two multiplexer rates, 50 and 150 Mbits/s assuming a mean rate $r = 1.8 Mbits/s$ (corresponding to the sequence in section 4). Figure 2 shows that multiplexing efficiency decreases with increasing peak rate and decreasing link rate. In the considered example, efficiency appears satisfactory (greater than 70%) for a peak rate of $2.6 Mbits/s$ on the $50 Mbits/s$ link and a peak rate of $3.2 Mbits/s$ on the $150 Mbits/s$ link.

It should be noted that for this multiplexing scheme, only a small buffer is needed to absorb cell scale congestion. This buffer introduces negligible delays,

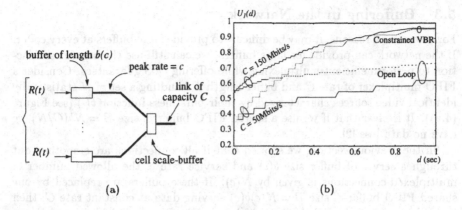

Figure 3: Buffers at the source: (a) Smoothing the peak rate, (b) link utilization

it is therefore ignored for the rest of the document. QoS requirements do not necessarily impose such performance and it is natural to ask if there is scope for greater efficiency at the expense of longer delays.

5.2 Buffering at the source

In this section we introduce a smoothing buffer between the source and the network. The intention is to reduce the source peak rate to improve multiplexer occupancy.

Assuming conformance to burstiness constraints $b(\cdot)$, each video source is smoothed out using a buffer of size $b(c)$ and constant service rate c (see Figure (3.a)).

The allowed number of multiplexed sources is then given by $N(c)$ as defined in (4). The delay experienced by each source is $D_1(c) = b(c)/c$ which is the only significant component in the end to end delay (since multiplexer delay is negligible). This delay is in a tradeoff with multiplexing efficiency, measured by the link occupancy. We can write the achieved link occupancy $U_1(d)$ as a function of the allowed QoS delay d:

$$U_1(d) = U(D_1^{-1}(d)) = \frac{rN\left(D_1^{-1}(d)\right)}{C} \tag{6}$$

Figure (3.b) shows that link utilisation increases with increasing d. We used the traffic of the video sequence described above when coded in both open loop and constrained VBR modes (curves of Figure (3.b) are derived using the two burstiness functions of Figure (1.a)). Clearly, there is a significant gain in link utilization resulting from reducing the traffic burstiness. We notice that satisfactory utilization (e.g, 0.7) can be achieved at the expense of reasonable delay (180 ms for the 150 Mbits/s link and 300 ms for the 50 Mbits/s link).

5.3 Buffering in the Network

For economical reasons, it may be difficult to provide large buffers at every coder. If the network can provide for large buffers, we can still use the burstiness function to perform efficient multiplexing while offering QoS guarantees. Consider a FIFO multiplexer of rate C and buffer length B handling a set of N statistically identical video sources characterized by their burstiness function $b(\cdot)$ (see Figure (4.a)). It is clear that if we use a shared FIFO buffer of size $B = Nb(C/N)$ we have no data loss [2].

In the previous section, we showed that if all connections are smoothed out through a server of buffer size $b(c)$ and service rate c, the allowed number of multiplexed connections is given by $N(c)$. If these buffers are replaced by one shared FIFO buffer of size $B = N(c)b(c)$ serving data at constant rate C, then the cell loss rate is bounded by ϵ (which is involved in the calculation of $N(c)$). This logical result is formally proven in [5].

In fact, the cell loss rate is expected to be much better than ϵ as a result of sharing one common buffer. While this improvement can be pointed out using statistical properties of multiplexed sources, it remains difficult to provide guarantees on it based on the proposed traffic descriptors.

The maximum multiplexer delay is $D_2(c) = N(c)b(c)/C$. It is slightly longer than the previous one (i.e. $D_1(c)$) since we have $C \leq cN(c)$. As a function of the maximum allowed QoS delay, the multiplexer link occupancy, noted $U_2(d)$ is:

$$U_2(d) = N(D_2^{-1}(d))r/C$$

In Figure (4.b) we plotted $U_2(d)$ and $U_1(d)$ for the constrained VBR coded video sequence and for two link rates. We notice that link utilization of the shared buffer multiplexer is slightly lower than the previous case but the difference is negligible. The maximum allowed number of multiplexed connection is then $N(f^{-1}(B))$ with $f(\cdot) = N(\cdot)b(\cdot)$.

6 Conclusion

An efficient traffic control in the B-ISDN should be able to define parameters that are both sufficient for traffic description and can be policed by the network. In addition, these parameters have to be useful for network operators to build up resource allocation mechanisms that allow optimized network utilization while maintaining the negotiated QoS. In this paper we presented a traffic control framework for VBR video traffic. We first defined the burstiness function as a set of burstiness bounds and established its relationship with the leaky bucket mechanism. We then proposed the use of the burstiness function as traffic descriptor and analysed the way it can be controlled in both the coder and the network. A closed loop rate control algorithm can be implemented in the video coder to ensure that its output traffic is compliant with a pre-defined burstiness function, thus avoiding cell discard at the network interface. We showed how the policing function can be implemented using a standardized mechanism only,

63

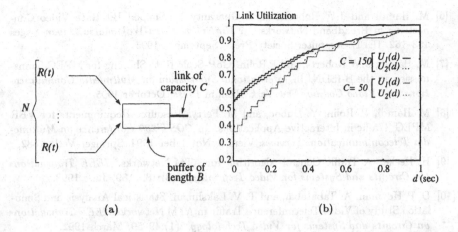

(a)   (b)

Figure 4: Buffering in the Network: (a) Shared Buffer, (b) Multiplexing Performance Comparison

i.e, the Generic Cell Rate Algorithm. There is a compromise between the UNI complexity and the precison of the policing function.

Based on the traffic contract, we proposed resource allocation schemes that guarantee the desired Quality of Service. For real time video, a worst case binomial distribution based on mean and peak rate was used. A satisfactory link occupancy is achieved for low values of the connection peak rate. If the application allows for longer delays, smoothing buffers are introduced at the source to improve network utilization. The results show that high link occupancies are achieved at the expense of reasonable delays. Finally, we show that the same dimensioning method can be used if individual source buffers are replaced by one shared FIFO buffer in the network.

References

[1] J. Beran, R. Sherman, M. S. Taqqu, and W. Willinger. Variable-Bit-Rate Video traffic and Long-Range Dependence. *Accepted for publication in IEEE Transactions on Communications*, 1992.

[2] R.L. Cruz. A calculus for Network Delay, Part I: Network Elements in Isolation. *IEEE Transactions On Information Theory*, 37(1):114–131, January 1991.

[3] D. Le Gall. MPEG: A Video Compression Standard for Multimedia Applications. *Communications of the ACM*, 4(34):305–313, April 1991.

[4] M.W. Garrett and W. Willinger. Analysis, Modeling and Generation of Self-Similar VBR Video Traffic. In *Proceedings of SigComm*. ACM, September 1994.

[5] M. Hamdi. Appendix to This Paper. In *Available at ftp://www.rennes.enst-bretagne.fr/ hamdi/proof.ps.* June 1995.

[6] M. Hamdi and J.W Roberts. QoS Guaranty for Shaped Bit Rate Video Connections in Broadband Networks. In *MmNet'95, Aizu-Wakamatsu, Japan*, pages 153–162. IEEE Computer Society Press, September 1995.

[7] M. Hamdi, J.W Roberts, and P.Rolin. GoP-Scale Rate Shaping for MPEG Transmission in the B-ISDN. In *International Symposium on Multimedia Communications and Video Coding, New York*. Plenum Press, October 1995.

[8] M. Hamdi, P. Rolin, Y. Duboc, and M. Ferry. Resource Requirements for VBR MPEG Traffic in Interactive Applications. In *COST'237 Conference on Multimedia Telecommunications Services*, Vienna, November 1994. Springer-Verlag 882.

[9] H. Heeke. A Traffic Control Algorithm for ATM Networks. *IEEE Transactions On Circuits and Systems for Video Technology*, 3(3):182–189, June 1993.

[10] D. P. Heyman, A. Tabatabai, and T. V. Lakshman. Statistical Analysis and Simulation Study of Video Teleconference Traffic in ATM Networks. *IEEE Transactions on Circuits and Systems for Video Technology*, 2(1):49–59, March 1992.

[11] ISO-IEC/JTC1/SC29/WG11. Coded Representation of Picture and Audio Information. In *MPEG Test Model 2*, July 1992.

[12] ISO-IEC/JTC1/SC29/WG11. Coding of Moving Pictures and Associated Audio for Digital Storage Media at up to 1.5 Mbits/s. In *DIS11172-1*, March 1992.

[13] ISO-IEC/JTC1/SC29/WG11. Generic Coding of Moving Pictures and Associated Audio: Video. In *Recommendation ITU-T H.262, ISO/IEC 13818-2*, November 1994.

[14] ITU-T. Traffic Control and Resource Management in B-ISDN. *I.371 Recommandation*, 1992.

[15] D.S. Lee, B. Melamed, A.R. Reibman, and B. Sengupta. TES modelling for analysis of a video multiplexer. *Performance Evaluation*, (16):21–34, 1992.

[16] S. Low and P. Varaiya. Burstiness bounds for Some BurstReducing Servers. In *INFOCOM'93*, pages 2–7. IEEE Computer Society Press Los Alamitos, California, March 1993.

[17] D.M. Lucantoni, M.F. Neuts, and A.R. Reibman. Methods for Performance Evaluation of VBR Video Traffic Models. *IEEE/ACM Transactions On Networking*, 2(2):176–180, April 1994.

[18] M. Nomura, T. Fujii, and N. Ohta. Basic Characteristics of Variable Rate Video Coding in ATM Environment. *IEEE Journal on Selected Areas in Communications*, 7(5):752–760, June 1989.

[19] I. Norros, J. Roberts, A. Simonian, and J. Virtamo. The Superposition of Variable Bit Rate Sources in an ATM Multiplexer. *To appear in IEEE JSAC special issue*, 9(3), April 1991.

[20] M.R. Pickering and J.F. Arnold. A Perceptually Efficient VBR Rate Control Algorithm. *IEEE Transactions On Image Processing*, 3(5):527–532, September 1994.

[21] A.R. Reibman and B.G. Haskell. Constraints on Variable bit rate video for ATM networks. *IEEE Transactions On Circuits and Systems for Video Technology*, 2(4):361–372, December 1992.

Implications of Self-Similarity for Providing End-to-End QOS Guarantees in High-Speed Networks: A Framework of Application Level Traffic Modeling

Bong Kyun Ryu

Department of Electrical Engineering and
Center for Telecommunications Research
Columbia University, New York, NY 10027
ryu@ctr.columbia.edu

Abstract. This paper is based on two novel research movements in the context of modeling, design, control, and management of future high-speed networks: (i) *end-to-end QOS guarantees* and (ii) *the self-similar (fractal) nature of real-world traffic*. The former pertains to building an integrated framework within which end-to-end QOS guarantees are fully supported in an efficient way while taking advantage of the statistical multiplexing. The latter concerns a salient characteristic observed in real-world traffic flow which may have critical impact on the former. However, little attention has yet been paid to determine which aspects of self-similar traffic would indeed affect control and management of future high-speed networks within an *integrated* framework of QOS provision end-to-end. The key objective of this paper is to address this issue by investigating (i) the distinctive characteristics of self-similar traffic from traditional Markovian traffic models, and (ii) how such characteristics would influence the three frameworks proposed in the literature. We provide evidences that the proposals considered in this paper did not properly take into account the self-similar nature of real-world traffic. As a result, a modeling methodology, called *Application Level Traffic Modeling* (ALTM), is proposed for more efficient QOS provision, which not only accounts for but also exploits the fractal nature of broadband traffic. An analysis of two high-quality Ethernet traces is presented supporting the ALTM. Several open research problems for further developing the ALTM are also presented.

Keywords: End-to-End QoS Provision, Self-Similarity, Traffic Modeling.

1 Introduction

It is expected that future B-ISDN (Broadband Integrated Services Digital Networks) will support a wide variety of services, such as voice, image, video, graphics, and CD quality audio, that exhibit diverse traffic characteristics, while guaranteeing the quality of service (QOS) of individual users. The key driving force behind this expectation is *multimedia* [27]. Because multimedia applications incorporate continuous media data types such as voice and video, the ability to *efficiently* provide isochronous applications with QOS guarantees is critical in such networks. Unlike the traditional circuit-switched networks in which sufficient bandwidth is assigned to an end user, the crucial objective of future B-ISDN is to benefit from the statistical multiplexing gain while maintaining reasonable network complexity. However, the *need* to provide end-to-end QOS guarantees while still taking advantage of the multiplexing gain remains an as yet largely unsolved problem and thus poses considerable challenges.

Significant effort has been, and is still being undertaken on achieving such need. Many approaches and proposals appear in the literature on providing end-to-end QOS guarantees in the context of packet-based or cell-based high speed networks. We identify three representative frameworks, among many other proposals, proposed by three different research groups: COMET research group of Columbia University (labeled as **CG**) [9, references therein]; Lancaster University in U. K. (**LG**) [1, references therein]; and TENET group of University of California at Berkeley (**TG**) [19, references therein]. Their approaches are novel and they all attempt to address key requirements of QOS provision both at the end user systems and at the edge and the core of the network with a broad range of perspectives and objectives [21].

In the meantime, recent statistical analyses of very large traffic data sets obtained from high-quality traffic measurements over various network environments have revealed that the statistical characteristics of real-world traffic may be best described by *self-similarity* (see Section 2 for its definition) [4, 7, 8, 14]. One of the main reasons for this result is that there exist various factors (*e.g.*, collision, flow control, and human response) that consistently affect traffic behavior during the entire lifetime of a connection, resulting in an extreme degree of fluctuation over a wide range of time scales [23]. Indeed, some recent studies of self-similar traffic report that self-similarity may critically affect the design, control, and management of future high-speed networks due to the drastical difference in nature between self-similarity and currently held Markovian assumptions [6, 14, 28]. These studies, however, have looked at somewhat isolated QOS-related problems such as experimental queueing analysis [6, 10], traffic modeling [14], and parameter estimation [5] with sometimes inadequate knowledge of the statistical characteristics of self-similar traffic. Little attention has yet been paid to (i) gaining a better understanding of the statistical behavior of self-similar traffic [23] and (ii) determining which aspects of self-similar traffic would indeed affect the control and management of future high-speed networks within an *integrated* framework providing end-to-end QOS.

This study indicates that currently proposed integrated frameworks for end-to-end QOS provision did not take into account this fractal behavior of broadband traffic observed over many currently working networks ranging from Ethernet to ISDN [6]. As a result, customers may face frequent service disruptions due to the potential underestimation of the impact of the bursty behavior of self-similar traffic on the ongoing connections.

We first provide the definition of self-similarity and stress its distinctive properties compared to traditional Markov-based traffic models in terms of *burstiness* and *fluctuation*. In Section 3, based on a proper understanding of self-similarity and current effort on efficient end-to-end QOS provision, we show how these two characteristics of self-similar traffic affect the three proposals. We also discuss several features of the proposals that might contribute to the prevalence of fractal nature of future broadband traffic not only at the edge but also at the core of the networks. In Section 5, issues discussed in the previous sections provide a set of motivations, based on which a new modeling methodology, called *Application Level Traffic Modeling* (ALTM), is proposed. It also includes an analysis of two high-quality Ethernet traffic traces supporting that the ALTM approach is realistic and appealing for more efficient end-to-end QOS provision in future broad-band networks. Several open research issues that arise in further developing the ALTM are also presented. Finally, conclusions are drawn in Section 6.

2 About Self-Similarity

2.1 Background

It was the work by Leland, Taqqu, Willinger, and Wilson [14] that brought out and applied Mandelbrot's idea, *Fractals* [17], to interpreting, understanding, and explaining the statistical behavior of Ethernet traffic. Their result is striking for two reasons: (1) it has brought out a clearly new perspective on teletraffic behavior; and (2) it has raised many controversial questions on evaluating various network-related performance, including queuing behavior and admission and congestion control, which are almost exclusively based on Markovian traffic source models. In what follows, we provide the definition of self-similarity and delineate the statistical properties that distinguish it from traditional Markovian traffic models, closely following [2, 14].

2.2 Definitions

Let $X = \{X_k : k \geq 1\}$ be a wide-sense stationary (WSS) process in the discrete-time domain with $\mu = \mathrm{E}[X_k]$, variance $\sigma^2 = \mathrm{E}[(X_k - \mu)^2]$, and an autocorrelation function $r(k) = \mathrm{E}[(X_n - \mu)(X_{n+k} - \mu)]/\sigma^2$ where X_k represents the number of packets/cells/bytes that have arrived during the k-th time interval of size T_s sec. Then, X is said to be a long-range dependent (LRD) process if its autocorrelation function $r(k)$ has the following form

$$r(k) \sim k^{-(2-2H)}, \qquad \text{as } k \to \infty, \tag{1}$$

with $0.5 < H < 1$. H is called the *Hurst parameter* and completely characterizes the relation (1).

This LRD property is one of many mathematically equivalent manifestations that the underlying traffic process is statistically second-order self-similar

[6]. Using the original process X, we define an aggregate WSS process $X^{(m)} = \{X_k^{(m)} : k \geq 1\}$, with an autocorrelation function $r^{(m)}(k)$, obtained by averaging the X_k's over non-overlapping blocks of size m, i.e., $X_k^{(m)} = m^{-1}(X_{(k-1)m+1} + X_{(k-1)m+2} + \cdots + X_{km})$. Formally, the process X defined above is called *asymptotically second-order self-similar* if for sufficiently large k the following condition holds;

$$r^{(m)}(k) \sim k^{-(2-2H)}, \qquad \text{as } m \to \infty, \tag{2}$$

implying a non-summable autocorrelation function, i.e., $\sum_k r(k) = \infty$. In contrast, X is said to be a short-range dependent (SRD) process if $X^{(m)}$ tends to second-order white noise, i.e., for all $k \geq 1$,

$$r^{(m)}(k) \to 0, \qquad \text{as } m \to \infty, \tag{3}$$

implying a summable autocorrelation function $\sum_k r(k) < \infty$. Typical Markovian traffic models considered in the literature have autocorrelation functions that decrease geometrically fast and thus are SRD processes [14].

2.3 Burstiness and Fluctuation

Indeed, self-similarity has rekindled attention to "burstiness". The significance of the burstiness of self-similar traffic is well described by the fact that there is no natural length of the bursts, i.e., traffic spikes riding on longer term ripples, that in turn ride on still longer term swells, known as burst-within-burst structure [13, 14, 15]. This implies that in the presence of self-similarity, currently used measures of burstiness such as *peak-to-mean ratio* and *coefficient of variation* (of interarrival times) become almost useless as they can acquire practically any value depending on the size of the averaging interval and the sample size [13, 15]. As an alternative, it has been suggested to use the Hurst parameter H as a measure of burstiness of self-similar traffic [14] as it completely describe the relation (1); however, it has been recently found that the Hurst parameter alone does not fully capture the essential behavior (burstiness) of various types of self-similar traffic [20, 23]. On the other hand, the burstiness of a typical Markov model such as MMPP is well defined by one quantity, e.g., mean burst length [26]. Later we show that some of the proposals use traditional parameters (mean burst length, average interarrival time, etc.) as traffic descriptors even for self-similar traffic, which may lead to frequent network congestion or service interruption.

The generic difference on burstiness between Markovian and self-similar traffic models appears to be caused by the high degree of traffic flow *fluctuation* that spans several orders of magnitude of time [23]. An important and practical implication of this is that detecting self-similarity of actual traffic naturally requires large number of measured samples, making it difficult to obtain correct information from short-time measurements. This has an immediate impact on the Comet group (**CG**) which uses an observation-based approach by means of measurement for obtaining cell-arrival statistics [21].

3 Implications of Self-Similarity

3.1 Three Proposals

We now turn our attention to issues of end-to-end QOS provision by briefly reviewing the three proposals. From the network provider's point of view, the following fundamental questions should be *explicitly* answered in order to build a complete framework for end-to-end QOS provision:

Q1 How do we allocate the network resources in such a way as to achieve the maximum statistical multiplexing gain while still guaranteeing QOS to end users with reasonable overall network complexity?

Q2 Given the current network load, can the requested connection be accepted by the network at its requested QOS, without violating QOS contracts of on-going connections?

The first question concerns the scheduling algorithms at switching nodes and the second the admission control algorithms at the edge of the network. Similar questions should be asked in the end systems such as multimedia workstations to guarantee full end-to-end QOS. Then the first question concerns the disk and memory scheduling of operating systems (**Q3**) and the second the system capacity (**Q4**).

The importance of the questions **Q1** – **Q4** can be illustrated by the following example. In [25], the performance of an experimental LAN-based videoconferencing application (Xphone) was investigated for various network load conditions. As expected, the frame rate of Xphone was reduced more than 30% when the network was highly loaded. End-to-end delay was also severely affected by high activity on the network. It is worth noting that the experiments were taken during the time when the host on which the Xphone was running was used solely by Xphone to achieve the best performance; if other users had shared the same host while taking the experiments, the performance of Xphone would have been degraded further correspondingly. As this simple experiment implies, a robust and efficient framework for end-to-end QOS provision must explicitly answer the above questions at the end systems as well as inside the networks.

In [21], three frameworks (the Extended Reference Model (XRM) of **CG**, the QOS Architecture (QOS-A) of **LG**, and the Dynamic Connection Management (DCM) of **TG**) proposed in the literature are extensively surveyed and compared by examining how they approach the problems mentioned above. Table 1 shows a comparison of various features of the three proposals.

3.2 Impact of self-similarity on three proposals

It would be ideal if traffic characteristics of a requested connection and its effect on existing traffic could be accurately estimated at any point in a network. Markov-based traffic models such as MMPP [26] have been exclusively used for this purpose mainly because of their analytical tractability and the preservation of the Markovian property under superposition. Unfortunately, there exist a

Group	Framework	Full end-to-end QOS provision	Scheduling Algorithm	VP	Suitability for large networks	Nature of QOS contract	Multicast Support
CG	XRM	Yes	MARS	Yes	High	Static	No
LG	QOS-A	Yes	EDF/WFQ	No	Low	Dynamic	Yes
TG	DCM	No	RCSP	No	Low	Dynamic	Yes

- XRM = Extended Reference Model
- QOS-A = Quality of Service Architecture
- DCM = Dynamic Connection Management
- MARS = MAGNET II Real-time Scheduling algorithm [11]
- EDF/WFQ = Earliest Deadline First / Weighted Fair Queue [1]
- RCSP = Rate Controlled Static Priority [19]
- VP = Virtual Path

Table 1. Comparison of the three frameworks based on the materials available at the time of writing.

variety of factors such as scheduling, flow control, and human response that affect traffic behavior during the entire connection lifetime.

The Comet Group (**CG**) uses an observation-based approach which relies on measurements of each of three cell-level traffic classes (see [11] for their definitions). Classes I, II, and III are defined based on the level of QOS requirements such as cell loss probability, delay, number of consecutively lost cells, etc., (I > II > III in the order of strictness; see [11]) and correspond to video, voice and data traffic, respectively, in a broad sense. For each class, cell-arrival statistics are obtained at each multiplexer (switch) from measurements. Such information, together with a given scheduling algorithm, is used to obtain the *schedulable region*: a set of the number of connections of each class a multiplexer can guarantee the pre-negotiated QOS at this node. This approach is limited, however, since measured data does not necessarily represent the actual traffic behavior of the class to which the measured traffic belongs. Aforementioned, in the presence of self-similarity large number of observations are required to infer correct information on the statistical nature of fractal traffic [16, 23]. As the accuracy of the schedulable region will be critically dependent of that of measurements, the development of new measurement methods on obtaining correct, essential statistics of traffic exhibiting self-similarity is required. The requirement that such techniques must also permit fast adaptation to the change of network status generally makes it difficult to apply measurement-based approach for fractal traffic.

On the other hand, both the QOS-A and the DCM schemes assume that traffic parameters of an application are known *a priori* and use them for accepting or rejecting a requested connection. The QOS-A scheme employs such parameters as peak and mean rates and mean burst size, while the DCM scheme uses minimum and average interarrival time with an associated interval, and maximum packet size. However, it was pointed out in [21] and mentioned earlier that those traffic parameters used in the QOS-A and DCM schemes are extremely inappropriate for accurately characterizing self-similar traffic. For example, the average interarrival time can acquire practically any value depending on the size of the averaging interval and the sample size [13, 15]. As the above traffic descriptors are employed as basic information for admission control of both frameworks, the resulting network status might be far from the expectation when self-similar traffic is present.

These somewhat discouraging results naturally arise when one considers that those proposals were developed before the self-similar nature of real-world traffic has been reported in the literature. It is only recent time that the notion of self-similarity has received tremendous interest from the networking community.

In the next section, we briefly discuss three factors, among many others, that might contribute to the prevalence of fractal nature of broadband traffic: flow/congestion control, scheduling, and human response.

4 Effect of Flow/Congestion Control, Scheduling, and Human Response on Traffic Behavior

A flow control mechanism is essential for providing end-to-end QOS guarantees. Its key functionality is to regulate the traffic rate depending on the states of the network. It can be exerted either by transport-layer protocol (*e.g.*, TCP sliding window) or by a network control function (*e.g.*, rate controller server of **TG**). It turns out that this mechanism is one of the major sources of self-similarity. For example, many examples of traffic that exhibit self-similarity employ TCP window-based flow control exerted during the entire lifetime of a connection [14, 25].

Indeed, the user side and the network side consistently influence each other through flow/congestion control. For example, an MPEG-coded VBR video source may have two parameters, N (interframe to intraframe ratio) and q (quantizer scale), which are adjusted to the network states as follows: (i) when the cell loss rate temporarily increases, N is reduced in order to increase the frequency of intraframes; (2) when congestion occurs, q is increased in order to reduce the output bit rate of the coder, at the expense of temporary loss of quality [18].

A well-known example is Ethernet in which all layers (possible exception of the physical layer) contribute to traffic randomness, giving rise to the self-similarity! It ranges from collision to flow control to error control (i.e., retransmission of TCP), even including human behavior and file size [3], all of which make self-similarity visible over an extremely wide range of time scales.

Based on the observation that the Internet traffic [3] is well explained by the self-similarity, one might expect that future B-ISDN will also exhibit self-similarity. However, we note one critical difference between the Internet and the B-ISDN: unlike the Internet, future B-ISDN requires *scheduling* at the end systems and the intermediate switch nodes for end-to-end QOS guarantees. Design and implementation of an efficient scheduling algorithm is perhaps the most critical requirement for satisfying **Q1** and **Q4**. All of the three frameworks incorporate scheduling algorithms such as the Magnet II Real-time Scheduling (**CG**), Earliest-Deadline-First/Weighted-Fair Queue (**LG**), and Rate-Controlled Static Priority (**TG**) into their frameworks for providing agreed end-to-end QOS requirements of accepted connections, see Table 1 or [21].

Our main interest lies on the effect of scheduling on departure processes, i.e., how scheduling changes the behavior of input traffic. The difficulty of analyzing an output process under a scheduling algorithm arises due to the complexity of the scheduling algorithm employed. However, we might be able to observe self-similar characteristic at the output of a multiplexer as a result of numerous interactions between traffic and control performed by a scheduling mechanism, as we did for Ethernet traffic. If this is the case, then the output process can be analytically represented as a self-similar process, so that end-to-end traffic modeling may be possible.

All of the previous discussions converge to the point that there exists an extreme amount of possibility of observing self-similar characteristics in future broadband traffic. The next section provides one framework exploiting the self-similar nature for more efficient end-to-end QOS provision.

5 Application Level Traffic Modeling (ALTM): A Framework

5.1 Motivation

An overview of the three frameworks given in [21], together with the self-similarity that appears to be prevalent in future B-ISDN, provides the following set of motivations by which the ALTM is driven.

M1 End-to-end QOS provision must be fulfilled on a *per-connection* basis.

M2 Different user applications result in statistically distinctive traffic characteristics, but the key statistical behavior of a user application may be invariant from one network to the other.

M3 Traffic from one source constantly interacts with traffic from other sources over a broad range of time scales by collision (Ethernet), scheduling, flow and congestion control, or human response to network states, resulting in an increased possibility of observing self-similarity not only at network access nodes but also at intermediate switch nodes.

M4 Under suitable conditions, self-similarity is preserved under superposition of both homogeneous and heterogeneous traffic sources.

M1 is obvious as it is pursued by all of the frameworks [21]. **M2** is based on the inherent traffic characteristics of some user applications due to protocol mechanisms and user behavior. For example, traffic behavior of an FTP application would be statistically different from that of a TELNET application. In what follows, a brief proof of **M4**, and two Ethernet traffic traces supporting the above motivations are provided.

5.2 Superposition

Consider two independent WSS stochastic processes $X = \{X_1, X_2, \ldots\}$ and $Y = \{Y_1, Y_2, \ldots\}$ with the following statistics:

$$\mu_X \overset{def}{=} E[X], \qquad\qquad \mu_Y \overset{def}{=} E[Y],$$
$$\sigma_X^2 \overset{def}{=} \text{Var}[X], \qquad\qquad \sigma_Y^2 \overset{def}{=} \text{Var}[Y],$$
$$r_X(k) \overset{def}{=} \text{Cov}(X_t, X_{t+k})/\sigma_X^2, \; r_Y(k) \overset{def}{=} \text{Cov}(Y_t, Y_{t+k})/\sigma_Y^2.$$

Let $Z = \{Z_1, Z_2, \ldots\}$ be constructed by the superposition of X and Y such that $Z_n = X_n + Y_n$. Then, obviously, Z becomes again a WSS process with mean $\mu_Z = \mu_X + \mu_Y$ and variance $\sigma_Z^2 = \sigma_X^2 + \sigma_Y^2$.

We consider two cases: (1) both X and Y are LRD processes; and (2) X is a LRD process but Y is a SRD process. The sharp difference in autocorrelation structures of LRD and SRD processes permits an analytical representation of the second-order statistics of Z in terms of those of X and Y. Here, we consider the autocorrelation function $r_Z(k)$ and the index of dispersion for counts function $IDC_Z(m)$ defined by

$$IDC_Z(m) \overset{def}{=} \frac{\text{Var}[Z_1 + \cdots + Z_m]}{E[Z_1 + \cdots + Z_m]}. \tag{4}$$

$IDC_Z(m)$ provides a measure of burstiness at time $t = mT_s$ sec where T_s is the counting time used for constructing the process Z. We note that $IDC_Z(\infty) = \infty$ if Z is an LRD (self-similar) process, while $IDC_Z(\infty) < \infty$ if Z is an SRD process [14]. Let $\gamma = \mu_X/\mu_Y$ and $\rho = \sigma_X^2/\sigma_Y^2$. Then, it is easy to show that $r_Z(k)$ and $IDC_Z(m)$ are given by

$$r_Z(k) = \frac{\rho}{\rho+1} r_X(k) + \frac{1}{\rho+1} r_Y(k), \tag{5}$$

$$IDC_Z(m) = \frac{\gamma}{\gamma+1} IDC_X(m) + \frac{1}{\gamma+1} IDC_Y(m). \tag{6}$$

From (5) and (6), it is clear that for $\rho > 0$ and $\gamma > 0$, in the presence of an LRD process, say $X = \text{LRD}$, $r_Z(k)$ will be also non-summable and $IDC_Z(\infty) = \infty$, making Z again an LRD process asymptotically.

5.3 Example

Two Ethernet traffic traces (LAN-I and LAN-II) from the two research labs (Systems and Image) of Center for Telecommunications Research at Columbia University have been recently collected by *tcpdump* [12] running on a SUN SPARC-station 10. Since only a few users have access to both LANs, we assume that the two traces are independent of each other. Both traces were collected for relatively long periods of time (93 consecutive hours for LAN-I and 66 hours for LAN-II) in order to observe traffic behavior over a broad range of time scales. Each trace

application	LAN-I		LAN-II	
	A (pkts)	B (sec)	A (pkts)	B (sec)
Aggregate	790,911	0.42	2,019,406	0.12
TELNET	327,474	1.00	1,066,854	0.22
NNTP	221,604	1.51	534,131	0.44
FTP	127,278	2.57	322,452	0.72
WWW	100,124	3.32	83,161	2.77
SMTP	14,924	22.31	12,808	18.19

Table 2. Statistics of traces LAN-I and LAN-II. (column A: number of packets observed; column B: average interarrival time)

consists of five TCP/IP-based applications: TELNET, FTP, NNTP, SMTP, and WWW. Table 2 briefly summarizes the sample statistics of each trace.

Figs 1 (a) and (b) show the aggregate traffic behavior of both traces during the entire measurement periods. We observe that LAN-II was about 3.5 times more active in terms of average arrival rate than LAN-I. Daily cycles are visible in both traces, and traffic fluctuation is indeed very high.

We use the IDC, among other equivalent techniques suggested in the literature [7], to detect self-similarity of each data set and estimate its Hurst parameter. For a self-similar process Z, the $IDC_Z(m)$ defined in (4) has the following form [14, 23]

$$IDC_Z(m) \approx 1 + am^D$$

with a positive constant a and $0 < D < 1$. D represents the slope of an IDC curve on a log-log plot and is related to H by $H = (1 + D)/2$. D is estimated using a simple least-squared-error fit on the log-log plot.

Figs 2 (a)-(f) show and compare the estimated IDC's of 12 data sets (two traces, six types of traffic for each trace). Ethernet traffic indeed manifests self-similarity as reported in [14, 25]. However, each type of individual application traffic is also self-similar. This is a new result. Indeed, these figures directly support M4; fractal behavior of individual application traffic results in the same behavior of the aggregate traffic. More surprisingly, except the possible exclusion of SMTP traffic, which appears from the fact that SMTP has been hardly used in both LANs, the estimated H for each type of traffic is within 0.1 for the two traces (M2). Also, every data set exhibits fractal behavior over the range of time scales that spans at least five orders of magnitude. In particular, fractal behavior begins to appear around or even before 10msec, which is within the range of time scales that may have significant effect on buffer performance [22]. These results may serve as strong evidence supporting M2 – M4.

5.4 Methodology

We consider a realistic scenario where a network provider serves many different network subscribers such as banks, universities, research centers, and companies. The modeling methodology of the ALTM can be summarized as follows:

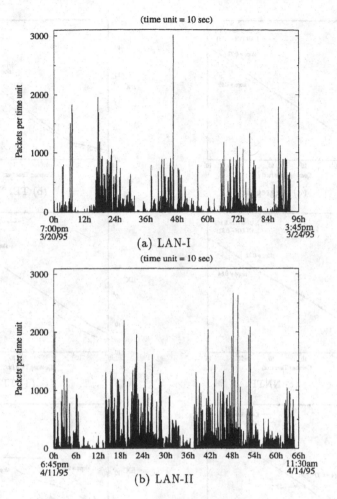

(time unit = 10 sec)

(a) LAN-I

(time unit = 10 sec)

(b) LAN-II

Fig. 1. Comparison of Aggregate Traffic Behavior of Two Traces.

1. A set of user applications of a subscriber is identified.
2. Each type of application traffic (i.e., TELNET, FTP) is characterized, if possible, by one of the well-studied traffic models at each node of the network.
3. The effect of a new connection on current network load along its path is evaluated.

The key purpose of the second step is to identify and estimate the parameters which are meaningful and necessary for evaluating network capacity that can guarantee full end-to-end QOS requirements of accepted connections. One method for this involves measuring a type of application traffic (target application traffic) for a relatively long period, estimating the representative statistics, finding a suitable traffic model for the target application traffic, and updat-

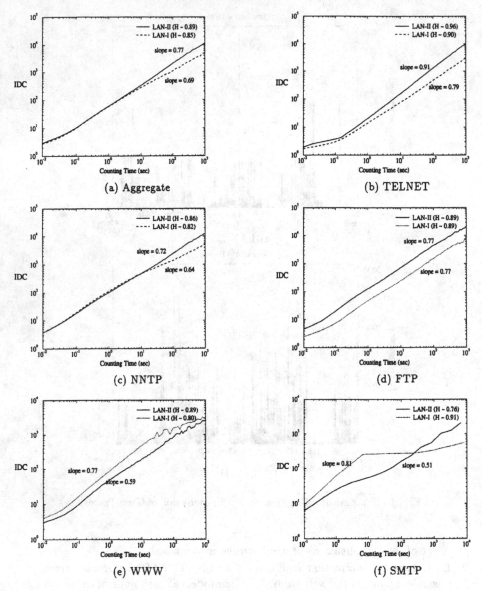

Fig. 2. Comparison of IDC Curves.

ing constantly based on ongoing measurements. As emphasized throughout this paper, self-similar models may play a major role in characterizing many types of application traffic. We note that the final step is fundamental to achieving our target objectives (**Q1 – Q4**) in that providing full end-to-end QOS on a per-connection basis would not be possible without it.

5.5 Remaining Work

Indeed, the ALTM approach is in its very early stage, giving rise to various open research issues such as:

- What is the impact of different types of new connections on existing traffic behavior?
- What is the impact of superposing heterogeneous types of self-similar traffic, possibly having different Hurst parameters?
- What is the impact of self-similar traffic on network capacity (required buffer size for given QOS parameters, number of connections that are allowed at each node, etc.)?
- What is the impact of scheduling and QOS parameters on departure processes? Is self-similarity observed from various types of application traffic after scheduling?
- Which traffic characteristics are crucial and meaningful in evaluating network capacity that can guarantee full end-to-end QOS requirements of accepted connections?
- What range of time scales is important for admission control? Is it sufficient to characterize traffic behavior in a range of *small* time scales only in order to build a simple and successful admission control?
- Are simple yet versatile self-similar traffic models available?

These questions, among many others not mentioned here, must be explicitly answered for the ALTM approach to be successfully employed for efficient end-to-end QOS provision. Studies on some of these issues are currently under way [22, 23, 24].

6 Concluding Remarks

Throughout this work, we have (i) reviewed current approaches on providing full end-to-end QOS guarantees [21], (ii) discussed impact of self-similar traffic on those approaches, (iii) proposed a new modeling methodology (ALTM) for more efficient end-to-end QOS provision, (iv) presented an analysis of two Ethernet traffic traces supporting that the ALTM approach is realistic, and (v) raised several open research issues regarding the ALTM approach. One of the key conclusion is that designing an integrated framework for end-to-end QOS provision must take into account the self-similar nature of real-world traffic due to its distinctive characteristics from well-known Markovian traffic models. The primary purpose of this work is to emphasize self-similarity (or fractal) as a promising candidate for modeling various types of traffic in B-ISDN which is likely to be extremely heterogeneous and complex. The fact that users and networks will constantly influence each other through numerous interactions *over a broad range of time scales* naturally gives rise to a fractal approach. Many measurement studies on current traffic behavior strongly suggest that fractal traffic will be ubiquitous in future high-speed networks as a result of those interactions. We believe that traffic modeling based on each type of application traffic (ALTM) with the fractal idea will considerably alleviate the burden required for achieving *efficient* full end-to-end QOS provision.

References

1. A. Campbell, G. Coulson, and D. Hutchison. A quality of service architecture. *Comp. Comm. Rev.*, pages 6–27, 1994.
2. D. R. Cox. Long-range dependence: A review. In H. A. David and H. T. Davis, editors, *Statistics: An Appraisal*, pages 55–74. The Iowa State University Press, Ames, Iowa, 1984.
3. Mark Crovella and Azer Bestavros. Explaining world wide web traffic self-similarity. Technical Report TR-95-015, Boston University, CS Dept, August 1995.
4. D. E. Duffy, A. A. McIntosh, M. Rosenstein, and W. Willinger. Statistical analysis of CCSN/SS7 traffic data from working CCS subnetworks. *IEEE JSAC*, 12:544–551, 1994.
5. A. Erramilli, J. Gordon, and W. Willinger. Applications of fractal in engineering for realistic traffic processes. In *Proc. ITC-14*, 1994.
6. A. Erramilli, O. Narayan, and W. Willinger. Experimental queueing analysis wiht long-range dependent packet traffic. submitted to IEEE/ACM Networking for possible publication, 1994.
7. A. Erramilli and W. Willinger. Fractal properties in packet traffic measurements. In *Proc. St. Petersburg Regional ITC Seminar*, 1993.
8. M. W. Garrett and W. Willinger. Analysis, modeling and generation of self-similar VBR video traffic. In *Proc. ACM SIGCOMM '94*, 1994.
9. COMET Research Group. Control and management of enterprises. Activity Report: 1991–1993, Center for Telecommunications Research, Columbia University, 1994.
10. D. P. Heyman and T. V. Lakshman. What are the implications of long-range dependence for VBR-video traffic engineering? preprint, 1995.
11. J. M. Hyman, A. A. Lazar, and G. Pacifici. Real-time scheduling with quality of service constraints. *IEEE JSAC*, 9:1052–1063, 1991.
12. Van Jacobson et al. *Tcpdump*, June 1989. manual.
13. W. E. Leland et al. On the self-similar nature of Ethernet traffic. In *Proc. ACM SIGCOMM '93*, 1993.
14. W. E. Leland et al. On the self-similar nature of Ethernet traffic (extended version). *IEEE/ACM Trans. Net.*, 2:1–15, 1994.
15. W. E. Leland and D. V. Wilson. High time-resolution measurement and analysis of LAN traffic: Implications for LAN interconnection. In *Proc. IEEE INFOCOM '91*, 1991.
16. S. B. Lowen and M. C. Teich. Estimation and simulation of fractal stochastic point processes. *Fractals*, 3:183–210, 1995.

17. B. B. Mandelbrot. *The Fractal Geometry of Nature*. W. H. Freeman, 1982.

18. P. Pancha and M. El Zarki. MPEG-coding for variable bit rate video transmission. *IEEE Comm. Mag.*, 32(5):54–66, May 1994.

19. C. Parris, H. Zhang, and D. Ferrari. Dynamic management of guaranteed-performance multimedia connections. *ACM/Springer Verlag Multimedia Systems*, 1:267–283, 1994.

20. M. Parulekar and A. M. Makowski. Buffer overflow probabilities for a multiplexer with self-similar input. to appear in *Proc. INFOCOM '96*.

21. B. K. Ryu. Implications of self-simialrity for providing QOS guarantees end-to-end in high-speed networks: A framework of Application Level Traffic Modeling. Technical Report CU/CTR/TR 401-95-07, Center for Telecommunications Research, Columbia University, March 1995.

22. B. K. Ryu and A. Elwalid. The importance of Long-Range Dependence of VBR video traffic in ATM traffic engineering. preprint, 1995.

23. B. K. Ryu and S. B. Lowen. Point-process approaches to the modeling and analysis of self-similar traffic: Part I - model construction. to appear in *Proc. IEEE INFOCOM '96*.

24. B. K. Ryu and S. B. Lowen. Point process approaches to modeling and analysis of self-similar traffic: Part II - queueing analysis. preprint, 1995.

25. B. K. Ryu and H. E. Meadows. Performance analysis and traffic behavior of Xphone videoconferencing application on an Ethernet. In *Proc. Third Int. Conf. Comp. Comm. Net.*, 1994.

26. N. B. Shroff. *Traffic Modeling and Analysis in High Speed ATM networks*. PhD thesis, Columbia University, 1994.

27. F. A. Tobagi, I. Dalgiç, and C. A. Noronha Jr. Multimedia networking and communications: Infrastructure, protocols, and servers. preprint, 1994.

28. W. Willinger. Traffic modeling for high-speed networks: theory versus practice. In *Stochastic Networks*, IMA Volumes in Mathematics and Its Applications. Springer-Verlag, 1994.

Multimedia Call Control: A Centralized Approach

Harald Müller

Lehrstuhl für Kommunikationsnetze,
Technische Universität München, 80290 München, Germany
E–mail: harald@lkn.e–technik.tu–muenchen.de

Abstract. To support multimedia services on a large scale, a powerful signaling system for broadband networks has to be provided. It has to enable arbitrary complex multimedia services with several participants, involving several media and heterogeneous terminal equipment. This paper describes a framework for an overall signaling architecture that fulfils those requirements. One of its basic principles is to perform call and resource control in a centralized manner, which is contrary to other advanced and traditional approaches. Another key feature of the proposed framework is an independent signaling system for connection control. One of the advantages of the new design is that no NNI signaling for call and resource control is required. Call control and the determination of special resources, like converters and conference bridges, is simplified. A cost optimization can be performed when locating the special resources. The architecture enables a straight forward inclusion of external providers of resource services. The autonomous connection control yields independence from the underlying switching and transmission technology, and it completely separates signaling for call and resource control from connection control. Hence it is feasible to have different providers for telecommunication services, resource services and bearer services.

1 Introduction

The wide–spread provision of multimedia communication services and applications is one of the most exciting topics nowadays. Multimedia is not simply an incremental improvement, but it will add a new dimension to communications and it will make a major contribution to the information age. It will enforce the integration of the traditionally separated domains of telecommunication, information technology and consumer electronics. As a result, it can be expected that a large number of heterogeneous multimedia terminals will develop, having different capabilities in terms of applications, supported media, presentation quality and coding algorithms.

The provision of a universal network access to different multimedia services for heterogeneous terminal equipment will be the key to a successful introduction and wide acceptance of multimedia communications. This requires a powerful signaling system that can handle multiparty calls, several connections (media) in a single call, asymmetrical connections, correlations among different media, heterogeneous terminals and dynamic change of call topology and connection parameters. Looking at the signaling systems that are currently available (e.g. Q.931 of narrowband ISDN), it is obvious that functionality has to be extended considerably.

The approaches in the standardization bodies for B–ISDN signaling started with the ISDN Q.931 protocol, taking into account the existing hardware and software investment and yielding a migration path from existing systems. Unfortunately, this incremental approach does not allow totally new concepts, and the functionality is moving very slowly towards the target described above. One example is the location of the call control functionality. Traditionally, call control is done in every exchange participating in the call. This was justified by the fact that a major part in traditional call/connection control was the link–wise setup of an end–to–end line. In future releases of ITU–T signaling and in other advanced signaling approaches, a functional separation of call and connection control will be introduced. This allows the different functions to be carried out at different physical locations. Several approaches take advantage of that fact and suggest to omit the processing of call control in transit exchanges (TEX). However, call control is still distributed over all local exchanges (LEX) participating in the call. Further, the separation of call and connection control is only partial, there are e.g. common signaling messages.

In this paper, a centralized approach to call control is proposed, where only one call control instance is active for each call. It will be shown that this yields several advantages against distributed call control. A further improvement is the introduction of an autonomous signaling system for connection control. The new functional model of a complete multimedia signaling system including centralized call control will be introduced and explained.

In Sec. 2 the evolution of signaling is outlined, starting with N–ISDN and leading to elaborate multimedia signaling approaches. The focus is on the applied call models and on call control. The drawbacks of distributed control of multimedia calls are explained. Section 3 introduces the idea of centralized call control. After explaining the basic operation, the advantages against distributed call control are pointed out. Solutions for different related tasks are given. An example of the processing of a multimedia call in Sec. 4 clarifies the overall operation of the proposed signaling architecture.

2 Evolution of Call Modelling and Call Control

2.1 Narrowband ISDN

The N–ISDN call model (Q.931, [1]) does not distinguish between call and connection. A call contains exactly one connection, which is bidirectional and symmetrical with respect to the connection parameters. The call state machines differ for the originating side and the terminating side of the call. Signaling protocols are asymmetrical, i.e. UNI and NNI protocols are not the same. The call model and the signaling protocols support only two–party calls. Although there is an end–to–end compatibility check, there is no negotiation mechanism.

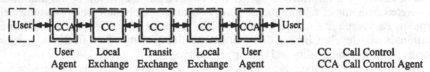

	User	Local	Transit	Local	User	CC	Call Control
	Agent	Exchange	Exchange	Exchange	Agent	CCA	Call Control Agent

Fig. 1. Functional model of N–ISDN signaling

Fig. 1 shows the functional model of N–ISDN call control. As there is no separation of call and connection control, call control is distributed, and it is executed in every switching node that lies on the communication path to be established. This model reflects the fact that the major tasks of the call setup procedure are dealing with the link–wise establishment of a communication channel. The more central, call–related functions, like e.g. compatibility checking or billing, seem less dominant and give no strong reason to separate call and connection control functionality. As a result, it is e.g. inevitable that an end–to–end communication channel (bearer) has to be set up, before an end–to–end compatibility checking can be performed.

2.2 ITU–T and ATM Forum

The current signaling approaches for B–ISDN of the ITU–T and the ATM Forum are very similar. Both bodies have used Q.931 signaling as a starting point for their first versions. The ITU–T signaling will be described first, and the few differences to the ATM Forum signaling will be listed afterwards.

The current standardization activities of ITU–T are considering the Capability Set 1 (CS 1) of a signaling protocol, which can be found in the draft Q.2931 [4]. Future releases (CS 2/3) will provide additional functionality [2]. CS 1 does not separate call and connection control functions. In principle, it is very similar to N–ISDN signaling with some additional bearer services. For CS 1, constant bit rate (CBR) synchronous and variable bit rate (VBR) connectionless bearer services with peak bit rate reservation are defined. They can be chosen at connection setup, modification is not possible later on. CS 2 will add VBR synchronous and VBR asynchronous bearer services with traffic parameters, allowing more elaborate reservation strategies [5]. Further, it will be possible to modify connection parameters during the lifetime of the connection.

In the Q.2931/CS 1 call model, a call contains a single bidirectional connection. Bandwidth parameters for both directions can differ. From CS 2, a call can have multiple connections [7]. In CS 1, connections can only be point–to–point. Only from CS 2, point–to–multipoint connections will be possible, parties can be added to or dropped from existing connections [6]. The addition of parties can only be initiated by the root of the point–to–multipoint connection. CS 1 will only allow end–to–end compatibility checking similar to N–ISDN signaling. The called terminal can only accept or reject the offered call, there is no possibility to negotiate parameters of the call. The negotiation functionality will be added in CS 2/3. As a result, communication using heterogeneous terminals is not possible at the moment.

The ATM Forum Phase 1 signaling [3] is equivalent to Q.2931, CS 1, with the following differences. A connection–oriented VBR bearer service is added. Unidirectional point–to–multipoint connections are possible. This is achieved by adding procedures on top of the basic Q.2931 protocol to add/drop parties to/from existing connections.

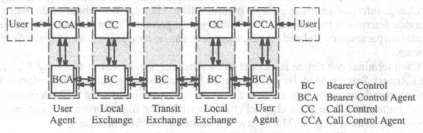

Fig. 2. Functional model of Q.2931 signaling (CS 2/3)

The functional model for Q.2931/CS 1 signaling is the same as for N–ISDN. Figure 2 shows the functional model of CS 2 and beyond, where a functional separation of call and connection will be introduced. The figure corresponds to a two–party call with two connections. Call control functions like e.g. end–to–end compatibility checking or negotiations can be carried out before setting up a connection (look–ahead). The execution of call control can be limited to the local exchanges, which saves processing effort and hence delay in the transit nodes. Several connections within one call can be handled in a straight forward manner. The different connections can be controlled and routed independent of each other.

Call control is still distributed since there is one call control instance for every participant, potentially located at different LEXs. The main task of bearer control is to establish an end–to–end communication channel consisting of a concatenation of virtual channels (VC). This includes, amongst others, resource reservation, connection routing and switching functions. It is obvious, that bearer control functions have to be carried out in every node that is part of the connection.

2.3 Advanced Approaches: MAGIC

There are several proposals for multimedia signaling that are ahead of the approaches in the standardization. One of them, called MAGIC [9,10], shall be described in more depth. It has a very elaborate call model that fulfils the before mentioned requirements. A similar call model will be used in the signaling architecture proposed in this paper. EXPANSE [11,12], another important signaling approach, also uses this call model, which therefore has a good chance to become a candidate for future releases of standardization.

Object–Oriented Call Model. MAGIC provides a flexible call model that allows the composition of arbitrary complex multimedia calls with heterogeneous terminals. Additional network functionality provides the necessary high–level functions, like conversions and mixing of information streams. These functions are carried out transparent to the end terminals. To support these functions, an additional level of abstraction has been introduced between the call level and the connection level, the resource level.

The call model is object–oriented. A call can be composed from basic elementary call objects. Each terminal can define its own access to the call independent of the other terminals. In this way, terminals with different media capabilities can participate in the same call. Each terminal has a local view of the call, i.e. it can only see those elementary call objects that are of interest for its own representation of the call. The local views that the terminals have of the same call differ from each other. In the signaling model, no global view of the call exists. It is introduced only to ease the understanding.

The example in Fig. 3 shows the global view of an audio/video call with three participants. An *Abstract Service* is an elementary call object that represents the possibility to exchange information using a single medium, like 'video'. An attribute of the Abstract Service determines the general information type (e.g. 'video'), but not the coding or presentation parameters. Related Abstract Services can be grouped into an *Abstract Service Group*.

Each terminal can define its access to an Abstract Service using *Mapping Elements* and *Access Service Modules* (ASM). An ASM represents a channel for information transport and corresponds to a bearer connection. The attributes of an ASM define lower layer parameters (like ATM channel characteristics) and higher layer information, like coding algorithms (e.g. 'G.711 A–law' for audio or 'H.261' for video).

The Mapping Elements define the flow of user information between ASM and Abstract Service. The Upstream Mapping Element defines how information is transferred from the terminal to the Abstract Service. The opposite direction is covered by Downstream Mapping Elements. Two different types of Mapping Elements exist in the downstream direction. The Composite Downstream Mapping Element represents the information of the Abstract Service as a whole. Party A in the example would receive the video information of the two Parties A and B merged into one video stream with a window for each party. The Component Downstream Mapping Element allows the receipt of information transmitted by a single user. In the example, Party B receives the video information of Party A only. Mapping Elements contain presentation attributes, as e.g. windowing or cropping parameters for video.

As is apparent from the example, there must be some functionality in the network to manage the merging of information, when several information streams are combined. If the terminals use different coding algorithms, the necessary conversion functions must also be managed. Therefore, a resource model has been defined, which is again object–oriented. MAGIC differentiates between ordinary resources and specialized resources. Ordinary resources represent the pure transmission capabilities of the network. They are handled by the bearer control (BC) entities. Specialized resources are concerned with higher layer functions like information conversion, merging and multicasting. Resource control (RC) handles these resources.

The bearer control (BC) is responsible for establishing and maintaining transmission channels. A bearer service is modeled as a bidirectional, symmetrical or asymmetrical two–party connection. Routing groups can be defined to assure differential delay limitation by performing common routing. Attributes of the bearer contain e.g. ATM layer and AAL capabilities, performance parameters (like delay and bit error rate) and channel identifiers.

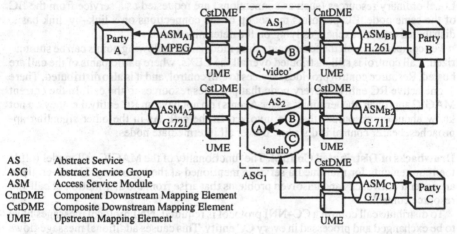

AS	Abstract Service
ASG	Abstract Service Group
ASM	Access Service Module
CntDME	Component Downstream Mapping Element
CstDME	Composite Downstream Mapping Element
UME	Upstream Mapping Element

Fig. 3. MAGIC call model: Global view of a three–party video/audio call [9]

Functional Model and Operation Principles. Fig. 4 shows the functional entities participating in MAGIC signaling. For reasons of clarity only a two–party, single connection call is shown. The relations between the different functional entities can be explained best by describing the normal operation during call setup.

A call is initiated by one party, which requests the desired elementary call objects via UNI signaling. The CC at the originating LEX interacts with the CC in the LEXs of the other participating users to offer the call and to negotiate their call objects. When the initial call configuration is complete the required resources can be determined. Using a mapping function the initiating CC entity maps the call structure to a set of resource objects.

The resource objects are requested from the RC entity of the same node. The service that RC provides to CC consists of implicit resource objects, i.e. the objects do not contain information about their physical location. The RC tries to find local physical resources with the required functionality. The local resources are reserved and set up. Where no local resource can be found, RC communicates with an appropriate RC in another node and propagates the remaining implicit resource object request. This requires an RC–NNI protocol.

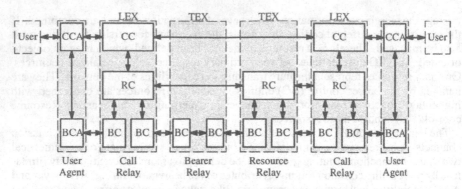

Fig. 4. Functional model of MAGIC signaling [9]

Local ordinary resources (such as connections) are requested as a service from the BC of the same node. BC performs the set up of the connections on a link–by–link basis, distributed over all exchanges lying on the determined routes.

For comparison with the former signaling models, the following points can be summarized. Call control is still distributed over all the LEXs, where participants of the call are hosted. Resource control provides services to call control, and it is also distributed. There is one active RC entity in every node that provides resources to the call. In the current MAGIC approach, the terminals (User Agents) do not have an RC entity, i.e. they do not know about the resources that are involved in the call. As in the other signaling approaches, bearer control is distributed over all intermediate nodes.

Drawbacks of Distributed Control. The functionality of the MAGIC call model fulfils the requirements for multimedia services mentioned at the beginning. However, there are several drawbacks and unsolved problems that arise from the distribution of call and resource control.

To distribute call control a CC–NNI protocol is required. NNI protocol messages have to be exchanged and processed in every CC entity. This causes additional message flows and signaling delays. The same disadvantages hold for the distributed resource control.

In order to perform the resource mapping function in CC, knowledge of the complete call is required. This is contradictory to the concepts of local call view and distributed call control where each CC entity should only be responsible for a part of the call. The solution for this is to let at least a single CC entity have the global view. However, this results in additional message flows. Further, a dedicated 'master CC' has to be selected which gathers the global view and requests resource objects. This could be the CC in the originating LEX.

The resource control has to map implicit resource objects onto physical resources. There are two possible ways in which this can be done. The first is by using only local resources. If some resource type is not available locally or is temporarily occupied, the complete call has to be rejected. As a consequence, all types of specialized resources have to be built into every node in sufficient quantity to avoid 'resource blocking'.

The second and much better solution is to distribute resource control with the effect that non–local resources can be used. This enables the supply of rarely used specialized resource types in only few nodes and the switching to resources at other nodes when local resources of a certain type are busy. However, even with this better solution, there are several drawbacks. An RC–NNI protocol is required and the problem of 'resource routing' has to be solved, i.e. the RC entity has to find out where the required resources are located and hence where the RC request has to be sent. In general, there will be several locations where the desired resource functions are available. They will have different costs that consist of the costs of the specialized resources themselves and of the costs of the bearer connections that are required to reach them. It will be necessary to perform some kind of 'resource optimization' to choose the least cost resources. However, the distributed control of the specialized resources makes it very difficult to apply such an optimization. Further, the change and release procedures for special resource configurations are more complicated with a distributed control.

Because all parts of resource control are embedded into the signaling entities of the exchanges, the special resources are coupled very tightly to the switches and hence to the network providers. The chosen architecture does not deal with the inclusion of external providers of specialized resource services. It can very well be foreseen that external providers would speed up the deployment of new specialized resources. Consider for example a vendor of a video codec hardware using an improved coding algorithm. He will probably have a strong interest in providing a conversion service from his format to more common ones in order to enable buyers of his hardware to communicate with other users that apply different formats.

The tight coupling of special resources to the switches also complicates the introduction of new resource types. RC entities in the exchanges as well as RC protocols have to be changed.

3 The Novel Approach: Centralization of Call Control

The key idea of the proposed signaling architecture is to perform centralized call and resource control in order to avoid the previously mentioned drawbacks. The way the terminals perceive the call and define it at the call level will remain almost unchanged. This includes the call model, which will be very similar to the one described in the previous section, and which is capable of supporting complex multimedia services. What will be changed is the processing of the call within the network which has an effect on the signaling entities and their relationships.

3.1 Functional Model and Basic Operation of the New Signaling Architecture

Figure 5 shows the functional model of the proposed signaling architecture. As in the previous section where the MAGIC model was explained only the instances for a rather simple call are depicted. The call that is shown here could represent one video connection between two parties and involve a converter between two coding algorithms. Again, the relationship between the entities will be explained by describing a normal call setup procedure.

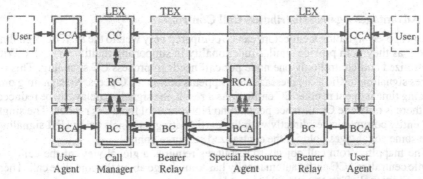

Fig. 5. Functional model of the proposed signaling architecture with centralized call control

The Call Control Agent (CCA) of the initiating user's terminal requests the desired call objects from call control via CC–UNI signaling. There is only a single instance of CC per call, which interacts directly with all terminals participating in the call. This instance belongs to the *Call Manager*. The Call Manager negotiates the call objects with the other participants and gathers a global view consisting of all call objects. Note that although the CC has a global view of the call, the call views of the terminals are still local.

When the global call configuration is known, CC determines the necessary resources and requests them from RC. The resources include special resources like converters and combiners, as well as multicast connections with appropriate parameters. More abstract QoS measures (like e.g. 'MPEG video, CIF resolution') have to be mapped to connection parameters like bandwidth or delay. The terminals do not have an RC entity, i.e. they are not aware of the existence or location of any resources that are necessary to provide the call. All resource and mapping functions of the network are transparent to the terminals.

There is only one RC entity per call. The separation from CC is only a functional one, and it does not have to be physical. Nevertheless, the separation yields an independent design of CC and RC, and later changes to these can be introduced easier. The main function of the RC is to locate the necessary special resources and to reserve them. This is done by using a database, where the resources, their functionalities and their availability is recorded. The best resources with respect to the costs are chosen. The special resource services are requested using an RC signaling protocol.

The special resources are connected to the network like ordinary terminals. Instead of a Call Control Agent, they have a Resource Control Agent, which is capable of handling RC signaling. The latter is used to request resource services and to set up the required functions of the special resource.

When the required special resources are located and reserved, the end points for the connections are known. The RC will request all required connections from Bearer Control. This includes also those connections where the local node is not included in the transmission path. BC and its signaling protocol are totally independent of the CC and RC functions and could also be used as a stand–alone protocol. As BC protocol, the Multicast Connection Management Protocol (MCMP) is used. MCMP has been further described in [13]. It shows the following properties:

- Connections can be multicast. *Note that in contrast to MAGIC, the multicast function is not considered as a special resource. It is a feature of the connection model.*
- Connections have dynamically changeable parameters (e.g. bandwidth, delay)
- Topology changes of connections during life time are possible
- 3rd party connection setup is handled

3.2 Advantages Against Distributed Call Control

The proposed signaling architecture can be compared very well to the MAGIC signaling model as they both provide similar functionality in supporting multimedia calls. With centralized call control only one node per call needs to process CC signaling. This reduces signaling traffic as no message flow appears between CC peers. No signaling processing time is spent in other CC entities. As a result, the signaling delay can be reduced. As there is only one CC instance per call, no NNI protocol for CC is needed. The single CC entity communicates directly with the participating terminals using UNI signaling. The same advantages hold for the centralized resource control.

The mapping from call objects to resources requires a global view of the call. The single centralized CC entity automatically has knowledge about the complete call. There is no additional effort to gather this knowledge.

The task of RC to locate the appropriate resources is simplified, as RC has knowledge of all available resources in the network, and it can choose the required ones. A network–wide optimal 'resource routing' can be performed. The resource routing algorithm can be implemented independent of any other protocols or signaling entities. This gives the chance to start with simple solutions and to introduce more elaborate methods later.

As special resources are connected to the network like terminals, external providers of special resource services can easily be included. How this can accelerate the deployment of new special resource types has been described above.

Bearer control has been totally separated from call and resource control. BC is the only part of the signaling system that strongly relates to the switching hardware, and hence depends on switch technology and vendors. Using MCMP makes bearer control service–independent, i.e. new multimedia services or special resource types can be introduced without having to change any instance or part of the protocol of BC at all. Having an external interface between CC/RC and BC yields the advantage that the two parts can be run on different platforms. Further, different switching and transmission technologies can be used without changing RC or CC, as long as the BC service interface is provided. These properties give the possibility to provide pure bearer services by an independent authority. Altogether, the proposed signaling architecture enables separate and independent providers of three levels of services, telecommunication services, special resource services and bearer services. This can be considered an important feature in the landscape of a telecommunication market, which is becoming more and more deregulated.

3.3 Implications

Changing the paradigm for call control in a signaling system can have two kinds of implications. New problems can arise that did not exist with the distributed call control, and already existing tasks might have to be solved in another way. One of the problems is how the Call Manager can get information about remote terminals (e.g. access rights). This information can be stored in a database system that is accessible by every CC entity. It is also possible to store another kind of (more abstract) addressing, which yields a name service or directory service. End users and terminals can be reached by alternatively giving an address in form of a hierarchical name system (like e.g. internet e–mail addresses). Most of the terminal information, like terminal capabilities, will be exchanged during the call negotiation phase.

It is obvious that every central solution bears the danger of traffic concentration. However, the CC function is not central to the whole network, one CC entity is active per call. Another call can have another CC entity in another node. It is straight forward to locate the active CC entity in the node where the call is initiated. Note that distributed solutions must always have a CC entity in the local exchange. The proposed solution does not impose more but fewer load on the LEXs. If an exchange tends to become overloaded, the centralized call control could apply another locating procedure, e.g. on a least load basis.

The single active CC entity must be able to exchange signaling messages directly with the remote terminals. This situation is new, since traditionally a signaling entity always exists in every local exchange, and signaling messages are transmitted over signaling VCs that are fixed or set up using metasignaling and that are terminated at the local exchange. A signaling transport network only exists between network nodes. To provide centralized call control, the signaling transport network has to be extended to the terminals. The connection control MCMP provides capabilities to send higher layer signaling messages to any addressable entity (ADDRESSED_DATA [13]). These messages will be sent over the signaling VC of connection control signaling, and it will be routed by the MCMP control entities. Another possibility is to establish virtual channel connections between the active call control entity and the terminals for the transport of call control signaling. Further studies will show which solution is preferable.

Call objects requested by the terminals and established bearer channels are related. One task of the signaling system is to provide an association between the two. For example, if a terminal requests an Abstract Service 'video' and an appropriate Access Service Module, the user plane of the terminal must know the VPI/VCI value of the established virtual channel connection to transmit any video data. In the classical approach, the association is done in the local exchange. This cannot be done when call control is centralized, as not every local exchange has an active CC entity. Therefore, the association has to be done by the terminal itself. To support this, each connection that is requested by the Call Manager contains a unique high layer identifier, which refers to the appropriate call object (ASM).

The resource routing problem is not restricted to the proposed architecture, it must be solved for distributed control, too. The RC entity receives a request to set up a set of special resources. It has to find out where these resources are located, which resources are available at the moment, and it has to choose the best solution. This can be supported by maintaining a transaction–based database that can be used by any RC entity, and that resolves concurrent access. This database contains information about every special resource in the network, including location, functionality, state (free, busy) and some cost criteria to enable resource optimization. In order to find a cost–optimal solution the costs of the connections between terminals and special resources have to be considered, too. The connection costs depend upon the location of the terminals and the special resources. The whole process of 'resource routing', i.e. the mapping of requested special resource services to physical resources, is a non–trivial task. It gets even more complicated when single special resource functions are mapped onto several physical resources. For example, a conversion from coding format A to B could be mapped onto one converter A–>C in series with one converter C–>B, when the converter A–>B is not available. Similar examples can be found for combiners with several inputs.

The last issue concerns the processing of calls spanning different network domains (e.g. in different countries). Although the centralization of call and resource control raises no additional problems, the different scenarios and possibilities shall be sketched. If the foreign network domain does not use the same signaling system some kind of interworking unit has to be installed. If the foreign network does not allow resource control, or if interworking with foreign resource control is not possible, only resources of the own network domain can be used.

In the remaining cases, the foreign network is supposed to use the same signaling system and protocols. Hence, there should also be a signaling transport system so that signaling messages can be sent directly to remote terminals and resource nodes. There should be no reason against centralized control at the call level, where only terminals are concerned. Centralized resource control requires access to a resource database. There could be reasons for network providers not to show their information about special resources, including functionality, cost and location. In this case, resource control could still use resources in the own domain to support the call which may not be optimal. However, given that special resource services would bring revenues, the provider will have an interest to make them available to a large area. If this is the case, a scenario is possible where, even for a call within a single domain, special resources from outside are used because they are cheaper or not available inside the domain.

4 A Multimedia Signaling Example

To convey further understanding of the control flows and procedures within the proposed signaling system, a detailed example will be given. It is assumed, that a call complying to Fig. 3 in Sec. 2.3 shall be set up. The participating control entities and the sequence of control steps are depicted in Fig. 6. For reasons of clarity, only the entities for the video part of the call are shown in this figure. The control of the audio part is analogous. In the following, the procedures and flows will be described in their temporal order.

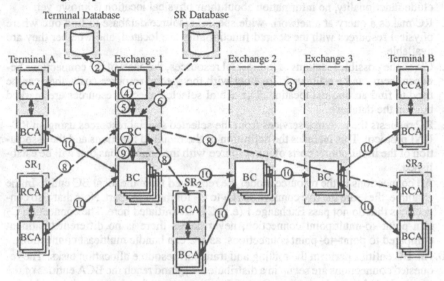

Fig. 6. Control procedures for the setup of the example call

1. The CCA of the initiating terminal A requests the call consisting of elementary call objects from the responsible CC entity. A defines a Party Set with Parties A, B and C, and the Abstract Services 'video' and 'audio'. Party A also specifies its own Composite Downstream Mapping Elements to both Abstract Services. The attributes of ASM_{A1} determine, that A wants to use MPEG for video coding, the resolution is also specified. The G.721 format is used for audio coding (ASM_{A2}).
2. The CC entity requests information about the participating terminals A, B and C from a terminal database. This includes terminal access rights and possibly addressing information for B and C.
3. The call control offers the call containing the Party Set and the two Abstract Services to the other participants B and C. The ASMs and Mapping Elements of each party are negotiated. In the example, Party B wants to use other coding algorithms than A for audio (G.711) and video (H.261). Party C does not have video capabilities, it only accepts the Abstract Service 'audio'. It is assumed that A has defined the participation of the other parties in the Abstract Service 'video' as optional. So the call can still be set up, even though C does not participate in the video component.
4. At this moment, all components of the complete call are known, and the CC entity can start to determine the appropriate resource configuration. The determination of the resource objects is not trivial as it is not a one–to–one mapping, and there can be several solutions. Figure 7 shows a possible configuration of resource objects, that can support the example call. The configuration consists of the special resources SR_1 to SR_6, connection end points in the terminals and resources, and unidirectional multicast connections with appropriate parameters (C_1 to C_{11}). The converters SR_1, SR_2, the combiner SR_3 and the connections C_1 to C_5 belong to the video part, which will be further discussed. If not already defined by the terminals or in any case for connec-

tions between special resources (like e.g. C_4), connection parameters like bandwidth or delay have to be derived from the more abstract coding attributes.

5. CC requests the special resource objects from RC. The requested objects do only include functionality, no information about their physical location is known yet.

6. RC makes a query at a network–wide special resource database to find out, where physical resources with the desired functionality are located, and whether they are available.

7. Given the possible locations of the special resources, the costs of the connections between them can be estimated. Together with the special resource costs, they can be used to find an optimal location. The actual selected special resources are marked busy in the database.

8. RC requests the resource services from the selected special resources using an RC–UNI protocol. This includes the definition of the required functions and the association of the input/output ports of the resource with the connections that will be established.

9. All connections in the resource model are requested from the local BC entity. In the example, these are the five connections C_1 to C_5 for the video part. Note that also connections that do not pass Exchange 1 (e.g. C_3) are initiated here. The connection C_2 is a point–to–multipoint connection, nevertheless there is no different treatment compared to point–to–point connections, as BC can handle multicast connections.

10. The BC entities perform the routing and transport resource allocation tasks. The requested connections are set up in a distributed way and reach the BCA entities of the terminals and special resources. One of the parameters of each connection is a higher layer identifier, which enables the association between the connection and the ASM. For example, connection C_1 carries an identifier of ASM_{A1} for party A.

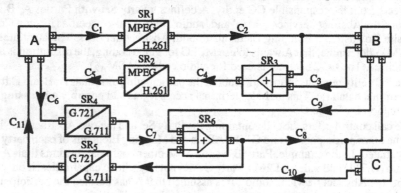

Fig. 7. Resource objects of complete example call

5 Conclusion

The short excursion into evolution of broadband signaling has revealed the following points. The current standardization approaches are not capable of supporting complex multimedia services. More advanced approaches offer the required functionality at the user–network interface. However, although they introduce a functional separation of the control plane into several levels (call, resource, bearer), they continue to perform control tasks in a distributed way. This leads to a number of problems, which have been pointed out.

A framework for a very powerful signaling system has been proposed. The object–oriented call model allows the configuration of arbitrary complex multimedia calls. Conversion and bridging functions that are necessary for multi–party services with heterogeneous terminals are supported by transparently including special resources. The two main architectural concepts are the centralized call and resource control and the use of an autonomous connection control system. The most important advantages are:

- no NNI signaling is required at the call and resource level
- signaling effort and delay is reduced
- location of special resources is simplified and network–wide optimal resource routing can be performed
- connection control and call/resource control can run on different platforms
- call and resource control is independent of switching and transmission technology
- services can be provided by independent authorities at three different levels: telecommunication services, special resource services and pure bearer services

Implications and new tasks that arise from centralized call and resource control have been described and possible solutions have been given.

References

[1] CCITT Recommendations Q.930–Q.940: Digital Access Signaling System, Network Layer, User–Network Management. *CCITT Blue Book,* Vol. VI, Fascicle VI.11, 1989.
[2] P. Blankers: Functional Modelling for Target B–ISDN Signaling. *Proceedings of IEEE Globecom '92,* 1992, S. 1166–1171.
[3] ATM Forum: UNI Signalling 4.0. *ATM Forum,* August 1995.
[4] ITU-T Draft Q.2931: B-ISDN, User-Network Interface Layer 3 Specification for Basic Call/Connection Control. *ITU-T Study Group XI,* Draft Q.2931, June 1994.
[5] ITU-T Draft Q.296x: B-ISDN, DSS2, Additional Traffic Parameters, Look Ahead Procedure, Negotiation, Connection Modification. *ITU-T Study Group XI,* Draft Q.296x, June 1994.
[6] ITU-T Draft Q.297x: B-ISDN, DSS2, Point–to–Multipoint. *ITU-T Study Group XI,* Draft Q.297x, June 1994.
[7] ITU-T Draft Q.298x: B-ISDN, DSS2, Point–Point Multiconnection. *ITU-T Study Group XI,* Draft Q.298x, June 1994.
[8] CCITT Study Group XI – A. Paglialunga: ISCP Baseline Document (Ver. 2.1). *CCITT Study Group XI,* Geneva, March 1992.
[9] RACE II MAGIC R2044: Stage 2 Specification. *RACE II Multiservice Applications Governing Integrated Control (MAGIC),* Deliverable 6, September 1993.
[10] RACE II MAGIC R2044: Protocols and Concepts of B–ISDN Signalling. *RACE II Multiservice Applications Governing Integrated Control (MAGIC),* Deliverable 10, October 1994.
[11] S. Minzer: A Signaling Protocol for Complex Multimedia Services. *IEEE Journal on Selected Areas in Communications,* Vol. 9, No. 9, December 1991, pp. 1383–1394.
[12] H. Bussey, S. Minzer, P. Mouchtaris, S.L. Moyer, F. Porter: EXPANSE Software for Distributed Call and Connection Control. *Int. Journal of Communication Systems,* Vol. 7, No. 2, April–June 1994, pp. 149–160.
[13] H. Müller: Flexible Connection Control for Multimedia Broadband Networks. – To appear in: *Proceedings of IEEE ICC '95,* June 1995.

A Generic Concept for Large-Scale Multicast

Markus Hofmann

Institute of Telematics, University of Karlsruhe,
Zirkel 2, 76128 Karlsruhe, Germany
Phone: +49 721 608 3982, Fax: +49 721 388097
E-mail: hofmann@telematik.informatik.uni-karlsruhe.de
URL: http://www.telematik.informatik.uni-karlsruhe.de/~hofmann

Abstract. Upcoming broadband networks offer a bearer service suitable for modern distributed applications. High bandwidth capacity and low transfer delay are two characteristics of such communication services. Beside these performance-oriented parameters, many applications have additional requirements in respect to quality of the used communication service. Computer Supported Cooperative Work, distributed parallel processing and virtual shared memory, for example, depend on error-free data exchange among multiple computer systems. Additional problems occur for the provision of a reliable multipoint service, where errors are more likely and the sender has to deal with numerous receivers. In order to meet the required reliability, powerful and scalable error control mechanisms are essential. Therefore, this paper presents a novel concept, named *Local Group Concept (LGC)*, for large-scale reliable multicast.

Keywords. Reliable Multicast, Large-Scale Networks, Local Groups, Implosion Control, Retransmissions, Dynamic Groups

1 Motivation

In the near future, global information exchange will become an essential resource in worldwide economies. Forthcoming computer applications will require reliable data transfer within large groups, whose members may be spread worldwide. To support reliable information exchange in such a scenario, protocols have to be changed to provide efficient and scalable multicast error control and traffic control.

One problem new protocols have to deal with is known as the *implosion problem*. As the number of communication participants becomes very large, a sender is swamped with return messages from its receivers. These messages may be generated as a result of status requests or as a result of data loss in conjunction with retransmission-based error control. The effect of implosion is twofold. Firstly, the large number of return messages results in processing overhead at the sender and, therefore, delays data communication. Secondly, a tremendous amount of control units may cause an excess of both, bandwidth and bufferspace, which in turn causes additional message losses. An optimal control scheme for multicast communication would reduce the number of status reports received by the sender down to one.

A second issue of great importance is the development of efficient *error correction* schemes for multicast communication. Common protocols use Go-Back-N or selective repeat to retransmit lost and corrupted data. Receivers do request missed data directly from the multicast sender without any consideration of network topology and actual network load. In case of group communication, it is also possible to exchange data with other receivers. It is preferable to request lost and corrupted data from a communication participant placed next to the end-system missing some information. An optimal error correction scheme would stimulate the retransmission of missed data units by the receiver located closest to the failing system. This would minimize transfer delay and network load. However, it seems to be very hard to develop an optimal control and error correction scheme. There will be always some kind of trade-off between complexity of protocol mechanisms and benefits to be achieved by an optimal solution.

Many protocols have been developed to support data exchange among various communication participants. Some of these protocols, such as MTP (Multicast Transport Protocol) [1] and RMP (Reliable Multicast Protocol) [2], implement a centralized control and error correction scheme to provide a totally ordered multicast delivery. The central instance controlling data transfer may become a bottleneck when dealing with numerous receivers. Therefore, these protocols will not scale well in respect of the group size. An approach based on the establishment of so called multicast servers within the network is given in [3]. Other protocols with centralized, sender-based control integrate special mechanisms to avoid acknowledgment implosion. A recent version of XTP (Xpress Transfer Protocol) [4], for example, defines damping and slotting algorithms to reduce implosion. Instead of returning control units exclusively to the multicast sender, receivers transmit them after a random delay to the whole group. Consequently, every group member receives this message and skips its status report if the incoming control unit corresponds to its personal state. This mechanism might reduce the number of acknowledgments to be processed by the sender. But in large-scale global networks, the usefulness of multicasting control units is very questionable [5]. A large number of participants together with the mechanism described above may cause a flood of control units all over the global network. SRM (Scalable Reliable Multicast) [6], which integrates a similar mechanism to implement a receiver-based error control, uses timers carefully set to avoid a flood of retransmission requests. However, the correct setting of timers will be very difficult for high dynamic networks with quickly changing load and frequently changing network structure.

This paper presents a novel error control mechanism suitable for global heterogeneous networks. It combines the advantages of both, sender-based and receiver-based control schemes. The concept aims at reducing transfer delay, network load and the acknowledgment processing overhead at the sending side. Section 2 and Section 3 present the proposed framework based on the Local Group Concept (LGC). Section 4 gives results of a performance analysis in respect to a simple example scenario. Finally, Section 5 draws concluding remarks.

2   The Local Group Concept (LGC)

The Local Group Concept (LGC) presented in this paper is based on a best-effort delivery model with multicast support. While these requirements are perfectly in conformity with IP and the current MBone, the Local Group Concept is not restricted to the Internet protocol family. Although this paper focuses on the use of the traditional IP multicast distribution model, the generic concept can also be integrated in an extended ATM Adaption Layer or in other protocol architectures with multicast support. Additionally, LGC avoids changes within internal network equipment, such as ATM switches or IP routers. The integration of implosion control into these systems requires protocol processing capabilities inside themselves. This implies more complex and less flexible intermediate systems. In addition, a router doing implosion control for hundreds of multicast connections will soon become a bottleneck. Therefore, the LGC approach focuses on implosion control on end-to-end basis.

2.1 Basic Principles of LGC

The problem with implosion is that a single system has to control a very large number of receivers. The basic principle of the Local Group Concept is to split the burden of acknowledgment handling and to distribute error correction as well as traffic control among all the members of the multicast group. To achieve better scalability of point-to-multipoint services in respect of group sizes, distances, and data rates, LGC divides global multicast groups into separate subgroups. These subgroups will combine communication participants within a local region, forming so called *Local Groups (LG)*. Each of them is represented by a specific communication node, named local *Group Controller (GC)*. These nodes support the provision of the following enhanced services:

- *Local Retransmissions*: Controllers of Local Groups are able to perform and to coordinate retransmissions of lost and corrupted data within their subgroup. This reduces delay caused by retransmissions and decreases the load for sender and network.

- *Local Acknowledgment Processing*: The integration of acknowledgment processing capabilities into local Group Controllers reduces the acknowledgment implosion problem of reliable multicast for large groups. Local Group Controllers evaluate received control units and inform the multicast sender about the status of the Local Group. This includes error reports as well as parameters to control data flow. Parallel processing of status reports and their combination to a single message per Local Group relieves the multicast sender as it reduces the number of control units to evaluate at the sending side.

In each local region, one of the receivers is determined to function as local Group Controller. The sender itself is defined always as a local Group Controller. The dedicated system has to collect status messages from all members of its subgroup and has to forward them to the multicast sender in a single composite control unit. Controllers of subgroups are also responsible for organizing local retransmissions. After evaluating received status messages a local Group Controller tries to transfer lost data units to all receivers that have observed errors or losses. To retransmit data units a local Group Controller can use either unicast or restricted multicast transmission. This decision may be static or dynamically based on the number of failed receivers. If a controller itself misses some data units, it will try to get them from another member of its Local Group. Therefore, a multicast sender has only to retransmit messages missed by all members of a subgroup. Local retransmissions lead to shorter delays and decrease the number of data units flowing through the global network.

An example scenario illustrating the basic idea and the advantage of the Local Group Concept is given in Figure 1. A multicast sender communicates over a satellite link with four receivers, which are connected to a common router. The satellite link is characterized by high transfer delay and high carrier fees. Therefore, it is desirable to reduce data traffic over this link. In this type of scenario it is useful to combine all four receivers into a single subgroup. One of the receiving end-systems has to function as the controller of the subgroup. In this case, local retransmissions do not traverse the satellite link. This reduces transfer delay and network load within the satellite link.

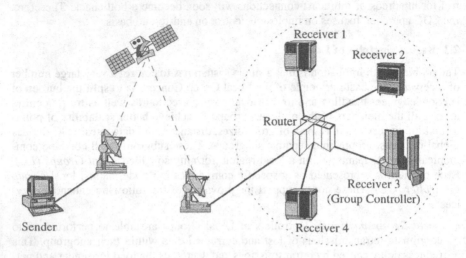

Fig. 1. Combining End-Systems to one Local Group

In this example, the decision to combine all four receivers into one single subgroup was based on the intention of minimizing transfer delay. The next Section deals with aspects that influence the creation and structure of Local Groups.

2.2 Metrics for Building Local Groups

The Local Group Concept is based on the combination of receivers within a 'local region'. Thus, the term 'local region' has to be clarified. Some kind of metric is necessary to determine the distance between two communication nodes. The suitability of different metrics, such as delay, bandwidth, throughput, error probability, reliability, carrier fees, or number of hops between two nodes, depends mainly on the application using a communication service. While an interactive application may wish to minimize transfer delay, a user transferring data files is interested in reducing financial costs of the transfer. However, there is a trade-off between complexity due to general distance metrics and efficiency when using simple metrics such as hop counts. A discussion about the usefulness of metrics such as transfer delay or hop counts can be found in [7] and [8]. The decision which kind of metric can be used also depends on the underlying network protocol. In some cases it is necessary that the routing protocol supports the chosen unit of measurement. IP, for example, considers metrics like delay, throughput, and reliability, while the ISO protocol IS-IS supports bandwidth, error probability, delay and financial costs [9]. Combinations of these mentioned metrics can also be taken into consideration. If the underlying network service supports no metric at all, the measured round trip time may be defined as the distance between two stations.

A simple solution for a similar problem has been proposed in [10]. The generation of local multicast regions as defined in the Designated Status Protocol (DSP) and the

Combined Protocol (CP) is mainly based on addresses and the geographical location of end-systems (for example, end-systems connected to a common switch). Such a static mechanism leads to problems supporting huge MANs (e.g., DQDB or FDDI) and mobile environments. The definition of local regions based on the geographic location of end-systems does not work very well in general, as mentioned in [7], because geographic information often does not correlate to network structure. Therefore, the Local Group Concept aims at flexible, dynamic and application-depending generation of Local Groups.

2.3 Selection and Placement of Group Controllers

An important detail in the design of a concept like LGC is the decision where to place the functionality of local Group Controllers. The question arises which type of communication system is suitable to be responsible for controlling a Local Group. Possible answers are, for example, end-systems, routers, or switches. If there are several suitable systems, which selection will fit best?

The CP protocol mentioned in [10] uses local exchanges (e.g., routers or ATM switches) to combine status messages of receivers within a local region. This solution implies modifications of local exchanges and restricts the Local Group Concept to operate in the network layer. By placing local Group Controllers into end-systems, the concept becomes 'layer-independent'. In addition, the end-system-based approach avoids the necessity of any changes in switches and routers. In this case, the protocol mechanism can be used in networks based on different protocol architectures. The placement of local Group Controllers in end-systems scales better for a large number of multicast connections. A local exchange doing retransmissions and acknowledgment processing for different multicast connections may soon become a bottleneck. It is advantageous to parallelize the job of handling different Local Groups and to distribute it to several communication nodes. Each multicast connection should have its own, separate local Group Controller. The end-system based solution allows for choosing different communication participants within the same local region for different multicast connections.

The Local Group Concept presented in this paper integrates the functionality of local Group Controllers into normal end-systems. In principle, every communication participant is able to perform the tasks necessary for controlling a subgroup. This strategy is similar to the selection of a monitor station within a Token Ring network [11]. One member per Local Group is designated as the active controller. The selected station represents the Local Group and performs local acknowledgment processing and local retransmissions.

Another important issue is the dynamic change of communication groups and network traffic. The structure of Local Groups and the placement of their controllers should always be adapted to the actual state of the network and the communication group. Therefore, it must be possible to move functionality of local Group Controllers during the lifetime of a connection.

The DSP protocol described in [10], for example, combines status messages in end-systems, which are selected at connection setup time. For that reason, no chance to reassign the role of a Group Controller is given. Careful selection and placement of local Group Controllers assist in optimizing network load and transfer delay. Characteristics such as processing capacity, memory size, and location within the network, determine the suitability of an end-system for the role as a local Group Controller. Other important issues concern network topology, group size, and group structure. Due to this wide range of parameters, there is no simple algorithm to find the optimal local Group Controller. Instead, near-optimal algorithms or other heuristics could be applied. A simple solution is to define the member next to the multicast sender as local Group Controller.

LGC defines local Group Controllers in the following manner. The initiator of a new Local Group always becomes the first controller of this subgroup. After a certain period of time or after joining of a new member, the local Group Controller initiates a reconfiguration process. During this process, every member of the Local Group reports its parameters relevant for the decision to the local Group Controller. Like the metric used to determine the distance between two systems, these values may depend on the application using the communication service. Possible parameters are distance to the multicast sender, performance, and buffer space of each end-system or the location of a receiver in relation to other members of the Local Group. Based on information obtained from its members, the local Group Controller determines the most suitable end-system to take over the role as local Group Controller in the future. If a change is advantageous, all necessary information migrates from the old to the new controller and all members of the Local Group are informed about this change. During this reconfiguration process, every join or leave request is queued and performed later.

2.4 Hierarchy of Local Group Controllers

To improve the benefits of LGC, Local Groups can be organized in a hierarchical structure. A hierarchy of Local Groups exists, if a local Group Controller LG_2 does not report directly to the multicast sender. Instead, it sends retransmission requests and status reports representing the whole Local Group to a local Group Controller LG_1 next to itself. This one treats LG_2 like a normal receiver. Therefore, all retransmission requests of LG_2 are redirected to and satisfied by LG_1. In this case, LG_1 is named the *parent* of LG_2, while LG_2 is the *child* of LG_1. It is obvious that this hierarchy will decrease transfer delay, network load, and implosion additionally.

Two changes have to be made to support a hierarchical structure. Firstly, every local Group Controller (except the multicast sender) has to subscribe to the nearest Local Group as a normal receiver. Secondly, every local Group Controller has to redirect its control messages to the next local Group Controller at an upper level (to its parent). Additionally, it may be necessary to increase timeout values of parents, because children have to collect all status reports of their members before answering to their parent. This timeout adaption could be done dynamically during a synchronizing handshake. The remainder of this paper does not consider the hierarchy in more detail. Nevertheless, all the algorithms described in the following Sections may be easily modified to support hierarchical Local Groups.

3 Protocol Mechanisms realizing LGC

The following Subsections give a more detailed description of the Local Group Concept. A complete specification including pseudo code for integrating the concept into existing multicast protocols can be found in [12].

3.1 Joining Group Communication

Before receiving any data, new communication participants have to subscribe to a nearby Local Group in respect to the chosen metric. A simple possibility to find an appropriate local Group Controller is to use a service similar to the Host Anycasting Service described in [13]. The search may be based on routing information, too, obtained directly from routing tables. If there is no support by the network layer at all, an expanding ring mechanism [7] based on round trip times can be used. The overhead produced by this search strongly depends on the mechanism applied to find an appropriate Local Group. An end-system using the Host Anycasting service just sends one anycast request and gets back a single answer within a period of time corresponding to the round trip time between itself and the answering Group Controller. In contrast, an expanding ring search will cause much more overhead. Depending on the result of the search, different actions have to be performed:

- If no Local Group is found within a defined distance according to the chosen metric, the joining end-system establishes a new subgroup, appoints itself as Group Controller and informs the multicast sender about this event. Hence, the multicast sender has knowledge of all Local Groups and their representative controllers.

- If an appropriate Local Group exists within a certain distance, the joining end-system registers itself at the corresponding Group Controller. Therefore, a local Group Controller always knows the identity of all its members. This is a prerequisite for providing a reliable multicast service.

After subscription to a Local Group, the end-system is allowed to participate in the group communication. The division of dynamic communication groups into local subgroups has to be updated after a certain time. Due to dynamic changes in group membership, it is necessary to modify the group structure. For example, it may be preferable to place the local Group Controller in a newly joined participant that is located in the center of the subgroup. Such a reconfiguration process could be implemented according to the mechanism mentioned in Subsection 2.3. A reconfiguration is also necessary, if the division into Local Groups is based on routing information, which can be incorrect for limited periods of time and which is not globally consistent. Periodic updates of the group structure correct errors caused by temporary inconsistent metric information.

Due to the knowledge of group membership, a multicast sender is able to guarantee the correct data transfer to all of its children. Local Group Controllers are responsible for correct transmission of user data to all members of their subgroup. This allows to provide a reliable multicast service that scales well for either long distances and a large number of receivers.

3.2 Transfer of Multicast Data

The sender multicasts data units (e.g., using IP-multicast) to all destinations. After a certain period of time, the sending station requests the status of all receivers. Instead of returning control units directly to the sender, receivers transmit them to the controller of their Local Group. The controller collects incoming status messages, processes them, and calculates the status of the whole subgroup. Applying the knowledge of received control units, a local Group Controller requests data units missed by itself from other members of the subgroup and performs all relevant local retransmissions.

In addition, the controller combines received status messages into a single acknowledgment and transmits it to the multicast sender. The acknowledgment includes the request for all data units missed by all members of the Local Group. This mechanism ensures that every data unit received successfully by at least one group member is not requested at the multicast sender anymore, even if the Group Controller itself misses this data unit.

After the reception of user data, receivers deliver the information to the service user. However, they are not allowed to release the corresponding buffer space. This is because the local Group Controller may miss these data and request it at a later stage. Therefore, receivers have to wait either for a sufficient amount of time, or for an explicit permission by the local Group Controller before releasing the receive buffer. The period of time data must be kept by receivers depends among other issues on the radius of the Local Group and the frequency of status requests.

3.3 Leaving Group Communication

If a normal receiver wants to leave group communication, it unsubscribes itself at the corresponding local Group Controller. After performing necessary retransmissions, the Group Controller confirms unsubscription and the receiver is allowed to leave.

If an end-system functioning as a local Group Controller wants to leave communication, first of all an appropriate successor has to be determined. The new controller obtains all relevant group management information, like status and member information, from its predecessor. Afterwards, all members of the Local Group are informed about the identity of their new controller. Henceforth, it is the new local Group Controller that deals with new data sent by the multicast sender and that receives status reports from group members. However, the leaving Group Controller has to ensure immediately that all outstanding retransmissions to its former members have been dealt with, before being allowed to leave.

If the leaving end-system is the only one within its Local Group, it just unregisters itself directly at the multicast sender or at its parent.

3.4 Fault Tolerance

With respect to fault tolerance it is interesting in which way the Local Group Concept handles failures of a local Group Controller. The question arises, how sending and receiving end-systems discover such a failure and what will be done in this case.

As mentioned above, local Group Controllers must periodically send messages to initiate a buffer release at receiving sides. If a local Group Controller fails, receivers will not get such messages any more. In this case, they infer a failure of their local Group Controller. The multicast sender or other controllers within a Local Group hierarchy detect a failing child from the missing answer in response to a status request. Every end-system reports such a failure to its parent.

The reaction to the failure of a local Group Controller depends strongly on the application given group semantic. If the application requires an all reliable communication service and at least one of the local Group Controllers fails, the connection will be aborted and an error will be indicated to the service user. If the failure of a receiver is acceptable and an active controller fails, another group member has to take its role. For example, this could be a replicated local Group Controller (similar to the solution presented in [14]) or a dynamically determined successor within the Local Group. It is also possible that all members of the failed Local Group search for a new Local Group according to the mechanism described in Section 3.1.

3.5 Integration of Local Groups in XTP

As an example of integrating the novel concept into existing transport protocols, the mechanisms described above have been adopted to XTP, version 3.6 [4]. Work is ongoing to integrate the Local Group Concept into an implementation of XTP 4.0 [15]. An additional packet type for migration of management information must be added. To implement a k-reliable multicast service, sender and Group Controllers must be able to distinguish between different receivers. Therefore, an additional field identifying individual source addresses has to be integrated into CNTL packets of XTP.

4 Performance Evaluation

Analytical methods were applied and simulations were performed in order to evaluate the benefits of the Local Group Concept. The simulations were based on BONeS/ Designer, an event-driven network simulation tool by Comdisco. During evaluation, two important measures were focused:

- *Transfer Delay*, which is the primary concern from the application's point of view.

- *Network Load*, the most interesting measure from the networker's perspective.

The simulation scenario consists of a sender and various receivers that are connected to a local area network. The sender is linked to the LAN over a wide area network. If the satellite link in Figure 1 is replaced by a wide area network, these two scenarios will be identical.

The evaluation examined the impact of group size on average transfer delay and network load. The values obtained for the Local Group Concept were compared to the corresponding results for two common sender-based techniques using multicast and unicast retransmission. Transfer delay was assumed to 20 ms for the wide area link, which is approximately the delay for transferring data between the East and West coast of the USA. Transfer delay within the local area network was fixed at 2 ms. Error probability was assumed to be 10^{-1} for message loss caused by buffer overflow or bit errors, which is not uncommon for highly loaded internetworks. This value was derived from many measurings to randomly selected internet hosts using *ping* command. Other simulation parameters included data rate, processing delay within communication systems, status request rate, and burst length.

The impact of group size on average transfer delay is given in Figure 2. The graph shows for all techniques an increase in the average transfer delay with increasing numbers of receivers. The sharp increase for the common, sender-based approach can be explained with the large number of acknowledgments that have to be processed solely by the multicast sender. The relative high values for average transfer delay are mainly caused by retransmissions over the wide area link. Local retransmissions, as proposed by the Local Group Concept, effect a less strong increase in transfer delay. The benefit for the Local Group Concept is extremely high for large communication groups, but transfer delay still increases with the number of receivers. The establishment of several Local Groups with a restricted number of receivers leads to a distribution of acknowledgment processing to different local Group Controllers. In the third simulation, the maximum size of Local Groups was defined to 50 members. Therefore, every Group Controller just has to process a maximum of 50 acknowledgments. This results in an soft upper bound for transfer delay, because every Local Group with more than 50 receivers is separated into diverse subgroups with fewer members. This division distributes the burden of acknowledgment processing to different controllers and avoids further increase of transfer delay due to acknowledgment processing overhead.

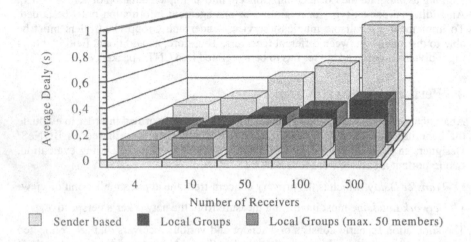

Fig. 2. Impact of Group Size on Transfer Delay

To clarify the difference between unicast and multicast retransmission for both of the common techniques, 50 receivers were added at the sending side of the wide area network for evaluation of network load. The result is illustrated in Figure 3.

For both common techniques, the ratio of retransmissions and total data traffic over the wide area link depends directly on the number of recipients. The Local Group Concept, in contrast, is not influenced by the number of participants and shows very low values compared with the results of the other techniques. The independence between network load and the number of receivers simplifies resource reservation for highly dynamic communication groups. The bandwidth required on the wide area link can be reserved regardless of the actual or future group size. Additional resource reservation during the data transfer phase due to joining participants is not required.

The Local Group Concept requires additional memory at receiving sides due to the delayed release of receive buffers. Simulations have demonstrated that the maximum amount of additional buffer space is independent of the error probability and of the group size. It corresponds directly to the number of data units received between two succeeding release messages. If every tenth data unit is followed by a release message from the corresponding Group Controller, the additional memory at receiving sides will not exceed the buffer space required to save ten data units. Therefore, the need for additional memory can be regulated by the frequency of release messages.

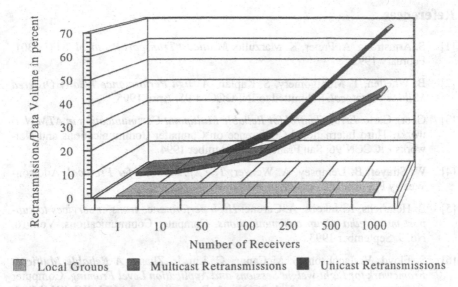

Fig. 3. Impact of Group Size on Network Load

Additional results and detailed information about the analytical methods, the simulation tools and the simulation models can be found in [12].

5 Conclusion

The general goal of the developed concept is the provision of scalable protocol mechanisms appropriate for multicast communication in global heterogeneous networks. Performance evaluation including simulation and analysis of network load shows for the Local Group Concept significant improvements compared to common multicast techniques. The benefits arise from replacing global retransmissions by local data exchange and additionally, from parallel preprocessing of acknowledgments within Local Groups. The performance benefits are achieved without the necessity to modify network equipment such as ATM switches or IP routers, and without a dramatically increased buffer overhead.

Careful placement of local Group Controllers assists in optimizing transfer delay and network load. Depending on particular group dynamics and network topology, it is advantageous to place controllers near or at the centre of a Local Group. There is no simple algorithm to find the centre of a dynamic group at all. Instead, some kind of heuristic has to be developed defining the right balance between simplicity and optimal placement of the controller.

6 Acknowledgments

I would like to thank my colleagues Torsten Braun and Georg Carle for valuable comments and suggestions on various subjects of this paper. In addition, thanks to Ralf Seufert, who performed some of the simulations presented in this paper.

References

[1]   S. Armstrong, A. Freier, K. Marzullo: *Multicast Transport Protocol*. RFC 1301, February 1992

[2]   B. Whetten, T. Montgomery, S. Kaplan: *A High Performance Totally Ordered Multicast Protocol*. Submitted to INFOCOM'95, April 1995

[3]   Georg Carle: *Error Control for Reliable Multipoint Communication in ATM Networks*. Third International Conference on Computer Communications and Networks - ICCCN'96, San Francisco, September 1994

[4]   W. Strayer, B. Dempsey, A. Weaver: *The Xpress Transfer Protocol*. Addison-Wesley Publishing Company, 1992

[5]   B. Heinrichs, K. Jakobs, A. Carone: *High performance transfer services to support multimedia group communications*. Computer Communications, Vol. 16, No. 9, September 1993

[6]   S. Floyd, V. Jacobson, S. McCanne, C. Liu, L. Zhang: *A Reliable Multicast Framework for Light-weight Sessions and Application Level Framing*. Computer Communication Review, Vol. 25, No. 4, Proceedings of ACM SIGCOMM'95, Cambridge, Massachusetts, August 1995

[7]   J. Guyton, M. Schwartz: *Locating Nearby Copies of Replicated Internet Servers*. Computer Communication Review, Vol. 25, No. 4, Proceedings of ACM SIG-COMM'95, Cambridge, Massachusetts, August 1995

[8]   M. Crovella, R. Carter: *Dynamic Server Selection in the Internet*. Third IEEE Workshop on High Performance Subsystems (HPCS'95), Mystic, Connecticut, August 1995

[9]   Radia Perlmann: *Interconnections: Bridges and Routers*. Addison-Wesley Publishing Company, 1992

[10] S. Paul, K. Sabnani, D. Kristol: *Multicast Transport Protocols for High Speed Networks*. Proceedings of International Conference on Network Protocols, Boston, October 1994

[11] W. Stallings: *Local and Metropolitan Area Networks*. Fourth Edition, Macmillan Publishing Company, 1993.

[12] Markus Hofmann: *Reliable Group Communication in Global Heterogeneous Networks*. Diploma Thesis at Institute of Telematics, University of Karlsruhe, August 1994 (in german)

[13] C. Partridge, T. Mendez, W. Milliken: *Host Anycasting Service*. RFC 1546, November 1993

[14] H. Holbrook, S. Singhal, D. Cheriton: *Log-Based Receiver-Reliable Multicast for Distributed Interactive Simulation*. Computer Communication Review, Vol. 25, No. 4, Proceedings of ACM SIGCOMM'95, Cambridge, Massachusetts, August 1995

[15] XTP Forum: *Xpress Transport Protocol Specification, XTP Revision 4.0*. XTP Forum, Santa Barbara, CA, USA, March 1995

On the Potentials of Forward Error Correction Mechanisms Applied to Real-Time Services Carried Over B-ISDN

Thomas Stock*
Alcatel STR AG
Friesenbergstrasse. 75
CH-8055 Zürich
E-Mail: thomas.stock@alcatel.ch

Xavier Garcia
EPFL/TCOM
ELD Ecublens
CH-1015 Lausanne
E-Mail: garcia@tcom.epfl.ch

Abstract

Variable bit rate real-time multimedia services like highly compressed video or audio tend to be very sensitive to transmission errors. This paper investigates cell and frame loss as well as delay and timing issues in an ATM-based B-ISDN environment that carries such services. The figures derived through simulation using artificial as well as real world sources motivate the introduction of forward error correction (FEC) mechanisms. We describe an approach that combines FEC mechanisms with extended loss recovery properties and cell interleaving and that is capable of dealing with variable bit rate services. If parameters and components are properly chosen the loss figures can be reduced significantly while still keeping the complexity of the FEC mechanism within acceptable bounds.

Keywords

ATM, B-ISDN, real-time multimedia services, forward error correction, timing regeneration

1   Introduction

Todays advanced transmission technologies feature very low bit error ratios. On the other hand it has been shown that in case of ATM-based switching on top of these transmission systems the cell loss ratios for variable bit rate (VBR) services might not be negligible. This problem is even more striking if the resulting frame loss ratios are observed (see, e.g. in [Stoc94] and [Rose94]). This is particularly true if the potential statistical multiplexing gain of the ATM network is to be exploited.

Unfortunately, variable bit rate real-time multimedia services expected in the B-ISDN, like highly compressed video or audio, tend to be very sensitive to such losses. For these service classes retransmission of data in case of cell loss and frame corruption in general is not an option, e.g. because of delay constraints or storage limitations. In such cases forward error correction (FEC) mechanisms promise to be a viable alternative. This paper tries to shed some light on the potentials and drawbacks of these mechanisms in the context of VBR real-time multimedia services carried over an ATM-based B-ISDN infrastructure.

We are faced with the fact that there is no generally valid network scenario for such an infrastructure. Neither do we know how the complete details of future real-time services' traffic characteristics will look like. Therefore we had to select some reasonable and — at least to some extend — representative environments to be simulated. As traffic under test we use a variable bit rate MPEG-1 encoded video stream which is sent through a sequence of ATM switching stages. The switches are loaded with various different background traffic streams.

---

*Contact person

Section 2 describes the basic network configuration simulated and the traffic streams considered. Section 3 presents some observations obtained by investigating the results of numerous simulations without applying FEC mechanisms. Section 4 describes possible FEC configurations and draws some conclusions concerning their dimensioning and usefulness on the basis of the simulation results of section 3. Briefly a method for timing regeneration in such environments is shown. Section 5 draws some general conclusions and provides an outlook of how this work will be continued.

2 Simulation scenario

2.1 Network environment

The aim of the simulation setup was to study both loss and timing aspects of a traffic under test (TUT) traversing an ATM network. The simulation scenario is depicted in figure 1. It consists of two end systems, e.g. multimedia workstations, communicating through a network of three ATM switches. Each switching node, implemented as a multiplexer with limited buffer, is loaded with background traffic from a number of different sources. The distance between the switches was assumed to be one cell length[1] which is reflected in the figures on cell transfer delay in section 3. All the links between the nodes as well as those to and from the end systems are assumed to be 34.368 Mbit/sec E3 links.

The TUT is generated by end system 1 and sent on to the first switching node. There the TUT is multiplexed with the background traffic and passed to the next node. Due to the limited amount of buffer of 53 cells in the multiplexing stage[2] — motivated by an earlier version of a cross connect product — some of the TUT or background cells may be lost. The background traffic is assumed not to interfere further with the TUT and thus is directly routed to a sink after leaving node 1. The situation at the subsequent nodes is analogous.

Statistics for the TUT were collected at each node to gain insight of the behavior and the evolution of main timing and loss parameters. See sections 3 and 4 for details.

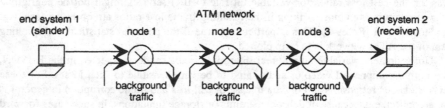

Figure 1: Simulation Setup

---

[1]It should be noted that the distance between nodes was chosen arbitrarily. Since there is no closed loop flow control the selection does not have any impact on the statistical behaviour of the system except for some constant additional factors in the cell transfer delays.

[2]This selection was motivated by the buffer size of an earlier realease of an Alcatel cross connect.

2.2 Traffic under test

For the TUT we concentrated on variable bit rate video traffic which features reasonable load and is sensitive with respect to timing and loss. Taking into account the bursty nature of VBR video traffic two approaches were considered: an unshaped source that transmits the cells back-to-back as soon as a frame is generated and a shaped source that does a per frame shaping. Bearing in mind that 24 new frames are generated per second, the latter source spreads all the cells belonging to a single frame equally along the frame duration interval of 1/24 seconds. Shaping over several subsequent frames, e.g. an MPEG 'group of pictures', has not been considered because of the excessive additional delays required (see [Rose94] where cell loss has been investigated for such an approach, named 'rate averaging'). Still our simple shaping approach significantly reduces the burstiness of the video source.

Accurate modeling of video sources has been proven to be difficult due to long range dependency statistics (see, e.g. [Garr93]). To get rid of model inaccuracy, we used the well known MPEG-1 frame size trace from the movie 'Star Wars' made available by M.W. Garret. A detailed description of the trace can be found in [Garr93]. The full trace being extremely long for simulation purposes (2 hours) we extracted a five minutes sequence to be used as a basis for the two different sources. We would like to state explicitly that this MPEG-1 stream is to be considered as one example of a real-time VBR stream. Other examples of 'real-world' real-time VBR stream can be envisaged.[3] This is the main reason why we will not consider the internal characteristics of the MPEG-1 stream, e.g. distinguish header and payload data, etc. in this paper.

frame size [bit]

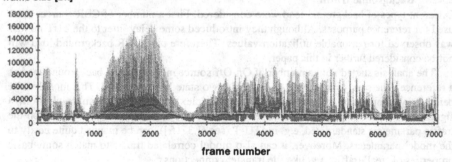

Figure 2: MPEG-1 frame size trace

Depicted in gray figure 2 shows the frame size trace of this selected sequence in bits versus the frame number. The instantaneous frame size averages over 24 successive frames, i.e. one second intervals, is depicted in black. It can be observed that the selected trace can be divided into three main parts of different activity. The starting phase of approximately one thousand frames shows several short bursts (scene cuts) but low average

---

[3]To some extent other examples have been considered yielding similar characteristics as the ones presented in the following.

load. The second phase of approximately one and a half thousand frames depicts a very active scene that results in high average and peak bit rates. The third phase again features a medium average bit rate of a more calm scene nevertheless containing a large number of rather short peaks of medium amplitude. These three main parts capture the variable and bursty nature of video traffic.

Table 1 shows some overall statistical values of the selected sequence of frames. In the literature these values are quite often referred to. Yet, observing the instant frame size averages over 24 successive frames, i.e. one second intervals, as depicted in black in figure 2, it can be easily seen that these instant averages vary significantly over time. Similar figures can be derived for the other parameters given in Table 1. Our conclusion is that these parameters are only of very limited value when describing real world variable bit rate traffic.

Mean frame size	Maximum frame size	Minimum frame size	Standard deviation
16558 bit	161730 bit	600 bit	20211 bit

Table 1: Statistical values of the MPEG-1 trace

Given that the source generates 24 frames per second, that the links simulated have a 'raw' capacity of 34.368 Mbit/sec and taking into account the overhead caused by the ATM cells and sequence numbering, the mean ATM link load generated by the video source is about 1.4% while for single frames peaks of about 14% are reached.

2.3 Background traffic

Different types of background load were considered. First a number of CBR sources were used for reference purposes. Although they introduced some delay jitter to the TUT no loss was observed for reasonable utilization values. Therefore pure CBR background load will not be considered further in this paper.

The analysis started with a number of On-Off source models for the background load as a reference. The sources are based on a simple two state Markov model. This model while being very simple can reflect quite well a multiplex of traffic that can result from a significant number of statistically independent connections in the network. In addition the traffic parameters standardized, e.g. in ITU-T (see [I.371fr]), can be mapped quite easily to the model parameters. Moreover, it can also model correlated traffic to match some basic properties of 'real' traffic, e.g. like file transfer connections.[4]

These were complemented by some mixed background loads. A number of video sources with selected sequences from the video trace mentioned above were combined with either On-Off and CBR sources. The video sequences were chosen from the same trace as the foreground traffic but such that there was no overlap between any of the subsequences.

The total background load applied was 80% and 85% leading to a mean utilization of the multiplexing stages of less than 82% and 87%, respectively, for the pure CBR and On-Off background sources. In the case of mixed source types the background load varied

---

[4]Note that we do not claim that this model exactly reflects the background traffic patterns to be expected in today's and future ATM networks.

between 60% and 75%. The video part accounted for about 14% the rest was due to CBR and uncorrelated On-Off sources (configuration 7). Table 2 summarizes the configurations studied.

traffic configuration	background source types
1	2 CBR
2	2 uncorrelated On-Off
3	2 correlated On-Off
4	8 uncorrelated On-Off
5	8 correlated On-Off
6	10 video, 4 CBR
7	10 video, 4 uncorrelated On-Off

Table 2: Background traffic configurations

3 Observations and problems

As mentioned already the following results are based on simulations each representing about 5 minutes of real time. All results presented in this section are based on the shaped video source as TUT. The use of the unshaped video source yielded figures with similar characteristics but higher amplitudes due to the significantly increased burstiness. See tables 3 and 4 for some results on unshaped video as TUT.

3.1 Loss behaviour

For various configurations the number of successively lost video cells and the absolute frequency of these 'burst losses' were calculated. Figures 3 and 4 depict these measures versus the number of video cells transmitted up until the loss and versus the burst loss size, respectively. Other configurations differ with respect to the absolute frequency of losses which strongly depend on the characteristics of the background sources. As can be expected 'well-behaved' ones, like CBR or uncorrelated On-Off sources, cause a lower number of losses than 'badly-behaved' ones, like correlated On-Off or video sources, (see tables 3 and 4 for more loss related figures). Note that yet the 'shape' of the corresponding figures was very similar for all configurations, just their 'density' varied.

In figure 3 sharp increases in the number of lost cells and consequently corrupted frames of the traffic under test can be observed when the background load of 8 correlated sources is increased from 80% to 85%. For a background load that includes video traffic figure 4 shows a similar behaviour at a load of 65% and 75%, respectively. See table 3 in the following section for the total number of lost cells and corrupted frames for these and other configurations. Although the x-axis values in figures 3 and 4 are given as the number of video cells transmitted while in figure 2 they denote the frame number, it can easily be seen that the major loss appears while the MPEG stream is in its very active second phase (as described in section 2). Once the load reaches a certain level (85% in figure 3 and 75% in figure 4) the shape of the burst loss size curve closely resembles the shape of the instantaneous frame size averages depicted in figure 2. That means: the larger the instantaneous frame size average the more and the larger burst losses can be observed.

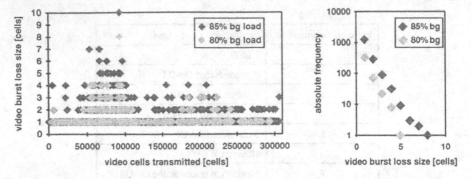

Figure 3: TUT cell loss for 8 correlated On-Off background sources

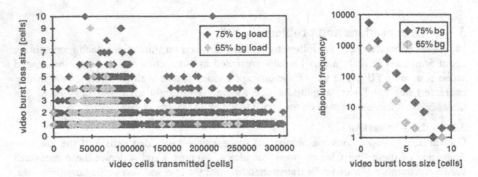

Figure 4: TUT cell loss for 10 shaped background video and 4 uncorrelated On-Off sources

Nevertheless even in the worst cases simulated more than 98% of the burst cell losses were no longer than 5 video cells. In the case closer to a reasonable operating point the ratio was even higher. This observation can be interpreted as a motivation for the introduction of a forward error correction mechanism which can cope with these limited length burst losses. Given the above figures such a mechanism might yield a decrease of the loss measures by almost two orders of magnitude if long burst losses do not follow each other, i.e. if losses are distributed rather equally over time and are not strongly correlated with the inter-loss duration (see [McAu90] and [Bier93]).

3.2   Transfer delay
Even in cases where no or limited loss was observed (80% background load in figure 5 and 65% in figure 6) the cell transfer delay (CTD) varied significantly. Although a large number of TUT cells only experienced a relatively small CTD a significant amount of TUT cells were delayed severely in all configurations investigated. Their CTD was in the range

of about 20 cells to the maximum possible, i.e. n times the ATM buffer size of 53 cells, plus the pure line transmission delay of one cell duration per switch output link. Here n denotes the index of the ATM node where the CTD was measured. In our scenario the maximum is $3.53+3 = 162$ cells or approximately 2 msec at the receiving end system, given that we use ATM on top of 34.368 Mbit/sec links. Note in particular the flat and long tails of the CTD distributions depicted in figures 5 and 6.

The above observations have some severe implications for the receiver of the video stream. Even in situations with relatively low loss — e.g. the 80% curves depicted in figure 5 where 35 frames are corrupted[5] during the second minute of transmission — the video is unacceptably disrupted at least during certain periods of time. In addition the significant fluctuation of the CTD will impose problems for decoders needing strictly limited jitter.[6] This problem will be even more prominent if additional nodes are introduced in the transmission path: figures 5 and 6 show that the CTD range generally increases with the number of nodes passed.

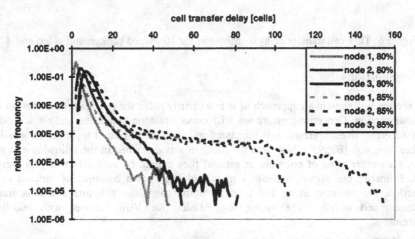

Figure 5: TUT cell transfer delay distribution for 2 uncorrelated On-Off background sources

The distinct local maxima of the third node's CTD distributions depicted in figure 6 are due to the high burstiness of the multiplex of the background video traffic at that node. In the 65% load case all the observed losses (see figure 4) appear at node 3. At 75% load there also appears some loss at nodes one and two but the majority of losses still is observed at the last node.

---

[5]Note that due to the particular type of data encoding used in MPEG this might imply errors in a whole group of subsequent video pictures shown of the receiver's monitor.
[6]In case of MPEG-II transport streams the delay jitter at the system layer has to be limited to 4 msec. At the AAL a jitter limit of 1 msec is under discussion.

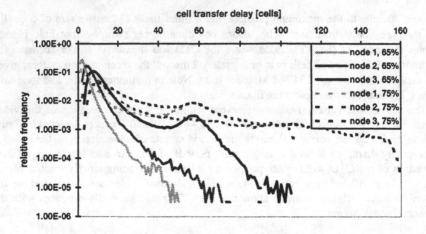

Figure 6: TUT cell transfer delay distribution for 10 shaped background video and 4

4   Solutions

This section will present an approach to at least partly solve the problems described in the previous section. Concerning errors we will concentrate on cell losses only since other types of error, like bit errors, cell misinsertions etc., are expected to be of far smaller number (see, e.g. [Bier93]). Note that the mechanisms described in the following can still detect such other types of error but in general they cannot be used effectively to correct them. Primarily our approach consists in combining known concepts of forward error correction, cluster interleaving and time stamping to reduce cell loss and thus frame corruption and to allow for timing regeneration for VBR services with real-time requirements.

4.1   FEC mechanisms and interleaving

Let d be the so-called interleaving modulus. Calculate $k < d$ FEC cells for each group of d-k subsequent payload cells and insert the k FEC cells into the original cell stream at the end of their respective group of payload cells. Put the resulting cell stream into the order depicted in figure 7, where each horizontal 'cluster' of d cells contains k overhead cells (note that [I.363dr] describes two similar special cases on the bases of octets for use in AAL1, i.e. CBR services). The first d-k columns hold the original data cells. Columns d-$k+1$ to d hold the corresponding FEC cells. All the cells are then transmitted columnwise.

Given some sequence numbering in all the cells, FECs can be designed such that with k FEC cells up to k payload cell losses can be recovered per cluster of d cells. See [McAu90] for a FEC based on Reed-Solomon codes which features these properties. For $k = 1$ such a recovery mechanism can easily be implemented by introducing a parity cell defined as the XOR product of the payload of the data cells; this might even be done in software. The case $k = 0$ is also covered by the above scheme. Note that we did not actually implement the FEC mechanisms.[7] Since we were working in a simulated environment we were able to calculate the results without the need for a real FEC implementation.

up to d-k payload cells per cluster					k FEC cells per cluster		
1	2	...	d-k-1	d-k	d-k+1	...	d
1	1	2	...	d-2	d-k	FECcells(1,...,d-1)		
2	(d-k)+1	...	...	2(d-k)-1	2(d-k)	FECcells((d-k)+1,...,2(d-k))		
...	...	...	...	...	...	...		
d-1	(d-2)(d-k)+1	...	...	(d-1)(d-k)-1	(d-1)(d-k)	FECcells((d-2)(d-k)+1,...,(d-1)(d-k))		
d	(d-1)(d-k)+1	...	...	d(d-k)-1	d(d-k)	FECcells((d-1)(d-k)+1,...,d(d-k))		

Figure 7: Interleaving scheme

The major problem of FECs as well as cell interleaving mechanisms when applied to real-time services is the delay they add to the 'pure' transmission and switching delays D_t and D_s. The mean additional delay Δ_i due to the interleaving process for the example shown in figure 7 is $(d-1)(d-k-1)D$, where D is the mean time between two successive (payload or FEC) cell transmissions (consider, e.g., payload cell d-k). Taking into account possible payload cell losses or errors the reception of the FEC cells corresponding to a particular cluster has to be awaited at the receiver. This results in a mean additional delay Δ_r of $d(d-1)D$ (consider, e.g., payload cell 1). Since Δ_r is larger than Δ_i the proposed mechanism implies a total additional delay of Δ_r.

In case of a VBR stream with a stringent delay limit of L and a reasonable large cluster size d, there might occur situations where $D_t + D_s + \Delta_r$ are larger than L for some cells. Then the payload cell area may be filled partially from top to bottom and left to right (medium grey elements in figure 7) as long as it is guaranteed that the delay limit L will not be exceeded. Unavailable elements are then filled with empty cells (light grey elements). For the calculation of the FEC cells (dark grey elements) the empty cells are assumed to contain, e.g., all zeros. The empty cells are not transmitted. Obviously, this requires some additional indications to be able to distinguish original data from overhead. This can be implemented by introducing — in addition to the sequence numbering scheme which is necessary, anyway — a label inside the cells to distinguish payload data from FEC cells.

The dimensioning of the parameter d may then be such that most of the time the clusters are completely filled, while in situations with exceptionally low traffic partially filled clusters are transferred. Partially filled clusters imply higher overhead. This increase in overhead can be kept low if care is taken that the number of partially filled clusters is significantly smaller than the total number of clusters transmitted.

---

[7]Meanwhile this has been done to be able to perform experiments in a 'real' environment.

Given the fact that the cells are to be stored in an intermediate buffer, they can even be transmitted in a 'shaped' fashion, i.e. in almost constant intervals thus reducing the burstiness of the traffic injected to the network.[8]

4.2   Loss behaviour

Tables 3 and 4 depict cell loss and video frame corruption figures for various configurations and compare the cluster loss figures under the assumption that an FEC as described above had been applied with and without interleaving (see table 2 for a description of traffic configurations). Table 3 shows the results for cluster sizes of 4 and 16 and loss recovery FECs which can regenerate a single lost cell per cluster. Table 4 presents the figures for cluster sizes of 16 and 64 and 'four cell loss recovery' FECs, respectively.

It can be concluded that the use of non interleaved 'single cell loss recovery' FECs does not provide significant improvements: In the case of a 4-cluster FEC the number of cells belonging to corrupted clusters is roughly equal to the number of lost cells while in the case of a 16-cluster FEC the number of cells in corrupted clusters is roughly four times the number of cells lost. The use of interleaving does not significantly improve the cluster loss figures neither with 4- nor with 16-clusters.

This changes when a 'four loss recovery' FEC is introduced. Using 16-clusters yields significant improvements: In particular for the 80% background load situations corruption and thus loss is reduced to almost zero. The use of lower overhead 64-clusters in general yields similar results, although for some higher load configurations it is overtaxed (consider, e.g. the correlated background sources with 85% utilization, background configurations 3 and 5).

TUT / background traffic configuration / background load	cells lost	corrupted MPEG frames without interleaving	corrupted 4-clusters without interleaving	corrupted 4-clusters with interleaving	corrupted 16-clusters without interleaving	corrupted 16-clusters with interleaving
shaped video/2/80%	0	0	0	0	0	0
shaped video/2/85%	316	25	81	51	79	68
shaped video/3/80%	143	35	33	18	33	16
shaped video/3/85%	1357	124	355	254	321	302
shaped video/5/80%	572	186	122	60	120	63
shaped video/5/85%	2306	636	472	262	528	344
shaped video/7/65%	1483	67	349	266	387	385
shaped video/7/75%	10421	1132	2545	2218	2497	2401
unshaped vid./2/80%	251	45	62	37	66	24
unshaped vid./2/85%	5712	444	1466	1180	1417	1332

Table 3: Loss figures for 'single cell loss recovery' FEC

[8]This was a major motivation behind the focus on shaped video TUT in section 3.

TUT / background traffic configuration / background load	corrupt 16-clusters	corrupt 16-clusters with i/l	corrupt 64-clusters	corrupt 64-clusters with i/l
shaped video/2/80%	0	0	0	0
shaped video/2/85%	16	3	28	1
shaped video/3/80%	6	0	10	0
shaped video/3/85%	82	4	112	20
shaped video/5/80%	16	2	32	1
shaped video/5/85%	81	4	136	22
shaped video/7/65%	73	16	135	38
shaped video/7/75%	688	300	746	785
unshaped vid./2/80%	10	0	20	0
unshaped vid./2/85%	397	40	484	189

Table 4: Loss figures for 'four cell loss recovery' FEC

4.3   Timing regeneration

The second main problem area we mentioned in the previous section is the need for timing regeneration caused by the buffering and shaping process as well as by the jitter introduced by the ATM nodes. Assuming that sender and receiver do have access to some relatively or absolutely synchronized clocks[9] we propose to use time stamps generated by the sending AAL/ATM entity when it receives the payload data. Several of these time stamps are to be carried in a single separate cell inserted into the payload cell stream in regular intervals. Note that these cells also have to be covered by the FEC mechanism. They can be distinguished from the payload or FEC cells either by assigning dedicated sequence numbers or by introducing an additional label to mark them. Carrying the time stamps in separate cells has the advantage that an application which would like to utilize the FEC option but does not need timing regeneration (or which implements its own timing regeneration facility) is not required to spend the necessary additional overhead.

Alternatively time stamps can be integrated into the payload data cells. Obviously they should also be covered by the FEC mechanism. Note that whatever option is actually used the basic method will cover not only CBR services as does AAL1, but shall be also applicable to VBR services with strict timing constraints.

5   Conclusions and Outlook

We have identified problems that VBR real-time multimedia services will be faced with in an ATM environment that tries to exploit the statistical multiplexing gain. On the basis of these observations we investigated a scalable and configurable concept which combines

[9]which by itself is not a trivial problem;

loss recovery mechanisms of limited complexity and a cell interleaving approach. A timing regeneration mechanism needed in conjunction with this concept was described briefly.

The examples we presented show an improvement of the loss figures between one and two orders of magnitude. Even better figures can be expected if less bursty video sources were used than those in our examples. Therefore the concept might be applied in particular in environments which are run close to their maximum operating point where it could avoid sudden QoS degradation in case of short term overload. We would like to state explicitly that the concept described is not intended to solve the problem of exploiting the potential statistical multiplexing gain of real-time VBR traffic by itself. In our opinion this problem is of a more fundamental nature where we do not envisage a 'simple' solution.

The applicability of the concept described is not restricted to real-time multimedia services. It might, e.g. also be used in cases where response time although not being strictly limited should be kept low. Consider, e.g. widely distributed database and corresponding query systems where retransmission although acceptable in principle is not desired because overall system performance might suffer. In addition there is no limitation to ATM cells. The mechanism can advantageously be applied in any loss prone transmission situation where data is transferred in fixed sized chunks.

6   References

[Bier93]   Biersack, E.W., "Performance evaluation of forward error correction in an ATM environment", IEEE JSAC, vol. 11, no. 4, May 1993.

[Garr93]   Garrett, M.W., "Contributions toward real-time services on packet switched networks", PhD thesis, Columbia University, 1993.

[Garc96]   Garcia Adanez, X., Basso, A., Hubaux, J.-P., "Study of AAL5 and a New AAL Segmentation Mechanism for MPEG-2 Video over ATM", Proc. 7th International Workshop on Packet Video, Brisbane, Australia, March 1996.

[I.363dr]   ITU-T SG13, "B-ISDN ATM adaptation layer (AAL) specification, types 1 and 2", draft recommendation I.363, Geneva, Sept. 1994.

[I.371fr]   ITU-T SG13, "Traffic Control and Congestion Control in B-ISDN", recommendation I.371, frozen version, Geneva, July 1995.

[McAu90]   McAuley, A.J., "Reliable broadband communications using a burst erasure correcting code", Proc. ACM SIGCOMM '90, Philadelphia, PA, Sept. 1990, pp. 287–306.

[Rose94]   Rose, O., Frater, M.R., "Delivery of MPEG video services over ATM" University of Würzburg, Institute of Computer Science Research Report Series, 1994.

[Stoc94]   Stock, Th., Grünenfelder, R., "Frame loss vs. cell loss in ATM concentrators and multiplexing units", EFOC&N´94, Heidelberg, 1994, pp 174–177.

Can self-similar traffic be modeled by Markovian processes ?

Stephan Robert   Jean-Yves Le Boudec
robert@di.epfl.ch   leboudec@di.epfl.ch

EPFL - DI - LRC
CH-1015 Lausanne, Switzerland

Abstract

In this paper, we compare high time resolution local area network (LAN) traffic with three different traffic models: Poisson, ON-OFF and 5-state Markov process. Due to the measured data's extreme variability on time scales ranging from milliseconds to days, it is difficult to find a model for it, especially a Markovian one. Recent studies show that conventional models do not capture the characteristics of the observed traffic. Fractal-based models have already been built to characterize such a traffic but they are not easily tractable tractability of them is not great. Through a new method which integrates different time scales in the model, we have tried to find a quite simple Markovian process having the same behavior as the measured traffic on the LAN. We show in particular that a simple 5-state Markov process integrating different time scales can reasonably model the behavior of measured traffic up to a certain time interval.

Keywords: LAN Traffic, self-similar, Markovian models, long range dependences

1   Introduction

Measurements on a network at Bellcore [Leland 91] have shown that LAN traffic behaves quite differently to what had previously been assumed. The traffic sources were generally characterized by short-term dependencies but characteristics of the measured traffic have shown that it is long term dependent (self-similar). New models have been developed [Norros 94, Willinger 94] but, in general, they are not very tractable although a lot of progress has been made [Norros 94]. The advantage of these models are that they give a good description of the traffic behavior based on few parameters. in this paper, we show through experiments that certain Markov processes (note that previous work has been done in this direction [Andersen 95]) can behave as self-similar processes over a limited but large time domain. The fact that this time domain is limited is not a drawback, since it is not reasonable to consider years-long measurements for network dimensioning.

2   Measured traffic

In this paper, we used the LAN traffic data collected by Leland and Wilson [Leland 91]. Many papers describe measurements made on local area networks (for example Gusella [Gusella 90], Jain and Routier [Jain 86], Feldmeier [Feldmeier 86]).

Software is also available for measuring LAN traffic: Tcpdump, which is distributed by Lawrence Berkeley Laboratory, for example. Tcpdump prints out the headers of packets on a network interface that match a Boolean expression. It is very easy to filter traffic out in order to watch specific traffic. This software tool is available via anonymous ftp from host `ftp.ee.lbl.gov` and contains some utilities for easier processing of the data. But the advantages of using the Bellcore data are: 1) A dedicated piece of hardware has been built for measuring each packet arrival. These measurements are performed without loss (irrespective of the traffic load) and the recorded time-stamps are accurate to within 100 µs.(Tcpdump precision depends on the workstation clock precision, which can be 0.5 ms) 2) A portion of the huge amount of collected data is available on the Bellcore server via anonymous ftp from `bellcore.flash.com`, directory pub/lan_traffic. Each file (pAug.TL, pOct.TL, OctExt.TL, OctExt4.TL) contains time-stamps (arrival time of packets) and length of packets (in bytes).

3 Characterization of traffic

The term "self-similar" was introduced by Mandelbrot [Mandelbrot 68, 69, 79] to describe behavior in the field of hydrology and geophysics. Intuitively, in the case of traffic behavior, self-similarity indicates that the behavior of a process is very similar (in a distributional sense) over different time-scales. Leland et al. [Leland 91] observed that Ethernet traffic seems to look the same whether the scale is hours, minutes, seconds, or milliseconds. A number of quantities have be evaluated to demonstrate the invalidity of the Markovian models:

3.1 Index of dispersion for counts

Here we consider the number of packets in an interval. We can define a function for packet counts: the index of dispersion (IDC) for counts at time t is given by the variance of the number of arrivals in an interval of length t divided by the mean number of arrivals in t: $IDC(t) = \frac{Var(Nt)}{E(Nt)}$, Nt is the number of arrivals in an interval of length t. The IDC has been defined in order that a Poisson process has the value of IDC(t)=1 for all t. The estimate of IDC of measured arrival processes is performed considering time as discrete, equally spaced instants τ_i. We see in [Leland 91] that IDC(t) increases monotically throughout a time span of 6 orders of magnitude. In contrast, all finite Markovian models have indices of dispersion that converge to fixed values over time scales on the order of the time constants of the models.

3.2 Coefficient of variation for counts

As in the preceding paragraph, we consider the number of packets in an interval. The coefficient of variation (CV) for packet counts at time t is given by the standard deviation of the number of arrivals in an interval of length t divided by the mean number of arrivals in t: $CV(t) = \frac{\sqrt{Var(Nt)}}{E(Nt)}$, Nt is the number of arrivals in an interval of length t. For Poissonian arrivals, it is easy to see that the coefficient of variation is equal to 1 for very small time intervals and converges to zero as the time interval increases. The observation of Ethernet traffic show that the coefficient of variation is much more than 1 for small time intervals.

3.3 Hurst Parameter, analysis of the variances

Let $X = (X_t : t = 0, 1, 2, \dots)$ be a covariance stationary stochastic process and define

$$X^{(m)} = \frac{1}{m}\left(X_{km-m+1} + \dots + X_{km}\right), \quad k \geq 1$$

The process X is called second order self similar with self-similarity parameter $H = 1 - \frac{\beta}{2}$ if for all $m = 1, 2, \dots$, $var(X^{(m)}) = \sigma^2 m^{-\beta}$ with $0 < \beta < 1$ (σ =variance for m=1). Estimating β, it is possible to deduce the Hurst parameter

H. β is given by the slope of the diagram $log_{10}\dfrac{var\left(X^{(m)}\right)}{\sigma^2}$ to $log_{10}(m)$ (here m=time interval)

3.4 Visual

Leland [Leland 94] considers pictures describing arrivals over time scales. He observed striking differences between measured datas behavior and batch-Poisson process behavior.

3.5 Others

There are other methods for determining the Hurst parameter, for example R/S statistics and methods based on periodogram. On other hand, Peak/Mean, packet losses and delays in a buffer [Leland 91] were compared for measurements and Poisson process.

4 Markov model

In this paper, we suggest the use of a Semi-Markov Process (SMP) for modeling data traffic. This process was introduced by Lévy [Lévy 54] and Smith [Smith 55] independently as a new class of stochastic processes. These processes are a generalization of both continuous and discrete parameter Markov Processes with countable state spaces. We consider here the discrete case. The Markov chain is modulated and can change its state at defined times: t, t+1, ... Between time t and t+1, the chain remains in the same state. In each state, the arrivals are generated according to a distribution ζ_i which is dependent on the visited state as well as the one to be visited next. The Special Semi-Markov Process (SSMP) is an SMP whose distribution ζ_i in a given state depends only on the present state. When ζ_i is a Poisson distribution, the SSMP is then called a Markov Modulated Poisson Process (MMPP). The kth moment of the random variable (RV) X_t is, by definition

$$E\left[X_t^k\right] = \sum_{x_l \geq 0} x_l^k \, pr(X_t = x_l)$$

X_t is the RV representing the number of cells arriving during the interval [t-1,t). X_t can be positive or equal to zero. In the case of an SSMP Process, the kth moment is

$$E\left[X_t^k\right] = \pi_t \Lambda_t^{(k)} e$$

with $\pi_t = \left(pr(Y_t = 1), \ \ldots \ , pr(Y_t = n)\right)$ which is the modulator's state probabilities vector, $(Y_t = i)$ is the modulator's state i, $i \in \{1,\ldots,n\}$ at time t. For the SSMP process, $\Lambda_t^{(r)}$ is equal to

$$\Lambda_t^{(r)} = diag\left(E\left(X_t^r \mid Y_t = 0\right),\ldots,E\left(X_t^r \mid Y_t = n\right)\right)$$

If the process is Wide Sense Stationary (WSS), the kth moment of X_t can be written $E\left[X^k\right] = \pi \Lambda^{(k)} e$ (e is the unity vector). The general expression of the autocovariance function of the RV X_t is [Papoulis 84]

$$cov\left(X_{t+\tau}^s, X_t^k\right) = E\left[X_{t+\tau}^s X_t^k\right] - E\left[X_{t+\tau}^s\right]E\left[X_t^k\right].$$

For an SSMP process $cov\left(X_{t+\tau}^s, X_t^k\right)$ becomes

$$cov\left(X_{t+\tau}^s, X_t^k\right) = \pi_t \Lambda_t^{(k)}\left(A(t+\tau,t) - e\pi_{t+\tau}\right)\Lambda_{t+\tau}^{(s)} e$$

with

$$\left(A(t+\tau,t)\right)_{kj} = a_{kj}(t+\tau,t) = pr\left(Y_{t+\tau} = j \mid Y_t = k\right) 1 \le k,j \le n$$

which are the elements of the Markov chain transition matrix, and if the process is WSS

$$cov\left(X_\tau^s, X_0^k\right) = \pi \Lambda^{(k)}\left(A^{|\tau|} - e\pi\right)\Lambda^{(s)} e, |\tau| \ge 1.$$

The index of variation is defined as $IDC(t) = \dfrac{Var(N_t)}{E(N_t)}$. After calculations:

$$IDC(t) = \frac{t\left(\pi_t \Lambda_t^{(2)} e\right) - t^2\left(\pi_t \Lambda_t e\right)^2 + 2\left(\sum_{i=0}^{t-1}(t-i)\left(\pi_t \Lambda_t A^{i+1}\Lambda_t e\right)\right)}{\left(\pi_t \Lambda_t e\right)t}$$

and the coefficient of variation

$$= \frac{\sqrt{t\left(\pi_t \Lambda_t^{(?)} e\right) - t^2 \left(\pi_t \Lambda_t e\right)^2 + 2\left(\sum_{i=0}^{t-1}(t-i)\left(\pi_t \Lambda_t A^{i+1} \Lambda_t e\right)\right)}}{\left(\pi_t \Lambda_t e\right)t}$$

5   Intuitive nature of the proposed model

The aim of the study is to generate a source model reflecting some properties of a real source. If we consider the measured LAN traffic as a source, we must take into account that many people are working on the network, having their own schedules. The human behavior has a big influence on the network utilization but the inverse is true too: the patience of the human being is not unlimited. The influence of the protocol used should be taken into consideration. Here we consider only data traffic in a first step: voice, sound and video traffic is ignored in our model (the measured Bellcore traffic is 99.5% IP, Internet Protocol). Data traffic consists of a large variety of service types (file-transfers, workstations communications, terminal communications, ...). IP can support different types of protocols, for example UDP (User Datagram Protocol), TCP (Transmission Control Protocol), ICMP (Internet Control Message Protocol),EGP (Exterior Gateway Protocol).

If we consider TCP/IP protocol [Comer 91], it has a very specific behavior. Manthorpe [Manthorpe 94] studied the influence of the transport layer on traffic modeling. This protocol uses a sliding window to ensure the efficiency of the transmission. The flow sent on the network depends on the window size but also on the network load and size. Maybe the most complex aspect in TCP is embedded in the way it handles time-out and transmission. Every time it sends a segment, TCP starts a timer and waits for an acknowledgment. If the timer expires before data in the segment has been acknowledged, TCP assumes that the segment was lost or corrupted and retransmits it. TCP is intended for use in an internet environment. In such an environment, a segment traveling between a pair of machines may traverse a single, low-delay network or it may wind across multiple intermediate networks through multiple gateways. Thus it is impossible to know a priori how quickly acknowledgments will return to the source. Furthermore, the delay at each gateway depends on the traffic. TCP monitors the performance of each connection and deduces values for time-outs and uses an adaptive retransmission algorithm which allows it to adapt itself if the performance of the connection changes. Furthermore, TCP reacts to congestion. When congestion occurs, TCP normally uses two techniques: slow start and multiplicative decrease of the window size.

Ethernet, when it has something to transmit, waits until the channel is free. Furthermore, it is able to detect collisions on the network. In other words, if two workstations sense the channel to be idle and begin to transmit simultaneously they will both detect the collision almost immediately. The two stations should abruptly stop transmitting as soon as a collision is detected. Therefore, our model for Ethernet will consist of alternating contention and transmission periods.

To resume, we roughly assume that LAN traffic measured on Ethernet can be examined at three major levels of behavior corresponding to a certain resolution of time:

- The connection level describes the human behavior. The connection duration is determined by the file sending time and the file length. The duration between calls on an Ethernet network is typically in the time range of 10 ... 1000 sec.

- The TCP/IP level describes the transport level. As we have seen before, the traffic sent on the network depends of an uncontrollable number of parameters but the major influences on it is the network behavior. The transmission duration of a TCP/IP packet typically varies from 0.01 to 10 sec.

- The Ethernet network level where the sent traffic depends essentially on the local traffic flowing on the network. The time between sending and not sending a frame is typically in the range 1 ... 50 msec.

At the human level, we consider two operational modes: sending or not sending a file. The change between the two modes depends uniquely on the human behavior and how he/she reacts when a congestion occurs on the network. At the TCP/IP level, the protocol is principally controlled by the network behavior. As seen before, the analysis of the jumps between the two states is difficult because of the dependencies between the protocol and the network. At the lowest level, Ethernet is waiting for the channel before transmitting data. The 5-states Markov model is based on this argument.

6   Characteristics of the models

Poisson

$$A = \begin{pmatrix} 0.99 & 0.01 \\ 0.01 & 0.99 \end{pmatrix} \qquad\qquad \Lambda = \begin{pmatrix} 0 & 0 \\ 0 & 1 \end{pmatrix}$$

ON-OFF

$$A = \begin{pmatrix} 0.999999 & 0.000001 \\ 0.000099 & 0.999901 \end{pmatrix} \qquad \Lambda = \begin{pmatrix} 0 & 0 \\ 0 & 1 \end{pmatrix} \qquad \text{Figures 1, 2 and 3}$$

$$A = \begin{pmatrix} 0.999 & 0.001 \\ 0.099 & 0.901 \end{pmatrix} \qquad \Lambda = \begin{pmatrix} 0 & 0 \\ 0 & 1 \end{pmatrix} \qquad \text{Figure 4(d)}$$

5-states Markov

$$A = \begin{pmatrix} 0.99 & 0.00001 & 0.0099889 & 0.000001 & 0.0000001 \\ 0.00001 & 0.5999889 & 0.4 & 0.000001 & 0.0000001 \\ 0.001 & 0.001 & 0.997999 & 0.000001 & 0 \\ 0.000001 & 0.000001 & 0.000001 & 0.999997 & 0 \\ 0.0000001 & 0.0000001 & 0.0000001 & 0 & 0.9999997 \end{pmatrix}$$

$$\Lambda = \begin{pmatrix} 1 & 0 & 0 & 0 & 0 \\ 0 & 1 & 0 & 0 & 0 \\ 0 & 0 & 0 & 0 & 0 \\ 0 & 0 & 0 & 0 & 0 \\ 0 & 0 & 0 & 0 & 0 \end{pmatrix}$$

7 Results

Here we compare measured traffic (internal and external Bellcore traffic) with different models: Poisson, 2-state ON-OFF source and a 5-state Markov integrating different time scales. Figure 1 shows a set of four curves. The index of dispersion of the arrival of the measured traffic on the Bellcore network: OctExt.TL which grows linearly over approximately 5 time scales. For comparison, the index of dispersion of a Poisson process is given here. It is very easy to deduce it analytically (For this process, IDC=1 for all time scales). Another example is given for comparison: The classical ON-OFF (2-state Markov process) process. We see here, as well as in [Leland 91] that the index of dispersion increases monotonically until an asymptote defined by the time scale order of the model. In addition, the index of dispersion of our 5-states is plotted.

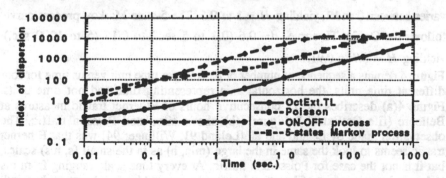

Figure 1 Index of dispersion (arrivals)

The coefficient of variation is plotted in Figure 2 for the external Bellcore traffic and the different considered models: Poisson, ON-OFF and 5-states Markov models. It is interesting here to notice that the ON-OFF process doesn't have a similar coefficient of variation as the one measured on the Bellcore network, as was the case with the index of dispersion. Exept for small times, the 5-states Markov model coefficient of variation behaves similarly to that of the measured data (OctExt.TL).

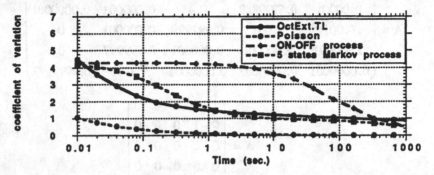

Figure 2 Coefficient of variation (arrivals)

An interesting phenomena is observed in Figure 3. In chapter 3, we have seen how to determine the Hurst parameter by the variances method. The "variance-time" plot is a graphical method for distinguishing between short range dependencies (limiting slope = β = 1) and long range dependencies ($0 < \beta < 1$) in a given record. As described in chapter 3, we plot the normalized variance of the aggregated series as a function of block m (which is equal to time in our case) on log-log coordinates. The Hurst parameter (H) is related to the slope β by H=1-β/2. Figure 3 shows that $\hat{\beta}$ (estimate of β) varies from 0.55 (0.05 to 1 sec.) to 0.17 (1...1000 sec.) yielding a Hurst parameter estimate $\hat{H}$ = 0.72 to 0.91 for the measured data (OctExt.TL). Note that for a Poisson process, β = 1 and H=0.5, which is typically a short range dependence process. For an ON-OFF process, $\hat{\beta}$ varies between 0 and 1, yielding $\hat{H}$ = 1 to 0.5. The 5-state Markov process curve follows the OctExt.TL curve. $\hat{\beta}$= 0.6 (0.1 to 5 sec.) to 0.17 (5 to 1000 sec.) yielding $\hat{H}$ = 0.7 to 0.91.

Figure 4 depicts a sequence of number of packets per time unit versus time for four different time units, the horizontal axis representing time and not time units. Figure 4(a) describes the 27 consecutive hours of Ethernet traffic measures at Bellcore (file OctExt.TL). Figure 4(b) shows simulated Poisson traffic. The observation made by Leland and al. [Leland 91, Willinger 94] was that Ethernet traffic seems to look the same in the large (min, h) as in the small (s, ms) scales, but it is not the case for Poissonian traffic. At every time scale ranging from ms to hours, bursts consist of bursty subperiods separated by less bursty subperiod. Note that the daily cycle is clearly apparent in this figure, which is not really compatible with the self-similar concept.

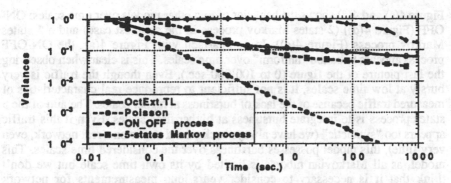

Figure 3 Method of the variances for checking the self-similarity property of the measured datas (OctExt.TL) and the different processes

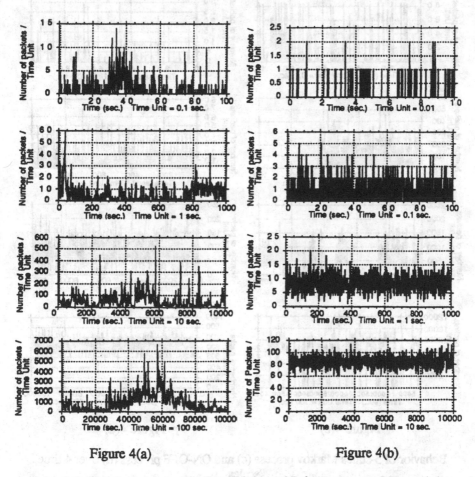

Figure 4(a)   Figure 4(b)

Behavior of measured datas (OctExt.TL) (a) and Poisson process (b) over 4 time scales

128

Figure 4(c) and 4(d) shows the behavior of two other models over time scales: ON-OFF (Figure 4(d)) (2 states Markov process in the simplest case) and a 5 states Markov process (Figure 4(c)). In comparison with Figure 4(a), the ON-OFF process appears to be too "uniform" over time scales. This is clear when observing the last picture of the figure (0 to 100'000 sec.). Even though the traffic is very bursty at low time scales, it is not sufficient to reproduce real characteristics of measured traffic because of the lack of burstiness at higher scales. The aim of the 5 states process is to integrate burstiness at higher time scales. Even if this traffic appears too "synthetic" (we have always background traffic on a real network, even very little), this model possesses burstiness over the considered time scales. This model, as all Markovian models is limited by its own time scale but we don't think that it is necessary to consider years-long measurements for network dimensioning (in this case let us consider quasi-"infinite" Makov chains).

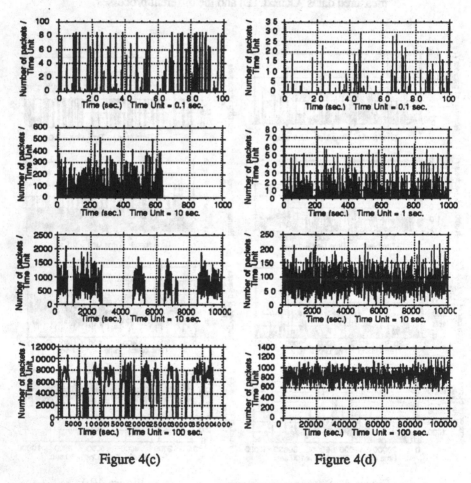

Figure 4(c)

Figure 4(d)

Behavior of 5-states Markov process (c) and ON-OFF process (d) over 4 time scales

8 Conclusion

In this paper, we have compared high time resolution local area network (LAN) traffic with three different traffic models: Poisson, ON-OFF and a 5-state Markov process. The comparison was based on the index of dispersion of arrivals, coefficient of variation of arrivals, Hurst parameter and "visual pictures". If Poisson and ON-OFF (2 states Markov chains) processes do not satisfy the requirements for modeling data traffic, we have seen that a simple 5-states Markov Model integrating different time scales can reasonably well approach its behavior up to a certain time interval. To confirm our approach, a certain number of other tests have to be performed (comparison of buffer behavior for example).

9 Acknowledgments

The authors would like to thank the Swiss Telecom PTT for their support.

10 References

[Andersen 95]   Allan T. Andersen, Alex Jensen and Bo Friis Nielsen, "Modelling of apparently self-similar packet arrival processes with Markovian Arrival Processes (MAP)", COST 242 technical document, Cambridge, 1995

[Comer 91]   D. Comer, "Internetworking With TCP/IP, Volume 1: Principles, Protocols, and Architecture, Second Edition", Prentice-Hall, 1991

[Feldmeier 86]   D. Feldmeier, "Traffic measurements of a Token Ring Network", Proc. of the IEEE Computer Network symposium, Washington, D. C. (November 1986)

[Grünenfelder 94]R. Grünenfelder and S. Robert, "Which Arrival Law Parameters Are Decisive for Queueing System Performance", ITC' 14, Plenary session, Antibes Juan-les-Pins, France, June 6-10, 1994

[Gusella 90]   R. Gusella, "A Measurement Study of Diskless Workstation Traffic on an Ethernet", IEEE Transactions on Communications, 38(9), Septembre 1990

[Jain 86]   R. Jain and S. A: Routier, "PAcket Trains: Measurments and a New Model for Computer Network Traffic", IEEE Journal on Selected Areas in Communications, SAC-4, Number 6 (September 1986)

[Leland 91]   W. E. Leland and D. V. Wilson, "High Time Resolution Measurements and Analysis of LAN Traffic: Implications for LAN Interconnection", IEEE Infocomm'91, paper 11D.3.1

[Lévy 54]   P. Lévy, "Systèmes semi-Markoviens à au plus une infinité
                   dénombrable d'états", *Proc. Int. Congr. Math.*, Amsterdam, vol.
                   2, 1954

[Mandelbrot 68]   B. B. Mandelbrot and M. S. Taqqu, "Robust R/S analysis of long
                   run serial correlation", in Proc. 42nd Session ISI, 1979, pp. 69-
                   99

[Mandelbrot 69]   B. B: Mandelbrot and J. W. Van Ness, "Fractional Brownian
                   motions, fractal noises and applications", SIAM Rev. vol. 5, pp.
                   228-267, 1969

[Mandelbrot 79]   B. B: Mandelbrot and J. R. Wallis, "Computer experiments with
                   fractional Gaussian noises", Water Resources Research, vol. 5,
                   pp. 228-267, 1969

[Manthorpe 94]   S. Manthorpe and X. Garcia, "TCP Performance Over ATM
                   Based LAN Interconnection Services", Interop'95, Engineers
                   Conference, April 28-30, 1995, Las Vegas, USA

[Norros 94]   I. Norros, "On the use of fractional Brownian motion in the
                   theory of connectionless networks", Technical contribution,
                   TD94-33, September 1994

[Papoulis 84]   A. Papoulis, "Probability, Random Variables and Stochastic
                   Processes", Second Edition, Mc Graw-Hill, 1984

[Smith 55]   W. L. Smith, "Regenerative stochastic processes", *Proc. Roy.
                   Soc.* (London), Ser. A, vol. 232, p. 6-31, 1955

[Willinger 94]   W. Willinger , W. E. Leland, M. S. Taqu and D. Wilson, "On
                   the self-similar nature of ethernet traffic (extended version).
                   IEEE/ACM Transactions on Networking, February 1994

Modeling and Analysis of MPEG Video Sources for Performance Evaluation of Broadband Integrated Networks

Nicola Bléfari-Melazzi

D.I.E. Dept.- University of Roma at Tor Vergata (Italy)
E-mail: blefari@eln.utovrm.it

Abstract

This paper consists of two parts. In the first one, we present a statistical analysis of experimental MPEG video sources and we test some literature models for the performance evaluation of a queueing system loaded with such bit streams. In the second one, we propose a new modeling approach that considers the particular structure of MPEG streams and gives extremely accurate results in a well-defined set of the queueing system parameters.

Our results show that the impact of correlations and the particular structure of MPEG sequences (the GOP) is very significant. As a consequence it does not seem possible to model MPEG sources with stochastic processes based on only one picture type, with "simple" bit rate distribution, and/or whose autocorrelation function is a sum of exponential functions.

The proposed model is validated with simulation results produced using many experimental video sequences (both MPEG1 and MPEG2); the sequences analysed are long enough to allow a significant statistical analysis.

Introduction

Video traffic is expected to be one of the major sources of loading of emerging high-speed packet media. Thus there is a need to find accurate video traffic models and to evaluate network performance.

The MPEG (ISO Moving Picture Expert Group) will be the major compression algorithm to be used in video applications [9]. The MPEG syntax supports a Group of Pictures (GOP) structure. An input video sequence is divided into units of group of pictures consisting of three types of pictures (frames): Intraframe (I) pictures, Predictive (P) pictures, Interpolative (B) pictures. A GOP is defined by its length, N (or the distance between I pictures) and by the distance between P pictures, M. An example of a GOP with $N=12$ and $M=3$ is [IBBPBBPBBPBB]. The coded bit stream is also characterized by a hierarchy, structured into layers; each video program is divided into the following layers (in order of decreasing size): scene, GOP, picture, slice, macroblock, block. A scene contains several GOPs and each of them contains N pictures. A given number of blocks forms a macroblock and an arbitrary number of macroblocks forms a slice [9].

Theoretically, modeling of an MPEG bit stream should be made at the level of each of the above layers, but this would be a very demanding and probably an unnecessary task. We chose to perform the analysis considering the layers higher than the slice one. Therefore in the rest of this paper we assume that video traffic is deterministically smoothed over the picture period. The reasons of this choice are discussed in [4].

Almost all video traffic models that can be found in the literature are based on matching statistical parameters of the sources. The MPEG traffic brings about additional problems due to the GOP structure (the video sequence has a pseudo-periodic behaviour). The bit rate distribution follows a complex behaviour. Normal, Gamma and Lognormal distributions do not fit empirical distributions; only a hybrid of a Gamma and of a "heavy-tailed" Pareto distribution provides good results [7]. Matching the autocorrelation is much more difficult. First, this function exhibits a pseudo-periodic behaviour that is already difficult to reproduce; second, MPEG streams are characterized by long range dependence (LRD). The autocorrelation function decays as a hyperbolic function and not as a negative exponential, as commonly assumed (see [7] and references therein). The presence of LRD has very serious effects: models that ignore LRD result in very optimistic estimates of performance.

In conclusion, it does not seem possible to analyse queueing systems loaded with MPEG streams by using models based on only one picture type, with "simple" bit rate distribution, and/or whose autocorrelation function is a sum of exponential functions (i.e. most of the commonly used stochastic models, see again the excellent analysis in [7] and § 1.2).

In this paper we face the study of a queueing system loaded with a superposition of video sources and with a single source. We propose a new approach that gives extremely accurate results in a well-defined set of the queueing system parameters.

Our model takes into account the three picture types and the GOP structure and succeeds in matching perfectly the bit rate distribution. This means that it offers a significant advantage over existing methods that are not able to capture the behaviour of the real sequences even for a restricted set of system parameters. The model will be validated with simulation results produced using directly experimental video sequences; the sequences analysed are long enough to allow a significant statistical analysis and the validation is done for a wide range of all possible or likely values of the system parameters. These points may seem obvious but they are sometimes overlooked. In particular it is very important to test the model *for different values of the utilization factor and of the buffer size.*

The performance evaluation will be carried out by assuming that the GOP is [IBBPBBPBBPBB] (N=12 and M=3) and will be focused on ATM/B-ISDN; thus we assume a payload of 48 bytes per packet (or cell in ATM terms). This is consistent with the use of AAL5 for the transport of MPEG. MPEG has also a scheme to do buffer smoothing, especially for use with CD-ROMs. The following analysis is applied to un-smoothed streams.

As for the organization of the paper, in the first part we present a statistical analysis of an MPEG video sequence (§ 1.1) and we examine some literature models (§ 1.2). In the second part, the proposed model is introduced (§ 2.1 and § 2.2) and used for the performance evaluation of a queueing system loaded with a single MPEG1 source (§ 2.3), a single MPEG2 source (§ 2.4) and a superposition of sources (§ 2.5). Finally, we draw the conclusions and discuss an extension of the model that allow to take into account the correlations, thus extending significantly its range of applicability (§ 3).

1 Statistical Analysis and Literature Models

1.1 Statistical Characteristics of MPEG Bit Streams

The results presented in this Section have been produced by using the experimental sequence "Star Wars", consisting of 174126 frames. The stream data are in the form of the number of *cells* generated per frame for each frame in the sequence. The frame rate is 24 frames/second. The mean bit rate is about 379 kbit/s while the peak bit rate is about 4451 kbit/s. A detailed statistical analysis has shown that the eight B frames are statistically similar to each other; the same happens for the three P frames. This means that we can safely simplify the description of the experimental sequence, using for the B frames, and for the P frames, only two sets of statistical parameters obtained by averaging on the 8 B frame types and on the 3 P frame types respectively [4,5]. In [4] we show the probability mass function of the frame size for the I, P and B frames, verifying that its complex behaviour can not be modelled by commonly used PMFs.

It is by now common knowledge that, to correctly model an arrival process, it is not always sufficient to keep into account only first order statistics. Correlation effects play indeed an important role in assessing the queueing performance. If a pre-buffering scheme is adopted, so that the cell transmission rate is constant during a frame, the correlations in an MPEG bit stream are due to:
- inter-frame, intra GOP correlations;
- inter GOP, long term correlations.

We will now investigate the amount and the effects of these correlations.

Figure 1 shows the first 2500 frames of the Star Wars sequence. It is clear that, in addition to the behaviour due to the use of a given GOP structure, there are strong and long term correlations between I, P and B frames, due to the content of the film.

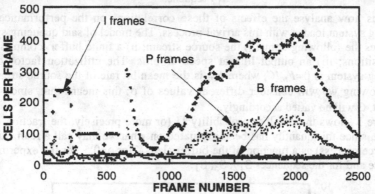

Fig. 1 - First 2500 frames of the experimental sequence.

In order to analyse the correlation properties of the experimental bit stream, and their effects on the performance of a queueing system, we generated three "artificial" sequences:
- S_{12}: it maintains the real GOP structure; the frame sizes are generated according to each of the 12 frame type distributions of the real sequence, but they are statistically independent;
- S_3: it maintains the real GOP structure; the frames are still statistically independent but all the B frames have the same distribution (obtained averaging on the 8 B-type distributions) and all of the P frames have the same distribution (obtained averaging on the 3 P-type distributions);
- S_1: it does not maintain the GOP structure; the frame sizes are generating according to a distribution obtained averaging on *all* frames.

Figure 2 shows the autocovariance function for the original experimental sequence and for the S_{12} sequence for lags among 0 and 2000. The autocovariance function of sequence S_{12} has obviously a periodic structure with a period of 12 frames, while the experimental sequence has a pseudo-periodic structure with the same period but is much more correlated. Thus the presence and the amount of correlations noted in Fig. 1 is confirmed.

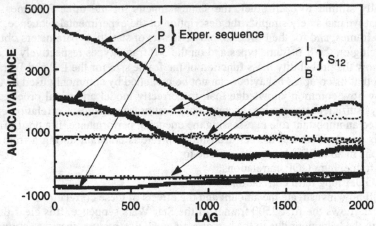

Fig. 2 - Autocovariance of the experimental sequence compared with the autocovariance of S_{12} sequence.

Let us now analyse the effects of these correlations on the performance of a queueing system loaded with this arrival process. The model of said queueing system envisages the following items: i) the source stream; ii) a finite buffer, comprising K cell positions; iii) an output link at speed C bit/s. The utilization factor of the queueing system is $\rho=F_m/C$, where F_m is the mean bit rate of the source stream. In the following we will consider different values of ρ; this means that, since F_m is constant, C will be varied accordingly.

Figure 3 shows the cell loss probability Π (or more precisely, the fraction of the cell lost, since this quantity has been evaluated on what we may call a realization of the source process) as a function of the buffer size, K for $\rho=0.5$, for the experimental sequence and for the sequences S_{12}, S_3, S_1.

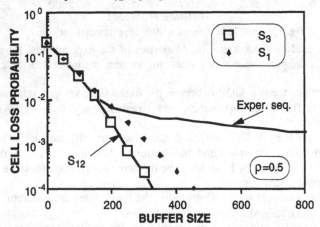

Fig. 3 - Cell loss probability Π as a function of the buffer size, K for $\rho=0.5$, for the experimental sequence and for the sequences S_{12}, S_3, S_1.

This figure enlightens very clearly the effects of correlations. For small values of the buffer size (i.e. $K<200$), the role of correlations is not important and the cell loss behaviour is well captured by a first order statistic (the distribution of the frame sizes);

in fact all the sequences have the same performance. This is in perfect agreement with known results [2 and references therein]. For the correlations to manifest their effects, a queueing system must have a large buffer size and a high load. For small buffer sizes and low loads the number of cells that interacts in the queue is not significant. Figure 3 shows also that there is no difference between the loss performance of the sequences S_{12} and S_3; this means that we can safely avoid to consider 12 different frame types and that we can characterize all the B frames with one statistical description and all the P frames with one statistical description.

If we ignore also the GOP structure (sequence S1) we obtain an overestimate of the performance of the sequences S_{12} and S_3. This is to be expected since in the latter sequences the I frames (that require more bits than P and B frames) are spaced by 12 frames and the P frames (that require more bits than B frames) are spaced by two or three frames; in the S_1 sequence there is not such a constraint (in other words the real sequence possesses also *negative* correlations, see Fig. 2). Finally Fig. 3 shows that, in analysing the loss behaviour as a function of the buffer size, we can define two operating regions: the "uncorrelated" region (for small buffer sizes), where the queueing performance depend only on first order statistics, and the "correlated" region, where correlations play their role and determine the queueing performance (for greater buffer sizes). The boundary between these two regions depends on the utilization factor of the queueing system: when the utilization factor increases, the system moves toward the burst region. In fact, the cumulative effect of the correlations among successive interarrival times becomes increasingly evident as the number of cells stored in the buffer increases, i.e. for large buffer sizes and for high loads [2,4,5].

This classification has a correspondence with the by now familiar "cell region" and "burst region" arising in the cell loss behaviour of a multiplexer loaded with a superposition of On-Off sources [e.g. 2,15]. Here the situation is very similar, with the difference that the arrival process consists of only one source. Thus the definition made above can be view as an integration and a generalization of the cell/burst region concept. In the uncorrelated region the cell loss rate decreases sharply when the buffer size increases, while, in the correlated region, increasing K does not result in a significant improvement of the loss performance.

A similar analysis has been performed also on many other MPEG1 sequences and MPEG2 sequences. The above conclusions apply to all the sequences analysed.

1.2 Analytical Modeling Issues

Almost all the approaches proposed for modeling video traffic can be roughly classified in four classes: i) TES models [13] and ad hoc models developed to generate complex arrival streams [7]; these are models that are very successful in reproducing complex statistical characteristics of the sources but have the disadvantage that it is very difficult to study a queueing system loaded with such arrival processes; the analysis must be performed via simulations; ii) autoregressive models; first order AR models proposed in [e.g. 12] have gaussian bit rate distributions and exponential autocorrelation functions; moreover they do not lead to a queueing analysis of manageable complexity; the sum of two AR processes has been used to catch a typical general behaviour of video autocorrelation function shown also by our sequence [14] (this function decreases quickly for short lag times and slowly for larger lag times); this model is a significant improvement with respect to models that exhibit pure exponential autocorrelation function, but even this model does not capture satisfactorily the behaviour of our real autocorrelation function; iii) MMPP, MAP [10], D-BMAP [6], Stochastic Fluid Flow [1]; these models lead to a tractable analysis of a queueing system; their autocorrelation function is a sum of exponential functions; the bit rate distribution is binomial, or can be set to a more complex distribution by increasing the dimensionality of the model [17 and references therein]; iv) simple models based only on first order statistics; these models do not keep into account the presence of correlation and are thus valid only for a restricted set of system parameters.

In addition to the difficulties encountered in modeling video sequence, a MPEG stream brings about additional problems. The bit rate distribution is quite complex [4] and the autocorrelation function is not a sum of exponential functions (Fig. 2). Approximating the real autocorrelation function by a sum of exponential functions is possible for a given time scale but there is always a time scale where the model breaks down. This means that it is difficult for the models discussed above (that ignore also the GOP structure) to accurately predict the MPEG queueing performance. We have substantiated this consideration by applying some significant literature models (analytically tractable) to the queueing analysis of our sequence.

Models based on first order statistics

A large part of this Section is based on results obtained in [17] since this work gives very useful insights in the behaviour of video traffic. In [17] some experimental sequences are analysed by means of the so-called Histogram Approximation (HA) and assuming that cells are transmitted during a frame according to a uniform smoothing mode; i.e. the cells are distributed in a frame randomly with a uniform distribution. The authors point out that this random smoothing is unlikely in a real system, but that the obtained results are similar to the case of deterministic smoothing when the cells are deterministically spaced, i.e. the case in which a pre-buffering scheme is used. It is also assumed that, on a frame-by-frame basis, the system behaves approximately like an M/D/1/K system. For each frame the arrival process can be then considered as Poisson with a rate λ. During the sequence λ assumes different values. These values are quantized in a number N of possible values (bins), hence the name Histogram Approximation. The Histogram Approximation ignores the transient behaviour and solves the M/D/1/K system as a function of λ and then conditions over the range of λ.

This approximation does not keep into account precisely inter-frame correlations; thus the HA will not predict correctly the system performance when correlations play their role. The agreement between HA and simulation results is good only for small buffer size (see Fig. 6 in [17]). The HA performs very well for the sequences analysed in [17] but we have found that gives poor results for the MPEG sequence even in the uncorrelated region since it does not consider the GOP structure. Figure 4 shows the buffer occupancy distribution for $K=50$ and $\rho=0.5$ comparing HA (for different values of the number of bins of the bit rate histogram) to simulation results and to an MMPP model that will be discussed in the sequel.

Some observations are in order. First while a small number of bins ($N=8$) has been found sufficient for the sequences in [17], in the case at hand we had to use 80 bins; in fact the curves relevant to $N=10$ and $N=20$ are very different from the curve that takes into account all possible values of the arrival rate during a frame (484) and only for $N=80$ we obtained results very similar to the curve relevant to $N=484$. This seems due to the form of the arrival rate distribution and to the fact that there are many bins for which the arrival rate is higher then service rate. More important, the agreement between approximation and simulation results is not good even for small values of the buffer size.

A significant performance measure is the cell loss probability. Figure 5 shows the cell loss probability as a function of the buffer size for $\rho=0.5$ comparing HA (for $N=80$) to simulation results and to an MMPP and to a Stochastic Fluid Flow model that will be discussed in the sequel. The cell loss predicted by HA is almost constant as a function of the buffer size and gives good results only for K near to zero. This is due to the presence of many bins for which the arrival rate is higher then service rate. As noted in [17] in this case the HA predicts results very similar to deterministic smoothing. For deterministic smoothing the histogram approximation gives an expression that does not keep into account the buffer size and explains the behaviour of the curve HA in Fig. 5.

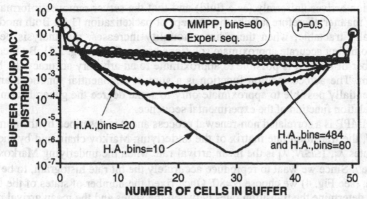

Fig. 4 - Buffer occupancy distribution for $K=50$; comparison between Histogram Approximation (for different values of the number of bins), the MMPP model and experimental sequence results.

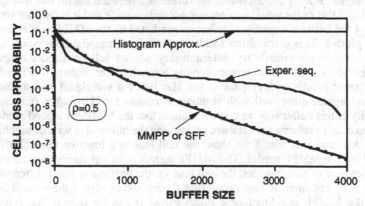

Fig. 5 - Cell loss probability as a function of the buffer size, K for $\rho=0.5$; comparison between experimental sequence, Histogram Approximation, MMPP and SFF model.

Here we find an example of the complexity inherent in modeling correlated traffic streams and of the various effects that correlations have; in fact it is true, as assumed in deriving the HA approximation, that the system quickly reaches the steady state, but, in our case, the buffer *does* have memory of past arrival rates (if we have two consecutive frames with a high number of cells, the second frame finds the buffer full and loses many cells; if high frame rates are spaced by low frame rates, the system loses fewer cells). The steady state is quickly reached during a frame because the arrival rate is much larger than the frame rate or, as explained in [17], because of correlations; the HA assumes that the arrival process changes infinitesimally slow, so in this sense, it would give the maximum degree of correlation possible; but the HA does not keep memory of past arrival rates and the same correlations that drive the system in steady state during a frame, thus suggesting a quasi-static approximation, make the same approximation not satisfactory in this case (it overestimates real loss performance).

Stochastic Fluid Flow and MMPP models

Let us now turn our attention to models of class iii) defined above. The MMPP is a doubly stochastic Poisson process whose mean arrival rate is a function of the state of a finite Markov process [10,3]. The Stochastic Fluid Flow (SFF) model assumes that information arrives uniformly (as a fluid) and that the server removes information in the same manner; therefore it lacks the concept of packetization [1,2]. Both models are analytically tractable; when their dimensionality increases, the analysis becomes cumbersome but accurate approximate solutions are available [e.g. 2,3]. By increasing the number of states it is possible to approximate to an arbitrary accuracy the bit rate histogram. The autocorrelation function is a sum of exponential functions and it is then potentially possible to approximate only to some degree the general form of the autocorrelation function of the experimental sequence.

The MMPP is a correlated non-renewal process and it is described by the number of states, N, the transition rate matrix of the underlying Markov chain and by the mean arrival rates λ_i, $1 \leq i \leq N$; λ_i is the mean arrival rate when the underlying Markov chain is in state i. Since we want to reproduce accurately the bit rate histogram, to be on the safe side, (see Fig. 4) we choose $N=80$. Once fixed the number of states of the MMPP we must determine the transition rates between the states and the mean arrival rates in each of the states. To determine these parameters we follow the approach proposed in [11 and 17], that is to say we use the mean arrival rates and the transition rates of the actual sequence. Having chosen $N=80$, the difference between the bit rate histogram of the model and that of the actual sequence *is not appreciable* and it is thus not shown.

In Fig. 4 the buffer occupancy distribution predicted by the MMPP model described above is plotted. This distribution follows correctly a bimodal behaviour, i. e. the buffer is, with high probability, alternatively almost full or almost empty; the remaining values of buffer occupancy having a much lower probability. The MMPP gives different results with respect to the HA being a correlated non memoryless system and agrees quite well with simulation results *for this buffer size*. So if we looked only at this buffer size we could conclude that the MMPP is a good model.

If we look at a performance measure by varying the buffer size we can highlight the limits of this model. In Fig. 5 we show the cell loss as a function of the buffer size predicted by the MMPP model. The MMPP agrees with simulation results only for buffer sizes near to zero; besides, the cell loss vs. the buffer size has a different decay rate than the experimental sequence. This figure shows also a theoretical limit of models like MMPP and Stochastic Fluid Flow. It can be shown that both these models have an asymptotic exponential decay rate of the cell loss probability as a function of the buffer size, while the behaviour of the same performance measure of a real sequence sooner or later drops to zero. The curve in Fig. 5 is labelled MMPP or SFF since in our case both these models give almost the same results. In fact we have tried to model the experimental sequence also with the SFF model following the same approach (more details can be found in [5]).

The difference between MMPP and SFF is twofold: first the SFF lacks the concept of packetization, second, during the sojourn in a state of the underlying Markov chain, the former emits information according to a Poisson process, while the latter emits information in a fluid, uniform way. Both these differences are more evident for small buffer sizes. For instance, when the SFF is applied to the analysis of a multiplexer loaded with a superposition of sources, it grossly underestimates the loss probability for small buffer sizes. In our case we have only one source, so the difference between the two models is a consequence only of the different emission characteristics in a state. Also this difference tends to decrease when the buffer size increases. The results presented above seem to show that commonly used models are not adequate to assess the performance of queueing systems loaded with MPEG sources.

Conclusions

Summing up, modeling queueing performance in the correlated region is a very demanding task. Anyway the point does not lie in the difficulties brought about by the performance analysis, which, after all, is the work of traffic engineers. The point is that a queueing system working *in the correlated region* has a number of nasty characteristics:

1): buffering is of little help in reducing cell loss since the decay of this quantity vs. the buffer size is much slower than in the uncorrelated region;

2): it is doubtful whether the average cell loss rate evaluated by means of stationary models is of significance in assessing the user perceived quality of service; the quality of service perceived by the user is somehow similar to the one of circuit switching networks: there are long periods during which there is no loss; but there are periods of time during which a single user loses almost all his cells, like if the connection fell (while others do not lose any cell); on the contrary, in the uncorrelated region, loss is more uniformly distributed. For instance, in the case of the Star Wars sequence, for a buffer size equal to 2000 cells and $\rho=0.5$, we have a cell loss rate of about $5.5 \; 10^{-4}$, but if we look at the temporal behaviour of the loss phenomena of the experimental sequence we find that loss is concentrated during only 52 frames over 174126 frames. In particular there are three short loss periods: one concentrated near frame number 87350 (3 frames) one near frame number 152710 (31 frames) and one near frame number 164600 (18 frames). During the rest of sequence there is no loss. Keeping in mind that the buffer size is equal to about four times the maximum frame size, this loss is evidently due (and it can be verified if we look at the frame size of the sequence in these periods) to three short series of *consecutive* high values of I, P and B frames. The buffer content accumulates because of transient and not memoryless phenomena. In [7] the authors remark that these loss periods are due to special video effects, e.g. "jump to hyperspace" or "Death Star explosion". From the point of view of user-perceived Quality of Service and of the validity of long term stationary performance measures, we remark that a user will not be happy to lose the Death Star explosion. It could then be argued that it is not possible for stationary processes to reproduce this behaviour. Nevertheless it could also be concluded that one has not found a satisfactory description of the process over all time scales [7];

3): in a ATM/B-ISDN traffic control framework, allocation rules become very complicated; moreover, when heterogeneous traffic mixes are considered, the long term cell loss rate seen by the different traffic sources is strongly biased by the most bursty sources in the mix; in other words, the cell loss performance are largely dominated by the "worst-behaved" traffic sources [2];

It can be objected that when dealing with many sources multiplexed together many of the problems enumerated above tend to disappear, but this happens only because, when the number of superposed correlated streams increases, the system moves toward the uncorrelated region (see § 2.5). Finally, buffer sizes in broadband network, are generally assumed to be in the order of 100-300 cells to limit delay and delay jitter; in our case, for such buffer sizes, the queueing system works in the uncorrelated region (especially if low values of the cell loss probability are to be achieved).

On the basis of the above points, we propose a simple renewal model that predicts very accurately the queueing performance in the uncorrelated region.

2 A New Modeling Approach

2.1 Renewal Approximation

In this paragraph we propose a Renewal Approximation (RA) to evaluate the performance of a queueing system loaded with MPEG bit streams. The RA approach takes into account the periodic GOP structure.

The buffer occupancy can be considered as resulting from the sum of two components; the first one is the one that we would have if the arrival process were a renewal one; the second one is due to correlations among interarrival times. For a given (non renewal) arrival process, the relative contribution of these two components to the buffer content (that we may call renewal and correlated components) varies as a function of the load and of the buffer size. For small buffer sizes and low loads the second component is negligible. The approximation that we are going to introduce takes into account only the renewal component (in fact this approximation gives *exact* results for the artificial, renewal sequences generated in § 1.1: S_{12}, S_3 and S_1).

Let K denote the buffer size, in ATM cells, and C the output link speed in cells/frame. The utilization factor of the queueing system is $\rho=F_m/C$, where F_m is the mean cell rate per frame of the source stream. The MPEG codifier generates one frame at a time; each frame is packetized and cell arrivals occur at the rate of the codifier. The time unit is the transmission time of a cell on the output link (slot).

The adopted model of the arrival process is described by the following parameters:

m: the number of frame types, or states (here $m=12$ since the GOP is [IBBPBBPBBPBB]);

T: the time spent in each state (in slots); the source has then a period of mT slots and every T slots changes its state, i.e. emits a different frame type;

$f_j(\cdot)$: the probability mass function (PMF) of the frame type j, i.e. $f_j(h)$ denotes the probability that the size of the frame type j of a GOP is h cells, $1\leq j\leq m$; $0\leq h\leq M_j$. These PMF's in our case are sampled by the real MPEG sequence.

λ_j: the mean number of cells generated during frame type j, $1\leq j\leq m$, i.c. the mean of the distribution $f_j(\cdot)$;

M_j: the maximum number of cells generated during frame type j, $1\leq j\leq m$;

The overall mean arrival rate is given by:

$$F_m=\frac{1}{m}\sum_{j=1}^{m}\lambda_j \tag{1}$$

The source then alternates periodically between m states; while in a given state (frame type) the source generates a number of cells according to an arbitrary distribution independently from state to state, hence the name Renewal Approximation. Note that, in this setting, T and C are the same quantity; the time (in slot) spent in a state (frame) is equal to the number of cells that can be transmitted in a frame on the output link. The model can be easily extended to consider a superposition of different sources, eventually with a different synchronization between sources. A similar source model has been proposed in [8] to analyze intra-frame correlations in a video source, i.e. at the slice level. The queueing system will now be analysed in discrete time; in particular we focus on evaluating the buffer occupancy distribution at the epochs at which a frame terminates. Other performance measures of interest, like loss probability and queue length distribution at arbitrary time instants, can be determined through established literature results.

Let q_k and x_k be respectively the queue length and the state of the source at time instants kT, $(k \geq 1)$; i.e. at the end of frame boundary; thank to the above assumptions, $\{q_k, x_k\}$ is a Markov chain. The state space is $\{(i,j): 0 \leq i \leq K, 1 \leq j \leq m\}$ where (i,j) is the current state of (q_k, x_k). Let $p(i,j,i',j')$ be the transition probability from state (i,j) at time k to state (i',j') at time $k+1$.

It can be shown that through the transformation

$$\{i,j\} \Rightarrow \{l,l'\} \text{ with } l=(j-1)(K+1)+i \text{ and } l'=(j'-1)(K+1)+i' \tag{2}$$

$p(i,j,i',j')$ is mapped into $p(l,l')$ and that the resulting transition matrix $\mathbf{P}$ with elements $p(l,l')$ is given by:

$$\mathbf{P} = \begin{bmatrix} 0 & \mathbf{P_1} & 0 & \cdots & 0 \\ 0 & 0 & \mathbf{P_2} & \cdots & 0 \\ \vdots & \vdots & \vdots & \cdots & \vdots \\ 0 & 0 & 0 & \cdots & \mathbf{P_{m-1}} \\ \mathbf{P_m} & 0 & 0 & \cdots & 0 \end{bmatrix} \tag{3}$$

where the matrices $\mathbf{P_j}$ ($1 \leq j \leq m$) are given by [5]:

$$\begin{bmatrix} \sum_{i=0}^{C} \tilde{f}_j(i) & \tilde{f}_j(C+1) & \cdots & \tilde{f}_j(C+K-1) & 1-\sum_{i=0}^{C+K-1} \tilde{f}_j(i) \\ \sum_{i=0}^{C-1} \tilde{f}_j(i) & \tilde{f}_j(C) & \cdots & \tilde{f}_j(C+K-2) & 1-\sum_{i=0}^{C+K-2} \tilde{f}_j(i) \\ \sum_{i=0}^{C-2} \tilde{f}_j(i) & \tilde{f}_j(C-1) & \cdots & \tilde{f}_j(C+K-3) & 1-\sum_{i=0}^{C+K-3} \tilde{f}_j(i) \\ \vdots & \vdots & \cdots & \vdots & \vdots \\ \cdots & \cdots & \cdots & \cdots & 1-\sum_{i=0}^{C-1} \tilde{f}_j(i) \end{bmatrix} \tag{4}$$

where C is the output link capacity, (note that C must be an integer) and

$$\tilde{f}_j(h) = \begin{cases} f_j(h) & \text{if } h \leq M_j \\ 0 & \text{otherwise} \end{cases} \qquad 0 \leq h \leq C+K-1.$$

With the transition probability matrix $\mathbf{P}$ at hand, it is conceptually straightforward to obtain the unique stationary probability vector of the considered Markov Chain $\pi = (b_{0,1} \cdots b_{K,m})'$ where $b_{i,j}$ is the stationary probability of $q_k=i$ and $x_k=j$, $0 \leq i \leq K$, $1 \leq j \leq m$. However the numerical complexity and the storage requirements can be significant in real applications. This problem can be overcome by using the following property [quoted in 8]:

Property: A finite irreducible Markov Chain with transition matrix (3) has a stationary probability vector π given by: $\pi = (\pi_1', \cdots, \pi_m')'/m$

where π_j is the unique stationary probability vector for $A_j = P_j \cdots P_m \cdot P_1 \cdots P_{j-1}$ $(2 \leq j \leq m)$ and $A_1 = P_1 \ldots P_m$ $(j=1)$.

Thank to this property it is sufficient to deal with matrices of order $(K+1)$ instead of $(K+1)m$. It is also important to note that, in this model, the arrival process has the

same bit rate distribution of the experimental sequence. If the number of values of the PMFs of the frame size is significant, i.e. M_j $(1 \leq j \leq m)$ assumes large values, a reduction technique can be used. Each of the PMFs can be approximated by a histogram with a much lesser value of M_j; we have found that 20 bins are sufficient to obtain the same results relevant to the original PMFs that have a maximum value (for the Star Wars) of M_j equal to 484. Let us now turn our attention to the evaluation of the queueing system performance measures. The following relations hold:
the buffer occupancy distribution at the epochs at which a frame terminates is

$$b(s) = \sum_{j=1}^{m} b_{s,j} \qquad\qquad 0 \leq s \leq K \qquad\qquad (5)$$

the cell loss probability of the j-th frame type Π_j is:

$$\Pi_j = \frac{1}{F_m} \sum_{s=0}^{K} b_{s,j-1} \sum_{h=C+K+1-s}^{M_j} (h+s-C-K)\, f_j(h) \qquad 2 \leq j \leq m \qquad (6a)$$

$$\Pi_j = \frac{1}{F_m} \sum_{s=0}^{K} b_{s,m} \sum_{h=C+K+1-s}^{M_j} (h+s-C-K)\, f_j(h) \qquad j=1 \qquad (6b)$$

(the second sum in (6a) and (6b) must be weighted with the buffer occupancy distribution evaluated at the end of the frame that precedes the one considered);
the overall mean cell loss probability is:

$$\Pi = \frac{1}{m} \sum_{j=1}^{m} \Pi_j \qquad\qquad (7)$$

The solution presented above allows to analyse the queueing system at hand for whatever values of T and for a wide range of m (in particular, $m=12$ gives no problem). However we must still operate on matrices of order $(K+1)$. Therefore for large values of the buffer size there could be numerical and storage problems. For such cases we have developed an ad hoc iterative solution that has not numerical instability and does not require to store large matrices [5].

2.2 Overflow Probability Approximation

In this paragraph we discuss a very rough model for the evaluation of the cell loss probability in the uncorrelated region. This approximation, called Overflow Probability Approximation (OPA) assumes that the last cell of each frame leaves the buffer empty, so that loss can occur only if the number of cells generated in a frame is greater than $C+K$. In other words the epochs at which a frame terminates are renewal points and the system evolves without any memory, on a frame by frame basis. The cell loss probability of the j-th frame type $\Pi_{OPA,j}$ is then simply:

$$\Pi_{OPA,j} = \frac{1}{F_m} \sum_{h=C+K+1}^{M_j} (h-C-K)\, f_j(h) \qquad 1 \leq j \leq m \qquad (8)$$

the overall mean cell loss probability is:

$$\Pi_{OPA} = \frac{1}{m} \sum_{j=1}^{m} \Pi_j \qquad\qquad (9)$$

The reasons behind this approximation lie in the GOP structure. As it will be shown in the sequel, loss is mainly due to I frames; since I frames (that require more bits than P and B frames) are preceded by two B-frames, it is most likely that the first cell of an I frame will find the buffer empty. The same will happen also for the P frames. In other words during B frames the buffer content will, with high probability, tend to decrease; thus at the end of two consecutive B frames the buffer will be empty most of the times and will lose memory of past arrival rates.

2.3 Performance Evaluation: single MPEG1 Sequences

Figure 6 shows the buffer occupancy distribution for $K=100$ and for different values of the utilization factor, ρ, comparing simulation results and Renewal Approximation (for the Star Wars sequence). For $\rho=0.3$ there is no appreciable difference between model and simulation results. The RA succeeds in following very precisely the complex behaviour of the experimental sequence. We point out that an erratic behaviour like this can be hardly captured by the models analysed in § 1.2. As the utilization factor increases, we move toward the correlated region and the agreement between model and simulation results worsen, but it is still very good for $\rho=0.5$.

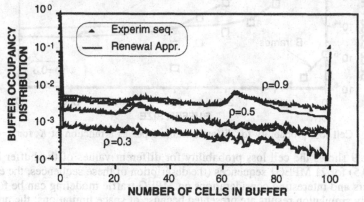

Fig. 6 - Buffer occupancy distribution for $K=100$ and for different values of ρ; comparison between experimental sequence and Renewal approximation.

The behaviour of the cell loss probability vs. the buffer size, for various values of ρ, depicted in Fig. 7, confirms our expectations. The RA and even the OPA are very good approximations in the uncorrelated region that, in the case at hand, comprises buffer sizes between 0 and about 300.

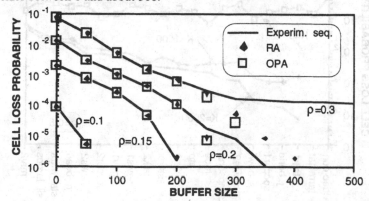

Fig. 7 - Cell loss probability Π as a function of the buffer size, K for different values of ρ; comparison between experimental sequence, RA and OPA.

In Fig. 8 we analyze the cell loss suffered during the different frame types. As expected, it happens, almost always, that I frames lose more cells than P frames and P frames lose more cells than B frames. However, in the correlated region, this it is not necessarily true; in fact the curves relevant to I and P frames cross. Another peculiar phenomenon depicted in this figure is that, sometimes, the cell loss concerning a particular frame type does not decrease as the buffer size increases. The overall cell loss does decrease as a function of K, as it must do, but not all of its components do. Both these phenomena are due to correlation and transient effects and confirm the difficulty of modeling satisfactorily these sequences.

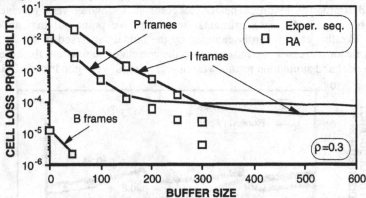

Fig. 8 - Cell loss probability for different frame types as a function of K, for $\rho=0.3$.

Figure 9 shows the cell loss probability for different values of the buffer size, K (and $\rho=0.5$) for 21 MPEG1 sequences (the description of these sequences, the encoder parameters and interesting considerations on MPEG traffic modeling can be found in [16]); only simulation results are presented because of space limitations; the proposed model has been found to be accurate in the uncorrelated region. The more important characteristic of this figure is the great difference between the cell loss experienced by different sequences. For instance, for $K=2000$, the "terminator" sequence does not lose any cells, while the "mtv2" sequence loses a percentage of its cells equal to 4.8 %.

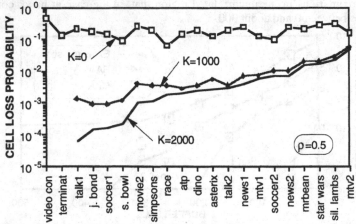

Fig. 9 - Cell loss probability for different values of the buffer size ($\rho=0.5$) for 21 MPEG1 sequences.

2.4 Performance Evaluation: single MPEG2 Sequences

To analyse the performance of MPEG2 bit streams we used two experimental MPEG2, MLGMP main profile, sequences. The first one is an Italy vs. Portugal soccer match; it consists of 2736 frames. The second is the Don Carlos opera; it consists of 2245 frames. These two sequences will be referred to, in the sequel, as IP and DC respectively.

We carried out a statistical study of these sequences similar to the one presented in §1.1 for the MPEG1 sequence and we obtained a similar qualitative behaviour. The relevant results are not presented for space limitations. A significant quantitative difference is, instead, that there are less inter-GOP correlations (maybe there are no special visual effects!). The system works in the uncorrelated region for a larger set of system parameters than in the case analysed in § 2.3; this means that the Renewal Approximation will perform very well.

Figure 10 shows the cell loss probability vs. the buffer size, for the DC sequence, for various values of ρ, and confirms our expectations. The RA is a very good approximation for the whole range of buffer sizes, (even for $\rho=0.7$ there is no appreciable difference in the plot between RA and simulations); only a small difference can be observed for large buffer sizes and for $\rho=0.9$. The OPA gives, for this sequence, the same result of the RA. This is because loss is due mainly to I frames that act as renewal points.

We have also analyzed the cell loss due to each of the frame types for the DC sequence, obtaining the same conclusions presented in the previous Section: cell loss is due predominantly to I frames; however the agreement between model and simulation is much better than that shown in Fig. 8 since in this case there are less inter-GOP correlations. Similar results have been observed also for the IP sequence. We have applied the proposed approach also to 12 very short MPEG2 sequences (e.g. Mobile & Calendar, Flower Garden and Tennis Table etc., that consist of only a few seconds of video). In this case there is no difference between simulation results and RA for every value of K and ρ [5]. Short sequences do not manifest important properties and are thus not very meaningful in studying the characteristics of VBR video.

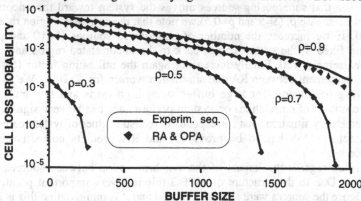

Fig. 10 - Cell loss probability Π as a function of the buffer size, K for different values of ρ for the DC sequence; comparison between experimental sequence, RA and OPA.

2.5 Performance Evaluation: Superposition of Sequences

In this Section we analyse the performance of a queueing system loaded with a superposition of MPEG sources. Both RA and OPA can be easily extended to this case since the Probability Mass Function (PMF) of the frame sizes of the aggregated stream can be obtained by convolving the individual PMFs (if the dimensionality of the problem increases, the reduction technique discussed in § 2.1 can be applied). We divided the long Star Wars sequence in S pieces ($S=5$ and $S=10$) obtaining S different sequence; each of them lasts about 24 minutes for $S=5$ and 12 minutes for $S=10$. The obtained sequences are long enough to allow a statistical analysis and can be considered uncorrelated from each other. Figure 11 shows the cell loss probability vs. the buffer size for $S=5$ and $S=10$ sources multiplexed together and for different values of ρ, comparing simulation results and RA. In this figure the sources are synchronized, i.e. they start at the same time (with an I frame). This figure is very useful to analyse the issue of time scales and correlations.

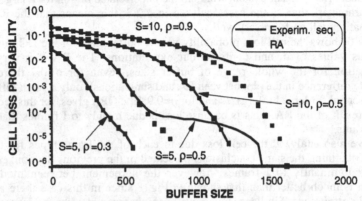

Fig. 11 - Cell loss probability Π as a function of K for a superposition of 5 and 10 sources: comparison between simulation results and RA.

For $S=5$ and $\rho=0.3$, the system works in the uncorrelated region for every K. The RA predicts *the same* results obtained by simulation. If we compare this curve with the curve relevant to only one source in Fig. 7, with the same value of the utilization factor, we see that superposing sources moves the system toward the uncorrelated region. If we increase ρ, ($S=5$ and $\rho=0.5$) we note that the correlated region re-appears for $K>600$. If we increase the number of superposed sources ($S=10$ and $\rho=0.5$) maintaining fixed ρ, the system comes back to the uncorrelated region and the RA performs extremely well. Finally increasing again the utilization factor ($S=10$ and $\rho=0.9$) the agreement between RA and simulation worsen for $K>1000$. We can draw the following conclusion: for large buffer sizes, high loads and few multiplexed sources the impact of correlations on system performance can be very significant; in the complementary situation (that corresponds to realistic values of system parameters, if a satisfactory QoS has to be provided) the RA predicts correctly the real performance.

A last issue regards the impact of the synchronization between sources on the performance. Due to the structure of MPEG this is a very important point. In the previous figure the sources were all synchronized and it is intuitive that this is a worst case; the I frames of the sources arrive simultaneously and the queue response deteriorates dramatically. We have analyzed also the performance of un-synchronized sources (i.e. sources that are offset by a multiple of the frame duration). As expected, the performance of un-synchronized sources are remarkably better; the loss probability is several orders of magnitude smaller. This means that a great improvement can be obtained if it could be possible to adjust source phase properly.

Finally, figure 12 depicts the cell loss probability for the superposition of 20 of the Rose's sequences [16] for two values of ρ as a function of K (synchronized sources). With 20 sources the system works, almost always, in the uncorrelated region; the proposed approximation is in very good agreement with simulation results, thus showing that when the number of superposed sources is significant (i.e. in realistic cases) the RA and the OPA could be useful to predict network performance.

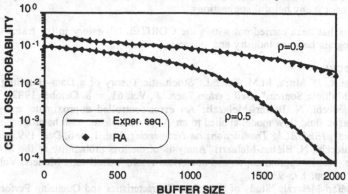

Fig. 12 - Cell loss probability Π as a function of K for a superposition of 20 of the Rose's sequences (for two values of ρ).

3 Conclusions

The approach proposed in this paper performs very well in the uncorrelated region. Anyway, the algorithm presented up to now does not give an assessment of the accuracy of the approximation introduced. In other words, if an arrival process is modelled with the RA, it is not possible to know a priori the closeness between model and real results.

In [5] we present an extension of the model presented in this paper that allow an a priori assessment of the extent of the uncorrelated region; the aim is to ascertain if a queueing system with given arrival process and system parameters, works in the correlated or in the uncorrelated region.

In order to accomplish this goal we have kept into account probability mass functions of the frame size of order greater than one and extended the OPA and the RA to analyze Markov process of order greater than one. In this way we are also able to evaluate the loss performance in the correlated region. In fact it can be shown that, by increasing the order of the Markov process that models the actual arrival process, the model results tend to the exact ones.

The results of the performance study presented are not very encouraging: i) correlations severely affect the performance and, in the correlated region, buffering is of little help in reducing the cell loss rate; so, constraining the cell loss rate under some very low threshold (e.g. 10^{-9}), implies a significant limitation of the output link efficiency; the multiplexing efficiency will be low unless a large number of sources are multiplexed together (i.e. unless the system works in the uncorrelated region); ii) loss affects mainly the I frames and it is concentrated in time; this has a negative consequence on the user-perceived QoS; iii) synchronization between sources heavily deteriorates the performance. Nevertheless our results show that is possible to squeeze more MPEG sources into a link than the number allowed by a peak allocation.

Finally we think that the approach proposed in this paper can be successfully applied also to model other complex and *real* arrival processes (e.g. LAN traffic).

Acknowledgements

We thank Mark Garrett (Bellcore, Morristown, NJ) for providing the MPEG1 Star Wars sequence and CSELT and Centro Ricerche RAI (Torino, Italy) for providing the MPEG2, Italy-Portugal and Don Carlos, sequences. We thank Oliver Rose (University of Würzburg) for the numerous MPEG1 traces that can be found on the ftp server: ftp-info3.informatik.uni-wuerzburg.de in the directory /pub/MPEG/. Finally, we thank the referees for many helpful suggestions.

This work has been carried out within the CORITEL laboratory in the frame of a joint research program between industry and university.

References

[1] S. Anick, D. Mitra, M.M. Sondhi: "Stochastic Theory of a Data–Handling Systems with Multiple Sources", *Bell System Tech. J.*, Vol. 61, n. 8, October 1982.

[2] A. Baiocchi, N. Bléfari-Melazzi: "An error-controlled approximate analysis of a stochastic fluid flow model applied to an ATM multiplexer with heterogeneous on-off sources", *IEEE/ACM Transactions on Networking*, Vol. 1, n. 6, Dec. 1993.

[3] A. Baiocchi, N. Bléfari-Melazzi: "Analysis of the loss probability of the MAP/G/1/K queue, part II: approximation and numerical results", Stochastic Models, Vol. 10, No. 4, 1994, pp. 895-925.

[4] N. Bléfari-Melazzi: "Study of Statistical Characteristics and Queueing Performance of MPEG1 and MPEG2 Video Sources", *IEEE Globecom '95*, Singapore, 13-17 November 1995, paper 13.7.

[5] N. Bléfari-Melazzi: "An error controlled algorithm for the performance analysis of MPEG video sources in a B-ISDN", submitted for publication.

[6] C. Blondia, O Casals: "Performance analysis of statistical multiplexing of VBR sources", *IEEE INFOCOM 92*, Florence 4-8 May 1992, pp. 828-838.

[7] M. W. Garrett, W. Willinger: "Analysis, modeling and generation of self similar VBR video traffic", *SIGCOMM 94- 8/94* London, pp. 269-280.

[8] R. Landry, I. Stavrakakis: "Multiplexing Generalized Periodic Markovian sources with an application to the study of VBR video", *ICC '94* New Orleans, May 1-5.

[9] D. Le Gall: "MPEG: A video compression standard for multimedia applications". *Communications of ACM*, 34(4), April 1991, pp. 46-58.

[10] D. M. Lucantoni, K. S. Meier-Hellstern, M. F. Neuts: "A single server queue with server vacations and a class of non-renewal arrival processes", *Adv. Appl. Prob.*, 22, 1990, pp. 676-705.

[11] D. M. Lucantoni, M. F. Neuts, A. R. Reibman: "Methods for performance evaluation of VBR video traffic models", *IEEE/ACM Transactions on Networking*, Vol. 2, No. 2, April 1994, pp. 176-180.

[12] B. Maglaris, D. Anastassiou, P. Sen, G. Karlsson, J. D. Robbins: "Performance models of statistical multiplexing in packet video communications", *IEEE Transactions on Communications*, Vol. 36, July 1988, pp 834-844.

[13] B. Melamed, D. Raychaudhuri, B. Sengupta, J. Zdepsky: "TES-based modeling for performance evaluation of integrated networks", *IEEE INFOCOM 92*, Florence 4-8 May 1992, pp. 75-84.

[14] G. Ramamurthy, B. Sengupta: "Modeling and analysis of variable bit rate video sources", *IEEE INFOCOM 92*, Florence 4-8 May 1992, pp. 817-827.

[15] J. W. Roberts: "Traffic control in the B-ISDN", *Computer Networks and ISDN Systems*, Vol. 25, N. 10, May 1993.

[16] O. Rose: "Statistical properties of MPEG video traffic and their impact on traffic modeling in ATM systems", *Proceedings of the 20th Annual Conference on Local Computer Networks*, Minneapolis, October 15-18, 1995.

[17] P. Skelly, M. Schwartz, S. S. Dixit: "A histogram-based model for video traffic behaviour in ATM multiplexer, *IEEE/ACM Transactions on Networking*, Vol. 1, No. 4, August 1993, pp. 446-459.

On the Scalability of the Demand-Priority LAN
A Performance Comparison to FDDI for
Multimedia Scenarios

Peter Martini, Jörg Ottensmeyer,
Dept. of Math. and Comp. Science, University of Paderborn, Germany
e-mail: {martini, otty}@uni-paderborn.de

Abstract
In this paper, we study the performance of the demand-priority LAN and compare it to the performance of FDDI. For different load models - one of them based on MPEG traffic - the demand-priority LAN is shown to come short of what it promises to be. In particular, it suffers from bad efficiency in case of large networks and small high priority frames. Most of our simulations address the effects that high priority traffic has on normal priority transmission. However, we also comment on the effectiveness of the high priority service. In some cases the network changes to a single priority operation where delays can no longer be bounded and where frame losses can no longer be avoided even for high priority traffic.

1 Introduction
In June 1995 the LAN/MAN Standards Committee of the IEEE finished a new standard defining the protocol and interconnection requirements for data communication via a central-repeater-controlled 100 Mbit/s LAN using the so-called demand-priority access method [IEEE95]. The ballot on the last draft resulted in 8 disapprove votes. One of the voters stated: "... the text seeks to provide low latency service for the support of multimedia applications by means of a two level priority access mechanism. There is no documentation to substantiate this claim. There are instead counter-examples to this capability...". The working group IEEE 802.12 rejected that view with reference to publications such as [SpGr94].

Our paper adds additional counter-examples to the capability of the demand-priority LAN to support multimedia applications as claimed by the standard. Again, it substantiates a request to change the protocol which was rejected by IEEE 802.12 in summer 1994; details on the proposed changes are discussed in documents submitted to IEEE 802.12 and in [MaOt95a].

This paper is structured as follows: In section 2 we give a brief overview of the most important aspects of the demand-priority LAN. In section 3 we present simulation results obtained for different scenarios and different load models. To put the results into the right perspective, we also show results obtained for FDDI in a similar configuration with identical load. The paper ends with a conclusion.

2 Topology and Protocol of the Demand-Priority LAN
This brief overview of the demand-priority LAN only addresses its most important characteristics as far as the performance in terms of throughput, delays, ... is concerned. There are several additional unique characteristics such as "filtering" or "training" which are irrelevant to our discussion.

Basically, the demand-priority LAN is a priority-based round-robin protocol where transmission is controlled by central network controllers called "repeaters". As far as the topology is concerned, a single-level demand-priority LAN looks very much like a FDDI network with concentrator: end nodes are connected to hubs by point-to-point twisted pair or optical fiber links. However, instead of waiting for a token an end node (DTE) with data ready for transmission sends a normal or high priority request to the network controller. This makes sure that the "right of way" is only given to stations really needing it. It also makes sure that high priority requests are served first.

Each repeater is configured with n down-link ports where one down-link port is required for each connected DTE. Optionally, a repeater can also have one up-link port which may be connected to the down-link port of a repeater on the higher level.

A large part of the standard describes the so-called "RMAC" which is the part of the MAC protocol belonging to the repeater. At any given time there may be at most one DTE transmitting a frame. This frame is forwarded to all repeaters. It is also forwarded to all DTEs explicitly addressed and to all DTEs connected to a port operating in promiscuous mode. Figure 1 shows a cascaded network where a unicast transmission is underway from DTE s to DTE k. The frame is also transmitted to the LAN Analyzer marked "LA".

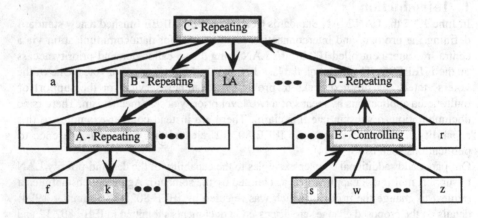

Fig. 1 Frame Transmission in a Cascaded Demand-Priority LAN

Our performance analysis presented in section 3 only studies a single-level network, i.e. a network where all DTEs are connected to the same repeater. In this case, the demand-priority access reduces to a local priority-based round-robin polling: DTEs send requests to the network controller, i.e. the repeater. The repeater keeps track of transmission requests by regularly checking each port. The standard does not require a specific method of checking for transmission requests. However, it requires each port to be checked at least once each frame time.

Normal and high priority requests are served in separate cycles. If the repeater is in the process of serving a sequence of normal priority frames and a high priority request is received, then it suspends service of the normal priority traffic to serve the high. However, it does not interrupt normal priority transmission in progress. When there are no more high priority requests pending, the normal priority cycle is resumed with the port following the last normal priority transmission. For both priority levels, ports are served according to a single frame service strategy, i.e. the DTE selected as the next sender may transmit one frame only before passing control back to the repeater. It should be noted that normal priority requests pending for more than approx. 250 ms are upgraded to high priority requests ("priority promotion").

The DTE selected by the repeater starts transmitting its frame immediately after detecting the "Grant" signal sent by the repeater. The repeater decodes the destination address of the frame currently transmitted and then forwards it to the DTE addressed and any promiscuous mode port (repeaters and specific DTEs) as soon as possible. Store-and-forward operation is only required for broadcast and multicast with 4-UTP link bundled cable. For the fiber optic networks studied in section 3 of this paper, frames may be forwarded on the fly.

For the transport of MSDUs, the demand-priority LAN uses either the 802.3 frame format or the 802.5 frame format. In addition to the MAC overhead (at least 18 byte), the IEEE 802.12 frames include preambles, start and end delimiters (260 bit for all physical media).

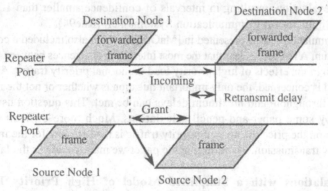

Fig. 2 Frame Transmission Space/Time Diagram

Figure 2 shows the data flowing through the repeater. The gap between outgoing frames is determined by two components:

- Firstly, by the time that elapses from selecting the next sender until the first bit of the frame is received. For a 100 m UTP link this is less than 1 μs, for a 2 km fiber link it is 20 μs.
- Secondly, by the "retransmit delay" which is the sum of the time that elapses until the destination address is completely received and the time needed for matching the destination address to the corresponding port. As opposed to draft 3.0, the standard now specifies a maximum time of 4.5 μs allowed for this operation.

Obviously, the gap between outgoing frames determines the efficiency of the network. It determines the difference between the data rate used on the links and the bandwidth really available to the end nodes.

3 Performance Analysis

A performance analysis of a newly proposed LAN does not make much sense if it does not include a comparison to existing standards and products. The obvious alternative to the demand-priority LAN is the FDDI multiple token ring. FDDI was the first available high speed local area network operating at 100 Mbit/s which was based on an international standard. The main advantages of the demand-priority LAN supposedly will be lower costs, use of the existing cabling and maintaining the frame format of Ethernet. Points like these have not been included in our performance studies. Instead, we compare the performance of FDDI to the fiber optic version of the demand-priority LAN.

A description of the well-known protocol FDDI [ISO9314] is not within the scope of this paper. For further details the interested reader is referred to publications such as [Jai94] or [Wit94]. It is well known that the performance of FDDI depends on the choice of the "operative Target Token Rotation Time" (TTRT) which is half the maximum token rotation time. Our choice of TTRT = 5 ms results in a maximum access delay for synchronous data of 10 ms which is a reasonable choice for the configuration studied, cf. [Spr94]. The bandwidth allocation for the synchronous service of FDDI is done as specified in [SMT7.3] based on the rate during a period of 125 µs. The actual allocation is scaled to the common TTRT value by the factor TTRT/125 µs.

For our simulations (run time: 45 s real system time, including 5 s initial part excluded from final evaluation, resulting in intervals of confidence smaller than 1 %) we used SimCom (Simulator for Communication Systems), cf. [Rue94].

The early simulation results presented in [MaOt95b] which also included a comparison to shared medium ATM made clear that the most interesting differences between the protocols may be seen in the effects of high priority load on normal priority traffic: As far as high priority load is concerned, the only important question is whether or not the requirements in terms of throughput and maximum delays can be met. This question usually can be answered by some paper-and-pencil calculations. Much more difficult to obtain is information on the price the normal priority traffic is made to pay for the integration of high priority transmission. Therefore, in the paper we mainly address the latter aspect.

3.1 Simulations with a Simplified Model of High Priority Load
In this section we study the demand-priority LAN (and FDDI) for a high priority load model which stays close to the VBR load model which was used in [Mar94] for a comparison of real-time communication strategies in DQDB. We show the sensitivity to both cable length and frame size.

The Load Model
Our load model includes both multimedia and conventional data traffic. The high priority load model is an on/off source generating fixed length bursts (16 ms) with isochronous frame arrival, cf. Fig. 3. The bursts are separated by idle periods of exponentially distributed length where the average length is 16 ms, too. Thus, the peak rate is two times the average rate. For each scenario studied, the average high priority load imposed to the MAC sublayer is chosen as follows:

average HP-load [Mbit/s]	0	2.048	4.096	6.144	8.192	10.24
frame size [byte]	-	512	1024	1536	2048	2560

The reason for choosing different frame sizes for different loads is to keep the frame spacing (1000 µs) identical within a given scenario. Of course, choosing frame sizes as multiples of 512 byte is somewhat arbitrary. Therefore, we also studied the impact of this assumption by changing the frame sizes to multiples of 48 byte resulting in a frame spacing of appr. 94 µs.

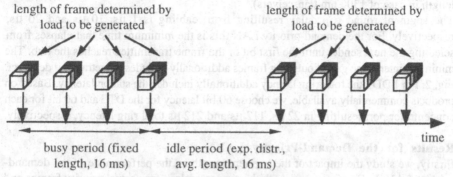

length of frame determined by
load to be generated

length of frame determined by
load to be generated

time

busy period (fixed
length, 16 ms)

idle period (exp. distr.,
avg. length, 16 ms)

Fig. 3 The VBR Load Model

Normal priority traffic (file transfer, telnet or e-mail) is assumed as Poisson arrivals of fixed size frames. We also choose 512 byte for this size. In [MaOt95a] we studied the impact of this assumption by choosing different frame sizes for normal priority traffic. The relevance in the scenarios studied here is very small.

It should be noted that the total load imposed to the network is higher because it additionally includes preambles, delimiters and 802.3 (or 802.5) MAC frame overhead for all normal and high priority frames. In the FDDI simulations included for comparison, we of course add the corresponding overhead instead.

The Configurations

We study a single-level fiber optic demand-priority LAN, i.e. a network where all DTEs are connected to one repeater. Obviously, this topology is the same as in a FDDI where all workstations are connected to the same concentrator.

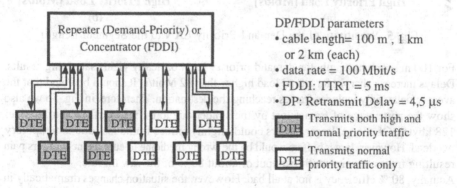

Repeater (Demand-Priority) or
Concentrator (FDDI)

DP/FDDI parameters
• cable length= 100 m , 1 km
  or 2 km (each)
• data rate = 100 Mbit/s
• FDDI: TTRT = 5 ms
• DP: Retransmit Delay = 4,5 µs

DTE   Transmits both high and
       normal priority traffic

DTE   Transmits normal
       priority traffic only

Fig. 4 The Configurations Studied

Figure 4 shows the configurations studied. Each of the 10 DTEs is assumed to transmit normal priority traffic (in case of FDDI using the asynchronous service). Five DTEs additionally send multimedia traffic at high priority (in case of FDDI using the synchronous service). The length of each cable connecting a DTE to the hub is chosen as 100 m (a typical working group LAN), as 1 km or as 2 km (maximum link segment length in case of 1300 nm transceivers).

The segment round trip delay resulting from cabling is 1 μs, 10 μs and 20 μs, respectively. For the demand-priority LAN this is the minimum time that elapses from selecting the next sender until the first bit of the frame transmitted reaches the hub. The minimum interframe gap of outgoing frames additionally includes the retransmit delay, cf. Fig. 2. For FDDI, the total ring latency additionally includes the station latency. Based on products commercially available, we choose 60 bit latency for the DTE and 60 bit for each concentrator port resulting in 22 μs, 112 μs and 212 μs total ring latency, respectively.

Results for the Demand-Priority LAN

Firstly, we study the impact of the segment length on the performance of the demand-priority LAN. In Fig. 5 we show both the average delays for normal priority frames and the frame loss characteristics versus the total high priority load (varying from 0 Mbit/s to 40.96 Mbit/s, for FDDI up to 51.2 Mbit/s). The normal priority (Poisson) load is set to 50 Mbit/s.

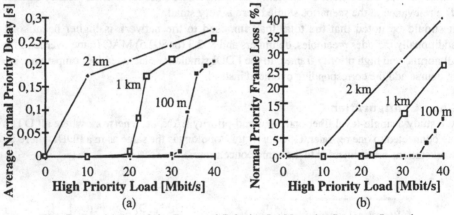

Fig. 5 Sensitivity of the Demand-Priority LAN to the Segment Length

For 100 m segment length, the demand-priority LAN basically yields satisfying results. Delays increase sharply for a total load higher than 82 Mbit/s. It should be noted that the average delay only includes frames reaching the destination. Therefore, in Fig. 5b we also show the percentage of frames lost by the sender due to buffer overflow (buffer size: 128 kbyte). Of course, larger buffers could slightly reduce frame loss due to temporary overload. However, larger buffers would be the wrong medicine to cure the networkers pain resulting from a maximum throughput of about 80 % of the raw bit rate.

Actually, 80 % efficiency is not at all bad. However, the situation changes dramatically in case of longer cables: For maximum length fiber segments (2 km with 1300 nm transceivers) everything is fine with the 50 Mbit/s normal priority load only, i.e. for 0 Mbit/s high priority load. With 10 Mbit/s high priority traffic the frame delay for normal priority changes to ≈ 170 ms. More important: ≈ 3 % of the frames are already lost. Frame loss increases linearly with increasing high priority load, i.e. normal priority yields to high priority load for these scenarios where the network is operated beyond its capacity (which is slightly below 60 Mbit/s for the load assumed here). For 1 km segments, the maximum throughput is between the other scenarios. The conclusion is that demand-priority users must keep their networks as small as possible.

As far as high priority traffic is concerned, our simulation model of the demand-priority LAN works fine: All frame delays are smaller than 10 ms, most of them considerably smaller. Frame losses do not occur for this traffic class. However, our model does not include the so-called "priority promotion" which makes normal priority frames interfere with high priority frames by upgrading normal priority requests pending for more than approximately 250 ms to high priority requests, cf. [IEEE95]. Obviously, the transmission collapses at both priority levels if priority promotion is implemented.

Results for FDDI

Next we discuss simulation results obtained for FDDI assuming the same cable lengths as for the demand-priority LAN. It has been pointed out by the designers of the demand-priority protocol that FDDI wastes a considerable amount of time passing the token around the ring even if all but one station are idle. This is due to the fact that the only way to come to know transmission requests is passing the token to the station. In contrast to this, demand-priority waits for the end stations to indicate their requests for transmission. Passive stations (even when far away from the repeater) do not harm the performance of the demand-priority LAN in any way. It should be noted that with a segment length of 2 km we are faced with quite a large FDDI network where the total cable length already contributes a latency of 200 μs. Thus, the time lost for traveling around the ring is 212 μs (including station latencies).

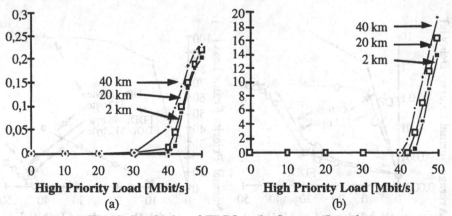

High Priority Load [Mbit/s]
(a)

High Priority Load [Mbit/s]
(b)

Fig. 6 Sensitivity of FDDI to the Segment Length

From Fig. 6 it comes clear that increasing cable length does not yield any significant performance degradation for FDDI. All simulation results are very close to each other. Frame loss does not occur even for 90 Mbit/s total load.

The major reason for the good results of FDDI may be seen in the fact that FDDI reduces token passing overhead in case of heavy load: For heavy load the average token rotation time approaches (and finally reaches) TTRT. For our simulations, TTRT was chosen as 5 ms. Thus, even with 2 km segments there basically still is 5 ms - 212 µs left for transmission in each token rotation.

The most important difference between FDDI and demand-priority is that after receiving the token an FDDI station is allowed to transmit several frames back-to-back (unless the token holding timer expires). In contrast to this, a demand-priority station must return the network control to the repeater after transmitting one single frame. Even if there is more data ready for transmission the station then must wait for the grant signal to be received again.

The simulation results for high priority traffic are not shown because they only confirm what is well known about FDDI: Both the average and the maximum MAC delays increase with increasing total load but are limited by TTRT and twice TTRT, respectively.

Comparison of Results

Summarizing and extending our discussion we now compare the simulation results for FDDI and the demand-priority LAN (DP). A change to logarithmic scaling allows us to examine the differences for a wide range of values (for the price of overemphasizing small values).

From Fig. 7a it comes clear that FDDI is the "winner". This should have been expected from our discussion in sections 3.1.3 and 3.1.4. However, the advantages of FDDI become even more obvious when assuming small high priority frames (48 byte as basic frame size, changing to 96 byte, 144 byte, 192 byte and finally 240 byte for increasing high priority load). For FDDI there is almost no performance degradation for this change; for demand-priority the results are hard to show even with logarithmic scaling.

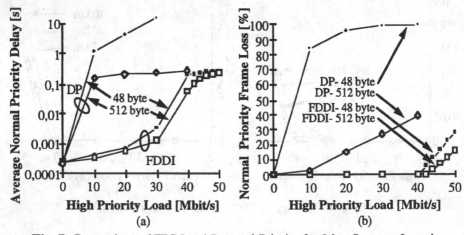

Fig. 7 Comparison of FDDI and Demand-Priority for 2 km Segment Length

The frame losses are compared in Fig. 7b. When comparing these results to the FDDI results presented in Fig. 6b it should be noted that the scaling had to be changed to accommodate the extremely bad results which have been obtained for demand-priority in case of small frames.

The most interesting curve in Fig. 7b definitely is the one for demand-priority and a basic frame size 48 byte. With 10 Mbit/s high priority load more than 80 % of the normal priority frames are lost! Basically, this means that even quite a low high priority load can dismiss the normal priority traffic from the network if priority promotion is not implemented. If it is implemented priority promotion makes sure that all data is transmitted at high priority which changes the two priority network into a single priority network. This clearly violates the major goals of the demand-priority LAN.

But even in a network without priority promotion nothing can be guaranteed for high priority frames if frames are small and the network is large. For 10 Mbit/s high priority load, the maximum delays measured in our simulations for small frames already exceed 66 ms. This can hardly be called a "low latency service". For 20 Mbit/s there is already loss of 1% of the high priority frames. For higher loads results become even worse.

3.2 Simulations with a High Priority Load based on MPEG Traffic

The results presented in the previous subsection were obtained for a simplified VBR model. Now we study the impact of this assumption by replacing the model for high priority load. MPEG is currently the most important video compression standard (for an overview see [LeGa91]). Thus, we use MPEG as the basis for a video load model. For our simulations with MPEG traffic, the segment length was set to 2 km resulting in a total ring length of 80 km.

The Load Model

As far as normal priority traffic is concerned we do not change anything when compared to section 3.1, i.e. we assume 50 Mbit/s Poisson traffic (512 byte frames).

For the high priority load we use a model of compressed video traffic. An uncompressed video is a sequence of frames with isochronous arrivals (33 ms spacing). With MPEG coding, consecutive frames are converted into a sequence of I-, B- and P-frames (intra coded, bidirectionally coded and predictive coded, respectively), each using specific compression techniques. A typical frame pattern is I-B-B-P-B-B with frame sizes of 10:1:2 for I:B:P frames [Ste93]. The result is a weakly regular stream of frames like the one shown in Fig. 8. The average load of a video stream is chosen as 1.5 Mbit/s.

We used such video streams as load models in our simulations. Of course, real video streams are more bursty due to scene changes resulting in different sizes for frames of the same type. For simplicity reasons we do not include this in our simulation model used here. It should be noted that we ran several simulations with real MPEG traces. As far as the aspects studied in this paper are concerned there were no significant differences except for much longer simulation times for small intervals of confidence.

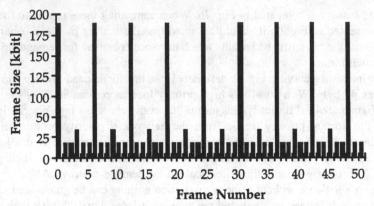

Fig. 8 Frame Structure of the MPEG Load Model

In section 3.1 we increased the load generated per station. Now the high priority load is increased by increasing the number of stations transmitting one video stream each. For each video sender the start of transmission is chosen independently (uniform distribution in a 200 ms interval). With a load of 1.5 Mbit/s per video stream and station, we increased the number of stations transmitting videos from 0 to 20.

When compared to the configuration studied in section 3.1, we now have 20 stations instead of 10. This has no impact on the performance of the demand-priority LAN since both the total load and the frame sizes remain unchanged. For FDDI, the change to 20 stations means twice as much station latencies resulting in more token passing overhead.

Simulation Results
A MPEG coder implemented according to the MPEG system standard [MPEG93] may generate two different output formats: a program stream (packets of variable and relatively great length) and a transport stream (packets of 188 bytes). When passing a program stream from the application to the transport system, packets have to be segmented (by the transport or network layer) according to the requirements of the network. For the demand-priority LAN we choose 1500 byte (which means CSMA/CD compatibility), 512 byte for comparison with the results of section 3.1 and 188 byte for the simulation of a transport stream.

The results presented in Fig. 9 confirm those presented in figure 7. Again, for a total load of 60 Mbit/s the demand-priority LAN reaches its limit. Obviously, the smaller the frames the worse the results. It could be argued that the bad results for the demand-priority LAN could result from possible synchronization of the video sources. Therefore, we also ran simulations where the isochronous frame arrivals were replaced by Poisson arrivals with the same average interarrival time. The effects on the results presented here are negligible. In fact, it would be impossible to see a difference in a graphical representation with scaling as Fig. 9.

For FDDI, delays were small (for light load slightly larger than for DP) and losses did not exist. Also, the segmentation has almost no impact on the results in the load range studied here. Of course, due to the additional PDU overhead differences could be seen for the load approaching the network capacity.

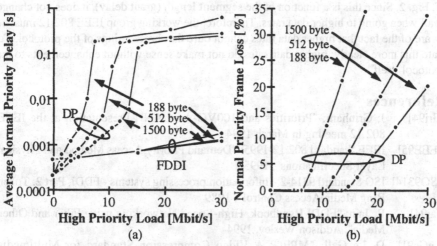

Fig. 9 Results for Load based on MPEG Traffic

The bursty high priority load assumed in this section results in increasing delays for this traffic class. This holds both for FDDI and DP. With priority promotion disabled, DP yields slightly smaller delays for high priority traffic than FDDI. In both DP and FDDI no high priority frames are lost. With priority promotion enabled both high and normal priority service collapse.

4 Conclusions

In this paper, we compared the performance of the demand-priority LAN to FDDI. Both networks claim to be able to support interactive multimedia applications and each includes priority mechanisms. Additionally, each of these networks provides normal priority traffic with the bandwidth reserved but momentarily not used by variable bit rate high priority traffic. The mechanisms are different and the effects on normal priority traffic are different, too. In all aspects covered by this paper, FDDI proved to be the much better choice.

In its overview section, the demand-priority LAN standard states its "purpose". In this part which was still called "goals" in draft 7, it claims that the protocol will provide a minimum data rate of 100 Mbit/s (purpose a). Our simulation study shows that "minimum" must have been mixed up with "maximum". In a configuration well in the limits imposed by the standard (purpose e: "allow topologies of 2.5 km and greater ...") even for large frames the usable bit rate was shown to be in the range of 60 Mbit/s. Priority promotion disables the provision of two priorities in case of heavy load which is a violation of "purpose h" of the standard. The statement "... low latency service ... for support of multimedia applications over *extended* networks" (purpose i) is somewhere between misleading and wrong. Here, it should be noted that the standard allows for cascading resulting in possibly much larger extensions than studied in this paper.

As far as the scalability of the demand-priority is concerned, only an increasing number of stations does not result in performance degradation. In contrast to this, smaller frames and larger distances throttle the available bandwidth seriously. The main reason for this behavior is the separation of frames leaving the repeater by at least the "incoming delay", cf. Fig. 2. Since this is a function of the segment length (grant delay), it does not change at all when going to higher data rates. Therefore, the working group IEEE 802.12 must be aware of the fact that higher data rates cannot cure the shortcomings of the protocol. To state this more clearly: higher data rates do not make sense without enhancements to the protocol itself.

References

[Gri94]   J. Grinham: "Priorities and 100VG-AnyLAN", Presentation at the IEEE 802.12 meeting in March 1994

[IEEE95]   IEEE Standard 802.12-1995: "Demand Priority Access Method and Physical Layer Specifications", 1995

[ISO9314]   ISO Standard 9314-2: "Information processing systems - FDDI, Part 2: Token Ring Media Access Control", 1989

[Jai94]   R. Jain: "FDDI Handbook. High-Speed Networking using Fiber and Other Media", Addison Wesley, 1994

[LeGa91]   D. Le Gall: "MPEG: A Video Compression Standard for Multimedia Applications", Communications of the ACM, Vol. 34, No. 4, pp. 46-58, April 1991

[Mar94]   P. Martini, "Connection Oriented Data Service in DQDB", Computer Networks and ISDN Systems, Vol. 26, Numbers 6-8, pp. 679-694, 3/1994

[MaOt95a]   P. Martini, J. Ottensmeyer "Performance Analysis of the Demand Priority LAN", Proc. of the "6th IFIP International Conference on High Performance Networking, HPN'95", pp. 255-266, Chapman & Hall, 1995

[MaOt95b]   P. Martini, J. Ottensmeyer, "Real-Time Communication in the Demand-Priority LAN - The Effects on Normal Priority Traffic", Proc. of the 20th LCN, pp. 350-357 , IEEE, 1995

[MPEG93]   ISO/IEC JTC1/SC29/WG11 N0601: "Coding of Moving Pictures and Associated Audio", Dec. 1993

[Rue94]   M. Rümekasten, "Simulation of Heterogeneous Networks", Proc. of the 1994 Winter Simulation Conference, pp. 1264-1271, IEEE, 1994

[SMT7.3]   ISO WD 9314-6: "FDDI - Station Management (SMT) Rev. 7.3a", June 1994

[SpGr94]   M. Spratt, J. Grinham, "Multmedia Applications and IEEE 802.12 Demand Priority", Proc. of the 12th Annual Conference of EFOC&N '94, pp. 47-50, 1994

[Spr94]   M. Spratt, et. al., "IEEE 802.12 - The local area network upgrade to 100 Mbit/s and beyond", Proc. of the 12th Annual Conference of EFOC&N '94, pp. 113-118, 1994

[Ste93]   R. Steinmetz: "Multimedia Technologie, Einführung und Grundlagen", Springer, 1993, (in German)

[Wit94]   R. Wittenberg, P. Martini, "Measurements in a FDDI Workstation Cluster", Proc. of the 19th LCN, pp. 52-58, IEEE, 1994

A Fast Switch Algorithm for ABR Traffic to Achieve Max-Min Fairness[*]

Danny H. K. Tsang[1], Wales Kin Fai Wong[1], Sheng Ming Jiang[1]
and Eric Y. S. Liu[2]

[1] Department of Electrical and Electronic Engineering, Hong Kong University of
Science and Technology, Clear Water Bay, Kowloon, Hong Kong
[2] Broadband and Data Strategies, Cable & Wireless, London, UK

Abstract. In this paper, a new rate-based switch mechanism is proposed for flow control of ABR traffic. By making use of the most up-to-date information from both the upstream and the downstream paths, the current bottleneck of the VC can quickly be found. The bandwidth allocation for different VCs can then be adjusted by using this new bottleneck information to achieve max-min fairness allocation [1]. We compare the proposed scheme to CAPC [2] and ERICA [3]. Simulation results showed that the transient response times of the sources are significantly reduced in the proposed scheme. Furthermore, the peak queue lengths of the switches are generally smaller.

1 Introduction

One of the challenges in developing Asynchronous Transfer Mode (ATM) is the support of connectionless traffic because connection-oriented ATM networks need to establish a connection before data can be transferred. Bandwidth negotiation is necessary such that appropriate resources can be allocated to the traffic to satisfy the user's declared Quality of Service (QoS). However, lacking of the prior knowledge of the characteristics of the connectionless traffic makes it very difficult to decide when and how much resources should be allocated. It is suggested that this kind of traffic should be supported by the Available Bit Rate (ABR) service. ABR traffic is used to fill in the bandwidth slack left by the scheduled traffic that has bandwidth and QoS guaranteed [4, 5].

Two flow control schemes for ABR traffic were under active discussion in ATM Forum [6]. They are credit-based flow control scheme [5] and rate-based flow control scheme [7]. After a prolong discussion, the traffic management group eventually adopted the rate-based approach. Since then, the original rate-based proposal was extensively modified [8]. Among the different approaches for rate-based flow control, the explicit rate (ER) approach requires the source to generate periodically resource management (RM) cells which traverse the path of the connection and eventually return back to the source. The switches along the

[*] Supported by Hongkong Telecom Institute of Information Technology grant HKTIIT93/94.EG01

path as well as the destination can use the ER field in the RM cells to carry congestion information about the path so that bandwidth allocation at the switches can be carried out. Upon receiving the RM cells, the source can then adjust its transmission rate based on the ER value in the received RM cells. Many of the proposed switch algorithms aim to achieve the max-min fairness allocation [1], a fairness criterion considered by ATM Forum. In this paper, a new switch mechanism which aims to rapidly achieve max-min fairness allocation is proposed. It is shown through simulations that the proposed scheme can significantly reduce both the transient response times of the sources and the peak queue lengths at the switches. In addition, the scheme is very simple and does not require any special parameters to be set.

The organization of the paper is as follows. In Section 2, two switch mechanisms previously proposed in ATM Forum are presented. In Section 3, a detailed description of the proposed scheme is given. Section 4 discusses the simulation results of both the transient response times of the sources and the peak queue lengths of the switches. Section 5 concludes the paper.

2 Previously Proposed Switch Mechanisms

2.1 Congestion Avoidance using Proportional Control (CAPC)

The basic idea of the Congestion Avoidance using Proportional Control (CAPC) [2] is to select a target rate, $R0$, at which the switch should operate. To achieve this, proportional feedback control is used along with the explicit rate approach.

The total input rate to the switch, $Rate$, is measured first. The rate adjustment factor, $delta$, is calculated as:

$$delta = 1 - Rate/R0 \qquad (1)$$

If $delta$ is greater than 0, the explicit rate for the switch, ERS, is increased as follows:

$$ERS = ERS \cdot \min(ERU, 1 + delta \cdot Rup), \qquad (2)$$

where ERU is the maximum increase of ERS (typically 1.5) and Rup is the proportional constant for rate increase (typically 0.025 to 0.01). Otherwise, ERS is reduced as follows:

$$ERS = ERS \cdot \max(ERF, 1 + delta \cdot Rdn), \qquad (3)$$

where ERF is the minimum decrease of ERS (typically 0.5) and Rdn is the proportional constant for rate decrease (typically 0.2 to 0.8).

When the switch receives a backward RM cell, the Explicit Rate (ER) field of the RM cell is updated to the minimum of the current value in the ER field and ERS. In addition, CAPC also marks the Congestion Indication (CI) bit of the backward RM cells when the queue length exceeds a threshold. When the source receives a RM cell, its allowed cell rate, ACR, is increased by the value of Additive Increase Rate (AIR) only if the value of CI is equal to zero

[2]. Furthermore, the final value of ACR is always set to the minimum of the current ACR, the peak cell rate (PCR), and the value of the ER field in the last received RM cell.

This scheme has several problems. First of all, the scheme requires the setting of many parameters. Incorrect setting of these parameters may lead to performance degradation. In addition, the use of queue length as overload indicator may lead to unfairness [8]. The scheme may also result in unnecessary oscillations [9].

2.2 Explicit Rate Indication for Congestion Avoidance (ERICA)

Instead of using queue length as the overload indicator, Explicit Rate Indication for Congestion Avoidance (ERICA) [3] uses the queue growth rate as the overload indicator. The switch measures the time T for N cell arrivals. If the available capacity of the link is C cells per second and the target utilization is U, the overload factor can be computed as follows:

$$Overload\_Factor = N/(T * U * C) \qquad (4)$$

At the end of the measurement interval of N cell arrivals, the switch computes the overload factor and informs all the VCs passing through it to adjust their rates according to the overload factor. The scheme also takes fairness into consideration and is achieved by ensuring that every VC gets at least a fair share of bandwidth, FS, which is computed as follows:

$$FS = Target\_Cell\_Rate/Number\_of\_Active\_VC, \qquad (5)$$

where $Target\_Cell\_Rate = U \cdot C$. The number of active VC is the number of distinct VCs that were seen transmitting during the last measurement interval of N cell arrivals.

By combining the two factors, the switch's recommended ER value for VC i can be computed as:

$$ERS(i) = \max(FS, CCR(i)/Overload\_Factor), \qquad (6)$$

where $CCR(i)$ is the current cell rate of VC i, which can be obtained from the most recently received RM cell of VC i.

When a backward RM cell of VC i is received at the switch, the switch first computes $ERS(i)$ and then updates the ER field of the RM cell to the minimum of the current value in the ER field and the computed $ERS(i)$. When the source receives a returning RM cell, its ACR is always set to the value of ER in the received RM cell.

3 Proposed Max-Min Scheme

The proposed scheme operates as follow. Each switch maintains an information table for all active VCs that pass through it (e.g., see Table 1). VCI denotes the VC identifier. ER_f and ER_b respectively denote the ER value of the most recent RM cell received in the forward and the backward directions. CA is the current allocation for the VC at the switch. *Constrained* is a boolean variable. When it is 1, the connection is a constrained one [10] and cannot achieve its fair share of bandwidth at this node because of the constraints imposed by its PCR or by the limited amount of bandwidth available at other nodes along its path. Similarly, when *constrained* $= 0$, this implies the bandwidth of the connection is only limited by the bandwidth available at the considered node. Denote N as the total number of active connections and M as the number of constrained connections at the switch. The number of active VC is the number of distinct VCs that were seen transmitting during the last measurement interval of N cell arrivals, as in ERICA [3].

Table 1. Information Table at the switch

VCI	ER_f	ER_b	CA	constrained
x	f1	b1	c1	0/1
y	f2.	b2	c2	0/1

When a RM cell is generated by the source, its ER field is set to Peak Cell Rate (PCR) as depicted in Figure 1. When the switch receives a forward RM cell of VC j with ER field equal to ER_RM, the switch will do the following:

    i. IF $ER\_RM = ER\_f(j)$ THEN GOTO step ix
    ii. $ER\_f(j) = ER\_RM$
    iii. IF $\min(ER\_f(j), ER\_b(j)) \leq CA(j)$ THEN
    $constrained(j) = 1$ and $CA(j) = \min(ER\_f(j), ER\_b(j))$
    ELSE
    $constrained(j) = 0$
    iv. For all unconstrained connections i, let $CA(i) = \Lambda$, where

$$\Lambda = \frac{Available\_Bandwidth - \sum_{constrained\_connection} CA(k)}{N - M} \quad (7)$$

    v. $changed = 0$
    vi. For all unconstrained connections i
    IF $\min(ER\_f(i), ER\_b(i)) \leq \Lambda$ THEN
    $constrained(i) = 1, CA(i) = \min(ER\_f(i), ER\_b(i))$ and $changed = 1$

vii. For all constrained connections k
IF $\min(ER\_f(k), ER\_b(k)) > \Lambda$ THEN
$constrained(k) = 0$ and $changed = 1$
viii. IF $changed = 1$ GOTO step iv
ix. END

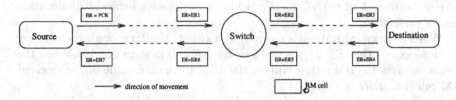

Fig. 1. Flow of RM cells

The algorithm works as follows. When a forward RM cell with $ER = ER\_RM$ for VC j arrives at the switch, the switch checks whether $ER\_f(j)$ is equal to ER_RM. If they are equal (step i), nothing needs to be done for this RM cell. Otherwise, $ER\_f(j)$ is set to ER_RM (step ii). If the minimum of the new $ER\_f(j)$ and $ER\_b(j)$ is less than $CA(j)$, this implies that the bottleneck of VC j is elsewhere along its path. Therefore, $CA(j)$ is reduced to the minimum of $ER\_f(j)$ and $ER\_b(j)$, and constrained is set to 1. Otherwise, constrained is set to 0 (step iii). For all unconstrained connections i, $CA(i)$ is updated to Λ (step iv) as in (7). Here, Λ is the new current allocation for all unconstrained connections and $CA(k)$ is the current allocation for constrained connection k. For all unconstrained connections i, Λ is compared to the minimum of $ER\_f(i)$ and $ER\_b(i)$ (step vi). If Λ is larger, $constrained(i)$ is set to 1 and $CA(i)$ is set to the minimum of $ER\_f(i)$ and $ER\_b(i)$. It is because if Λ is larger, the bottleneck is in fact elsewhere and thus the connection should be classified as a constrained one. The change in Λ is due to either the change in the available bandwidth at some switch or the change in the number of active VCs in the network. Similarly, if the minimum of $ER\_f(k)$ and $ER\_b(k)$ is larger than Λ for some constrained connection k (i.e., the bottleneck for the VC is not elsewhere but at the current switch), constrained is then set to 0 (step vii). If $changed = 1$ after steps vi and vii, further calculation of Λ is necessary because of the change of some VC's constrained status. Therefore, steps iv to viii are repeated until changed is 0 at the end of step vii. It is shown in [15] that the max-min calculation requires at most two iterations to converge.

The update of the ER field is now discussed. As depicted in Figure 1, let $ER1$ be the ER value in the RM cell when arrived at the switch and CA be the newly computed current allocation for the VC at the switch. The new ER value for the outgoing RM cell, $ER2$, is simply set to CA since the computation of CA in the algorithm has already taken both fairness and bottleneck information into account.

When the RM cell reaches the destination, it is turned around by the destination and the ER value of the returning RM cell is reset to the minimum of PCR and the destination's supported rate (i.e., $ER4$ in Figure 1). The resetting of the ER value at the destination is important since it permits the independent flows of upstream and downstream congestion information via the RM cells to the switches. By making use of the most up-to-date congestion information from both the upstream and the downstream paths, the switches can know of the current bottleneck of the VC more quickly so that better bandwidth allocation can be performed.

When a backward RM cell is received at a switch, similar procedures as above are done except that $ER\_f(j)$ is replaced by $ER\_b(j)$ in steps i and ii. When the source receives the RM cell, it will set its ACR to the ER value in the received RM cell (i.e., $ER7$ in Figure 1).

When either the number of active VC or the available bandwidth at the switch changes, steps iv to viii of the above pseudocode must also be executed in order to determine the new allocation. When a VC is terminated, its entry in the information table at each of the switches involved must be deleted. On the other hand, when a new VC is established, a new row needs to be created in the information table at each of the switches involved. The initial values of ER_f and ER_b are set to PCR while the initial constrained status is set to 0. The values of CAs for all VCs passing through the switches are recomputed using steps iv to viii of the above pseudocode.

In summary, by resetting the ER field of the RM cells at the destination, the proposed scheme can significantly reduce the response times of the sources because the new congestion information carried by the forward RM cells can immediately be used at the switch to calculate the new CAs, which are then quickly carried back to the sources by the backward RM cells. With this, the sources can adjust to the max-min fairness allocation in much shorter times.

4   Performance of Max-Min scheme

Figure 2 shows the simulation model [11] which is implemented by using the simulation package BONeS [12]. In this network, there are two multi-hop VCs (VC2 and VC4) while the remaining VCs are single-hop. The source end system (SES) behavior is based on [4]. However, since no NI field is used in CAPC, the operation based on NI in the SES is disabled. Similarly, since no NI and CI fields are used in ERICA and the proposed scheme, the SES is modified such that the operations based on NI and CI are not carried out.

4.1   Simulation Settings

The values of the common parameters for the SES [4] are shown in Table 2. The one-way propagation delay between the source/destination and its attached switch is $5\mu s$ while the one-way propagation delay between two switches is $50\mu s$ (as suggested in [13] for LAN separation). The sources we used are staggered one (i.e., the sources become active one by one). VC1 starts at $0ms$. Ten starting

167

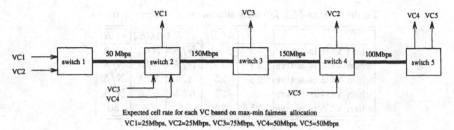

Expected cell rate for each VC based on max-min fairness allocation
VC1=25Mbps, VC2=25Mbps, VC3=75Mbps, VC4=50Mbps, VC5=50Mbps

Fig. 2. Simulation Model

times are tested for each subsequently active VC. The mean time of becoming
active for VC2, VC3, VC4 and VC5 are $5ms$, $10ms$, $15ms$ and $20ms$, respectively.
The ten starting times are equally spaced and cover an interval of width equal
to Nrm cell times of the previously started VC. The reason is to take into
account of the different arrival times of the RM cells. The sources remain active
after startup until the end of simulation. Each switch attempts to fully utilize
the total available bandwidth (e.g., $150Mbps$ for switch 2). Different initial cell
rates, $ICRs$, are used for comparison. The values of the parameters used in
CAPC are based on [14] and are shown in Table 3. For ERICA, the counting
interval N is 30 cells [3].

Table 2. Setting of Common Parameters for SES

PCR	MCR	Nrm	RDF	TOF
150Mbps	PCR/1000	32	1024	2

Table 3. Setting of Parameters for CAPC

AIR	Rup	Rdn	ERU	ERF	interval	Qthreshold
PCR	0.25	1.0	1.5	0.5	1ms	100 cells

4.2 Performance Comparison

Table 4 shows the max-min fairness allocation for different sources at different
times. The values of ACR for the different VCs are shown in Figures 3-5 when
the VCs become active one by one. Figure 3 shows the values of ACR for CAPC
for three different values of ICR. The figures show that, for most cases, the
$ACRs$ of the VCs cannot converge to the steady-state values before the next
VC becomes active. After all five VCs are active, it takes approximately 15 to
$20ms$ for the VCs to achieve the max-min fairness allocation. The scheme has
the longest response time when compared to the other two schemes.

Table 4. Max-Min fairness allocation at different times

	ACR1	ACR2	ACR3	ACR4	ACR5
when VC1 is active	50	N/A	N/A	N/A	N/A
when VC2 is active	25	25	N/A	N/A	N/A
when VC3 is active	25	25	125	N/A	N/A
when VC4 is active	25	25	62.5	62.5	N/A
when VC5 is active	25	25	75	50	50

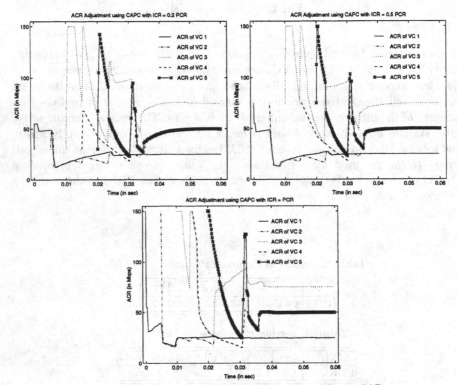

Fig. 3. ACR adjustment using CAPC under different ICRs

Figure 4 shows the values of ACR for ERICA under three different ICRs. It shows that the response times of the sources in ERICA are better than those in CAPC. However, ERICA sometimes cannot converge to the max-min fairness allocation (e.g., in Figure 4 after VC4 becomes active in the interval between 15 and $20ms$). According to the max-min fairness allocation, after VC4 becomes active, ACRs for VCs 2, 3 and 4 should be $25Mbps$, $62.5Mbps$ and $62.5Mbps$ respectively.

The convergence problem of ERICA can be explained as follows. Before VC4 becomes active, ACRs for VCs 2 and 3 are $25Mbps$ and $125Mbps$, respectively.

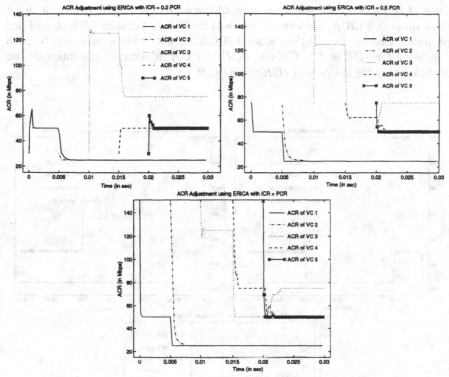

Fig. 4. ACR adjustment using ERICA under different ICRs

This means VCs 2 and 3 can fully utilise the output link of switch 2 (Figure 2). Therefore, once VC4 becomes active at $15ms$, the overload factor at switch 2 is over 1. For the case of $ICR = 0.2PCR$ (i.e., ACR for VC4 is $30Mbps$ at $15ms$), only ACR for VC3 is reduced because (i) ACR for VC4 is allowed to increase to at least $50Mbps$, which is the fair share value suggested by (5), and (ii) since the bottleneck for VC2 is at switch 1, ACR for VC2 is thus maintained at $25Mbps$. Therefore, the overload factor at switch 2 continues to be above 1 since only ACR for VC3 is reduced. The reduction ends when ACR for VC3 drops to $75Mbps$ and ACR for VC4 increases to $50Mbps$. For the case of $ICR = PCR$, ACRs for VCs 2, 3 and 4 should all be decreased since the overload factor at switch 2 is over 1. However, since the bottleneck for VC2 is at switch 1, ACR for VC2 is maintained at $25Mbps$. Moreover, ACR for VC3, a single-hop connection, decreases faster than that for VC4 because VC3 reacts faster to network feedback [11]. ACR for VC3 continues to decrease until it reaches $50Mbps$ (the fair share allocation) and ACR for VC4 is reduced to $75Mbps$. At that time, the overload factor becomes one and steady state is achieved.

For the proposed scheme, max-min fairness allocation is always achieved and the response times of the sources are the shortest among the three schemes examined in this paper (see Figure 5). Moreover, the performance of the proposed scheme is independent of ICR and is approximately the same for the different cases of *ICR*s. Since the response times of the sources in CAPC are much larger than those in ERICA, our comparison will focus only between ERICA and our proposed scheme. In addition, since ERICA cannot achieve max-min fairness allocation for $ICR = 0.2PCR$ and $ICR = PCR$ in certain time intervals, we compare only for the case of $ICR = 0.5PCR$.

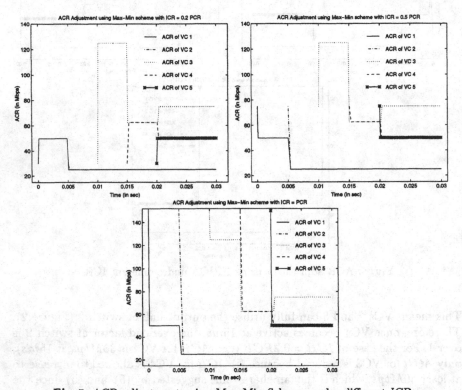

Fig. 5. ACR adjustment using Max-Min Scheme under different ICRs

Tables 5 and 6 show the transient response times of the sources for the case of $ICR = 0.5PCR$ for ERICA and the proposed scheme, respectively. They show that the response times of our scheme are much faster than that of ERICA.

In Table 7, the peak queue lengths at different switches for the two schemes are shown. It shows that a significant reduction in peak queue length is achieved by the proposed scheme. Small queue length is important for local area networks (LANs) because the buffer size of LAN switch is usually small. Better control of queue length can reduce the number of cell loss and therefore minimizes the performance degradation due to cell loss.

Table 5. Transient Response Time in μs for ERICA

	ACR1	ACR2	ACR3	ACR4	ACR5
when VC1 is active	405±0	N/A	N/A	N/A	N/A
when VC2 is active	531.7±268.4	3403.1±350	N/A	N/A	N/A
when VC3 is active	0±0	0±0	2791.6±25.8	N/A	N/A
when VC4 is active	0±0	0±0	2165.8±854.8	2161.8±886	N/A
when VC5 is active	0±0	0±0	1933.5±342.6	1463.3±514.2	484.7±84.6

Table 6. Transient Response Time in μs for Max-Min Scheme

	ACR1	ACR2	ACR3	ACR4	ACR5
when VC1 is active	134±0	N/A	N/A	N/A	N/A
when VC2 is active	159.2±77.8	408.9±1.6	N/A	N/A	N/A
when VC3 is active	0±0	0±0	128±0	N/A	N/A
when VC4 is active	0±0	0±0	69.9±23.4	338.5±1.1	N/A
when VC5 is active	0±0	0±0	322.3±46.3	227.3±54.6	131.2±1.2

Table 7. Comparison of Peak Queue Lengths in cells

	Switch 1	Switch 2	Switch 3	Switch 4
ERICA	131±4	53.8±2.8	2±0	54.2±7.6
Max-Min Scheme	66.9±4.6	22.6±3.5	2±0	21.9±1.6

5 Conclusion

A new switch mechanism that can quickly achieve the max-min fairness allocation is proposed. Max-min fairness is the fairness criterion considered by ATM Forum. In addition to always achieving the max-min fairness allocation, the proposed scheme also has several other advantages over the previously proposed schemes. One is the significant reduction of the transient response times of the sources. Another is the reduction of peak queue lengths at the switches. Furthermore, the proposed scheme is very simple and does not require any special parameters to be set. Therefore, the performance will not be degraded due to the improper setting of parameters.

References

1. Bartsekas, D., Gallager, R.: Data Networks. Prentice Hall, 2nd Edition, 1987
2. Barnhart, A.W.: Explicit Rate Performance Evaluation. ATM Forum 94-0983R1
3. Jain, R.: A Sample Switch Algorithm. ATM Forum 95-0178R1
4. Sathay, S.: ATM Forum Traffic Management Specification Version 4. ATM Forum 95-0013

5. Kung, H.T., Morris, R., Charuhas, T., Lin, D.: Use of Link-by-Link Flow Control in Maximising ATM Networks Performance: Simulation Results. Proc. IEEE Hot Interconnects Symposium, '93
6. Saunders, S.: ATM Forum Ponders Congestion Control Options. Data Communications, March 1994, page 55-60
7. Ramakrishnan, K. K., Jain, R.: A Binary Feedback Scheme for Congestion Avoidance in Computer Networks. ACM Transaction on Computer Systems, Vol. 8, No. 2, May 1990, page 159-181
8. Jain, R.: Congestion Control and Traffic Management in ATM Networks: Recent Advances and A Survey. Invited submission to Computer Networks and ISDN Systems
9. Jain, R., et al.: Rate Based Schemes: Mistakes to Avoid. ATM Forum 94-0882
10. Bonomi, F., Fendick, K.W.: The Rate-Based Flow Control Framework for the Available Bit Rate ATM Service. IEEE Network Magazine, March/April 1995, page 25-39
11. Chang, Y., Golmie, N., Su, D.: Comparative Analysis of the Evolving End System Behavior (Simulation Study). ATM Forum 95-0395R1
12. COMDISCO System Inc.: BONeS DESIGNER Core Library Guide. June 1993
13. Wojnaroski, L.: Baseline Text for Traffic Management Sub-Working Group. ATM Forum 94-0394R5
14. Barnhart, A.W.: Providing improved Explicit Rate Performance for the LAN. ATM Forum 94-1111
15. Charny, A.: An Algorithm for Rate Allocation in a Packet-Switching Network with Feedback. MIT/LCS/TR-601. April 1994.

Virtual Partitioning by Dynamic Priorities: Fair and Efficient Resource-Sharing by Several Services

Debasis Mitra[1] and Ilze Ziedins[2]

[1] AT&T Bell Laboratories, Murray Hill, New Jersey 07974, U.S.A.
[2] Dept. of Statistics, University of Auckland, Private Bag 92019, Auckland, New Zealand

Abstract. We propose a scheme for sharing an unbuffered resource, such as bandwidth or capacity, by various services. The scheme assigns a nominal capacity to each service class and implements a form of virtual partitioning by means of state-dependent priorities. That is, instead of each class of traffic having a fixed priority, as in traditional trunk reservation schemes, the priorities depend on the state of the system. An approximate method of analysis based on fixed point equations is given. Numerical results are obtained from the approximation, exact computations and simulations. The results show that the scheme is robust, fair and efficient.

1 Introduction

This paper considers the problem of sharing an unbuffered resource, such as bandwidth or capacity between several services both "fairly and efficiently". We restrict ourselves to looking at the single link and a single rate for the services.

We assume a single link, with capacity C, has K call classes arriving at it. Calls of class k, $k = 1, 2, \ldots, K$, arrive as independent Poisson streams of rate λ_k. All call classes have exponential holding times with the same mean $1/\mu$. Without loss of generality, assume that $\mu = 1$. The problem is to guarantee a given grade of service for each class of call (the "fairness" constraint), while at the same time ensuring that the resource is not underused (the "efficiency" constraint). The fairness constraint implies that some capacity should be allocated to each call class, while the efficiency constraint would suggest that some sharing of the allocated capacity should be allowed. The policy studied in this paper is based on forecasted traffic rates. Nominal allocations of capacity for each call class are precomputed in such a way that a given grade of service can be met, that is, there is an "engineered design". The key feature of our scheme for sharing the resource is the role of dynamic state-dependent priorities for each class. That is, instead of each class of traffic having a fixed priority, as in traditional trunk reservation schemes, the priorities may depend on the state of the system, and thus will vary dynamically with changing traffic patterns.

The problem of sharing resources between competing demands has been much studied. The simplest policies, and in some sense extremal policies are complete sharing (CS) and complete partitioning (CP). Complete sharing allows all services to share the resource indiscriminately. Complete partitioning divides the resource between the services, allowing each service exclusive use of its allocated capacity. Under complete sharing, one class of service may overwhelm all the others, so that the "fairness" requirement is not met. Under complete partitioning, the resource may be underused, so that the "efficiency" requirement is not met.

A scheme that falls partway between these two is one that we refer to as physical partitioning (PP). In this scheme, as in complete partitioning, the capacity is subdivided between the competing services. Calls of class k are assigned capacity C_k, where $\sum_{k=1}^{K} C_k = C$. Thus instead of having just one trunk group, one can think of there being K trunk groups, the kth having capacity C_k. Each trunk group is assigned a trunk reservation parameter, and calls of each class have priority at their allocated trunk group. A call arriving at its trunk group is accepted if there is spare capacity on it. If there is no spare capacity, then the call chooses, in accordance with some discipline, another trunk group where the amount of spare capacity exceeds the trunk reservation parameter, and is carried there. If no such trunk group exists, the call is lost. The simplest discipline picks the overflow trunk group at random. By allowing state-dependence the discipline for selecting the overflow trunk group may be generalized.

Physical partitioning is an attractive scheme, in that it offers some protection to each class, while at the same time making more efficient use of the resource, because, unlike complete partitioning, it allows some sharing. However, if we consider the sample-path behaviour we see that it has some undesirable characteristics. To see this, consider the case $K = 2$. Suppose a call of class 1 arrives. Its own trunk allocation may be occupied by predominantly class 2 calls (if, for instance, there has been a sudden burst of these), and yet it may be blocked in the second trunk group. Thus even though the number of class 1 calls in the system is low, a call of that type may be rejected. On the other hand, an arrival of class 2 that finds the trunk groups in exactly the same state will be accepted, thus leading to a state where usage is even more unfairly distributed.

To overcome the undesirable behaviour that the physical partitioning scheme exhibits, we propose a scheme which we call virtual partitioning (VP) and which is implemented by means of state-dependent priorities. As with physical partitioning, the kth class of call is assigned a nominal capacity C_k, where $\sum_{k=1}^{K} C_k \geq C$ and C_k is sufficiently high to ensure the desired grade of service. Let n_k denote the number of calls of class k in progress. An arriving call of class k is always accepted if $n_k < C_k$ and $\sum_{j=1}^{K} n_j < C$. If $n_k \geq C_k$, the call is accepted if $\sum_{j=1}^{K} n_j < C - r_k$, where r_k is a trunk reservation parameter chosen to protect the remaining classes against overloads of class k. Thus, depending on whether $n_k < C_k$ or $n_k \geq C_k$ the priority status of class k calls is either high or low.

In addition to overcoming the undesirable sample-path properties mentioned above, we would also expect this scheme to be very robust to changing traffic patterns, since it allows the priorities of traffic streams to vary dynamically

175

with changing traffic patterns. We observe that in the case $K = 2$, the virtual partitioning scheme is just physical partitioning with repacking. For $K > 2$, however, that is not the case, and we would expect the blocking probabilities to decrease by virtue of the pooling of resources that occurs.

Some of the authors who have considered related problems are Foschini et al. (1981), Gopal and Stern (1983) and Kraimeche and Schwartz (1984). Foschini et al. (1981) consider only admission policies that induce product-form equilibrium distributions and show that a threshold policy is optimal. Gopal and Stern (1983) apply Markov decision theory to the analysis of threshold policies, while trying to maximise the average occupancy of a link. Kraimeche and Schwartz (1984) consider a class of restricted access policies, with the aim of minimizing blocking probabilities. The work of Ash et al. (1991) on class-of-service routing deserves particular note. Implementations aimed at balancing fairness and efficiency are given. See also the work of Mason, Dziong and Tetreault (1992). Kelly (1991) and Ross (1995) are excellent sources of information on fixed point approximations for loss networks.

There are several ways in which the problem we have outlined could be formulated. Suppose that calls of class k attract revenue w_k. Then one formulation of the problem is to maximize the revenue earned (if $w_k = 1$ for all k this is just the number of calls carried) subject to certain grade of service requirements being met.

Another formulation of the problem arose from the observation that while we would normally wish to admit as many calls as possible, we do not wish to do so if other calls are then constrained to use less than their allocated capacity and hence rejected. It is therefore natural to consider a problem formulation where some penalty is incurred when undesirable behaviour is exhibited. We add a penalty structure to the problem by assuming that a penalty v_k is paid for each call of class k that is rejected when $n_k < C_k$ (in the numerical work we set $w_k = 1$ and $v_k = 2$ for all k). We can then seek the admission policy that maximises the revenue earned per unit time. The second formulation lends itself more easily to an analysis using Markov decision processes and it is the one used in this paper.

Although in the examples discussed in this paper, we have always assumed that $\sum_{k=1}^{K} C_k = C$, there are many cases of interest where it is more natural to have a strict inequality. For instance, it may be desirable to have some capacity that is completely shared between the types of call, with no trunk reservation operating on it. Or there may be classes of traffic that have high priority and that we wish to accept under any circumstances (e.g. some kinds of overflow traffic), in which case they could have nominal capacity C. And complete sharing is a special case of virtual partitioning, obtained by setting $r_k = 0$, $(k = 1, 2, \ldots, K)$. The fixed point approximation described below can be adapted to deal with these changes.

In this paper we investigate the optimality of the suggested policy, as well as developing a fixed point approximation for the blocking probability. We give numerical examples that show that the suggested virtual partitioning policy exhibits the desired properties of fairness and efficiency, as well as being fairly robust against changes in traffic patterns.

2 Fixed Point Approximations

To study the virtual partitioning policy further, it is necessary to consider the equilibrium distribution of the process. Unlike several other policies that have been discussed in the literature, this does not induce a product-form distribution on the state-space, and finding the exact solution is a non-trivial task, even for the simplest case with $K = 2$. It was therefore natural to consider an approximation, and in particular, some form of fixed point approximation. The fixed point approximation that was developed arose from the observation that the behaviour of a single link with a virtual partitioning policy was exactly the same as that of a certain circuit switched network.

Consider a network consisting of $K + 1$ links, as illustrated in Figure 1. The kth link, $1 \le k \le K$ has capacity $C - r_k$, and the $K + 1^{st}$ has capacity C. We call the $K + 1^{st}$ link the α link, and the kth link, β_k, for $1 \le k \le K$. Let n_α denote the number of calls in progress on link α, and n_k denote the number of calls in progress on link β_k. Calls of class k arrive at Poisson rate λ_k, and request a unit of capacity simultaneously from link α and link β_k. They are accepted if *either* $n_k < C_k$ and $n_\alpha < C$ *or* if $n_k \ge C_k$ and $n_\alpha < C - r_k$. Note that we are implicitly assuming here that $C - r_k > C_k$. The sequence of acceptance/rejection decisions for this network is indistinguishable from that of the single link with virtual partitioning, that is, if both start with the same numbers of type k calls in progress, and are offered the same sequence of arrivals, then both will accept and reject the same calls. The capacity of link β_k was set equal to $C - r_k$ because under virtual partitioning there can never be more than $C - r_k$ calls of class k in the system. We will refer to the characterization of the virtual partitioning policy in this form as the alpha/beta network.

For the alpha/beta network there is a natural fixed point approximation that can be constructed by assuming link independence. Let $\pi_k = (\pi_k(0), \ldots, \pi_k(C - r_k))$ be the marginal distribution for link β_k, and $\pi_\alpha = (\pi_\alpha(0), \ldots, \pi_\alpha(C))$ the marginal distribution for link α. Define $B_\alpha = \pi_\alpha(C)$, $b_\alpha^{(k)} = \sum_{i=C-r_k}^{C} \pi_\alpha(i)$, $B_k = \pi_k(C - r_k)$ and, finally, $b_k = \sum_{i=C_k}^{C-r_k} \pi_k(i)$. Then, if we assume link independence, the state-dependent transition rates on link β_k are given by

$$q_k(i, i+1) = \lambda_k[(1 - b_\alpha^{(k)}) + (b_\alpha^{(k)} - B_\alpha)I_{\{i<C_k\}}]$$

for $0 \le i < C - r_k$ and $q_k(i, i-1) = i$ for $1 \le i \le C - r_k$. The transition rates on link α are given by

$$q_\alpha(i, i+1) = \sum_{k=1}^{K} \lambda_k[(1 - b_k) + b_k I_{\{i<C-r_k\}}]$$

for $0 \le i < C$ and $q_\alpha(i, i-1) = i$ for $0 < i \le C$. The upward rate for, link α, for instance, is given by the observation that a call of class k will always be offered

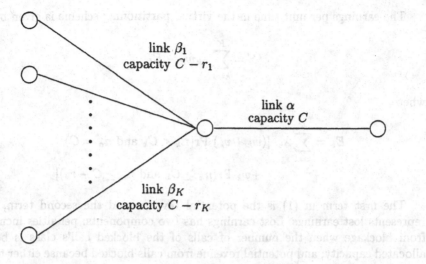

Fig. 1. The alpha/beta network

to link α if $n_k < C_k$, and this happens with probability $1 - b_k$. On the other hand, if $C_k \le n_k$ (which happens with probability b_k), the call is offered to link α only if $n_\alpha < C - r_k$. Now sum over all possible classes k, $1 \le k \le K$.

Under the assumption of link independence, the marginal distributions π_k and π_α are those of truncated birth and death processes with state-spaces $\{0, 1, \ldots, C - r_k\}$ and $\{0, 1, \ldots, C\}$, respectively, and transition rates as given above. Thus

$$\pi_\alpha(i) = \pi_\alpha(0) \prod_{j=0}^{i-1} \frac{q_\alpha(j, j+1)}{q_\alpha(j+1, j)}$$

for $1 \le i \le C$, where $\pi_\alpha(0)$ is a normalizing constant. Similarly,

$$\pi_k(i) = \pi_k(0) \prod_{j=0}^{i-1} \frac{q_k(j, j+1)}{q_k(j+1, j)}$$

for $1 \le i \le C - r_k$, where $\pi_k(0)$ is a normalizing constant.

Using this approximation the probability that a call of class k is lost is given by

$$
\begin{aligned}
L_k &= Pr(n_k < C_k \text{ and } n_\alpha = C) + Pr(n_k \ge C_k \text{ and } n_\alpha \ge C - r_k) \\
&= Pr(n_k < C_k)Pr(n_\alpha = C) + Pr(n_k \ge C_k)Pr(n_\alpha \ge C - r_k) \\
&= (1 - b_k)B_\alpha + b_k b_\alpha^{(k)} .
\end{aligned}
$$

The earnings per unit time in the virtual partitioning scheme is given by

$$E = \sum_{k=1}^{K} \lambda_k w_k - E_L \tag{1}$$

where

$$E_L = \sum_{k=1}^{K} \lambda_k \ [(w_k + v_k) \ \Pr(n_k < C_k \text{ and } n_\alpha = C)$$
$$+ w_k \ \Pr(n_k \geq C_k \text{ and } n_\alpha \geq C - r_k)] \tag{2}$$

The first term in (1) is the potential revenue and the second term, E_L, represents lost earnings. Lost earnings has two components: penalties incurred from blockage when the number of calls of the blocked call's class is below allocated capacity; and potential revenue from calls blocked because either there is no capacity or available capacity is reserved to forestall future penalties.

In the fixed point approximation,

$$E = \sum_{k=1}^{K} \lambda_k [w_k(1 - b_k b_\alpha^{(k)}) - (w_k + v_k)((1 - b_k)B_\alpha)] \tag{3}$$

To implement the approximation, it is necessary to use an iterative procedure, starting with some estimate of the marginal distributions π_k, $1 \leq k \leq K$, and π_α; using these to calculate the transition rates for the single link distributions given above; recalculating estimates for the marginal distributions using these transition rates; and continuing until the procedure converges.

3 Numerical Results

The aim of virtual partitioning is to protect traffic streams against one another. We examined the extent to which this was achieved by looking at two examples, one with two classes of traffic, and the other with four. In both examples the arrival rate for class 1 traffic was increased, while holding the arrival rate for the remaining classes constant. We also examined the adequacy of the fixed point approximation developed in Section 2 for these two examples.

The tools developed for evaluating performance were:

(i) the exact numerical solution of the equilibrium equations for a Markov chain using Gauss-Seidel iteration,

(ii) a discrete-event simulator, and

(iii) a fixed-point solver.

The first example had $K = 2$, $C = 120$, $C_1 = 80$, and $C_2 = 40$. The second had $K = 4$, $C = 240$, $C_k = 60$, $1 \leq k \leq 4$. We set $w_k = 1$ and $v_k = 2$ for all k.

3.1 Accuracy of Fixed Point Approximations

Gauss-Seidel iteration was used to find the equilibrium distribution for the case $K = 2$. In the standard example, even with a favourably chosen initial distribution, the number of iterations necessary to obtain convergence to 10^{-7} in the L_∞ norm was 100 or 200. In higher dimensions, the number of iterations necessary would be even greater, and this underscores the importance of developing an adequate approximation for the blocking probabilities.

The numerical results are presented in Tables 1 and 2, and indicate that the approximation suggested above performs reasonably well in many cases of interest, and aids in assessment of qualitative behaviour. When compared with the exact distribution, the percentage error of the blocking increased as the trunk reservation increased. The error is around 5-6%, and occasionally as high as 15%.

The approximation was also compared with simulation results for the second example with $K = 4$. Initially it was assumed that $\lambda_k = 55$, $r_k = r$, $1 \leq k \leq 4$, and then λ_1 was increased in 10% increments, up to a level of 82.5. Again the percentage error increased with the trunk reservation parameter. With $r = 2$ the percentage error was no greater than 6.8%, and mostly around 5% or less. With $r = 5$, the percentage error in blocking for calls of class 1 was less than 7%, whereas for the other classes it could be as high as 15%.

Table 1a. Accuracy of fixed point, compared with exact calculations $K = 2$, $C = 120$, $C_1 = 80$, $C_2 = 40$, $\lambda_1 = 73$, $\lambda_2 = 37$

(r_1, r_2)	Blocking class 1 Exact	Fixed point	Blocking class 2 Exact	Fixed point	Lost revenue Exact
(0,0)	.02781		.02781		5.93
(1,1)	.02882	.02728	.03069	.03070	4.58
(2,2)	.02971	.02738	.03553	.03373	4.13
(2,3)	.02873	.02570	.04064	.03900	3.97
(3,2)	.03185	.02956	.03489	.03152	4.22
(3,3)	.03096	.02792	.04001	.03688	4.06
(3,4)	.03041	.02648	.04433	.04203	4.04
(4,3)	.03279	.03016	.03970	.03488	4.18
(4,4)	.03232	.02877	.04402	.04014	4.15
(5,5)	.03368	.02983	.04773	.04347	4.31
(2,4)	.02814	.02440	.04497	.04390	3.94

Table 1b. Accuracy of fixed point, compared with exact calculations $K = 2$, $C = 120$, $C_1 = 80$, $C_2 = 40$, $\lambda_1 = 80.3$, $\lambda_2 = 37.0$

	Blocking class 1		Blocking class 2		Lost revenue
(r_1, r_2)	Exact	Fixed point	Exact	Fixed point	Exact
(0,0)	.0568	.0568	.0568	.0568	12.03
(2,2)	.0677	.0628	.0513	.0540	8.32
(2,3)	.0667	.0606	.0563	.0606	8.02
(3,3)	.0707	.0658	.0549	.0552	8.17
(3,4)	.0701	.0640	.0589	.0612	8.06
(4,4)	.0735	.0687	.0583	.0571	8.28
(5,5)	.0760	.0715	.0613	.0594	8.48
(2,4)	.0661	.0591	.0602	.0659	7.92
(2,5)	.0657	.0580	.0633	.0702	7.91
(3,5)	.0698	.0627	.0621	.0660	8.05

Table 1c. Accuracy of fixed point, compared with exact calculations $K = 2$, $C = 120$, $C_1 = 80$, $C_2 = 40$, $\lambda_1 = 87.6$, $\lambda_2 = 37.0$

	Blocking class 1		Blocking class 2		Lost revenue
(r_1, r_2)	Exact	Fixed point	Exact	Fixed point	Exact
(0,0)	.0924	.0924	.0924	.0924	19.52
(2,2)	.1154	.1089	.0619	.0693	13.64
(3,3)	.1198	.1143	.0641	.0673	13.40
(4,4)	.1236	.1190	.0666	.0675	13.54
(2,4)	.1140	.1053	.0691	.0808	13.10
(2,5)	.1137	.1044	.0714	.0843	13.05
(2,6)	.1135	.1039	.0732	.0870	13.05
(1,5)	.1056	.0961	.0784	.0949	13.65

Table 2. Accuracy of fixed point, comparison with simulation results (with 95% confidence intervals) $K = 4$, $C_k = 60$, $1 \le k \le 4$, $\lambda_2 = \lambda_3 = \lambda_4 = 55.0$, $r_1 = r_2 = r_3 = r_4 = 2$

λ_1	Blocking class 1		Blocking classes 2,3,4	
	Simulation	Fixed point	Simulation	Fixed point
55.0	.0128 ± .0004	.0124	.0128 ± .0004	.0124
60.5	.0266 ± .0007	.0273	.0191 ± .0004	.0182
66.0	.0475 ± .0010	.0491	.0256 ± .0005	.0241
71.5	.0715 ± .0009	.0752	.0318 ± .0005	.0300
77.0	.0993 ± .0015	.1030	.0388 ± .0006	.0362
82.5	.1269 ± .0019	.1320	.0451 ± .0006	.0425

3.2 Fairness and Robustness

Figures 2 and 3 give changes in the blocking probabilities as class 1 traffic increases its arrival rate. The data is drawn from the tables given in the section above. So for the case $K = 2$, the blocking probabilities are exact, whereas for the example with $K = 4$, the blocking probabilities are based on simulation results. For the latter example, therefore, a confidence interval for the blocking probabilities is indicated on the graph.

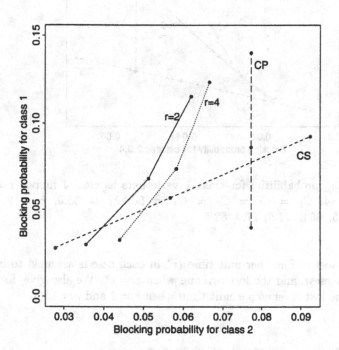

Fig. 2. Blocking probabilities for class 1 vs. class 2 as class 1 arrival rate increases. $K = 2$; $C_1 = 80$, $C_2 = 40$; $\lambda_2 = 37.0$; $\lambda_1 = 73.0$, 80.3, 87.6

Four cases are compared – complete sharing (CS), complete partitioning (CP), and virtual partitioning with two choices of trunk reservation parameter (the reservation parameter is the same for all traffic classes).

For both examples we see that the blocking probability for all classes increases as the arrival rate for class 1 increases, except, of course, for complete partitioning, where it remains unchanged. However, virtual partitioning with $r > 0$ does have the desired property of substantially reducing the effect of this increase on those classes with constant arrival rate.

We now examine the change in lost revenue as λ_1 increases for the example with $K = 4$. In Table 3 we give, for each value of λ_1, the choice of r, r^*, which

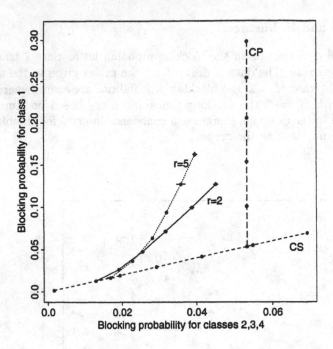

Fig. 3. Blocking probabilities for class 1 vs. others as class 1 arrival rate increases. $K = 4$; $C_1 = C_2 = C_3 = C_4 = 60$; $\lambda_k = 55.0$, $k = 2, 3, 4$; $\lambda_1 = 55.0, 60.5, 66.0, 71.5, 77.0, 82.5$

minimizes the lost revenue per unit time (r^* in each case is assumed to be the same for all classes), and the lost revenue when $r = r^*$. We also give, for each arrival rate, the lost revenue per unit time when $r = 2$ and $r = 5$.

Table 3. Lost revenue $K = 4$; $C_k = 60$, $1 \leq k \leq 4$; $\lambda_2 = \lambda_3 = \lambda_4 = 55.0$

λ_1	55.0	60.5	66.0	71.5	77.0	82.5
r^*	5	6	7	7	7	7
Lost revenue						
$r = r^*$	4.453	18.22	38.18	60.97	85.53	111.3
$r = 2$	5.063	21.34	44.99	71.68	99.70	128.4
$r = 5$	4.453	18.30	38.54	61.71	86.46	112.2

We see that lost revenue is relatively insensitive to the choice of trunk reservation parameter under the engineered loading. However, it appears to be important not to have too low a trunk reservation parameter, particularly in heavily overloaded conditions.

3.3 Optimality

When a penalty structure is attached to the problem, it is possible to find the optimal policy maximising the expected reward per unit time. It can be found numerically by using value iteration (see e.g. Ross (1983) or Tijms(1986)). Since the state space is finite, the optimal policy will be stationary, and can thus be simply described as an acceptance region for each class of service.

The example discussed in this section has $K = 2$, $C = 24$, $C_1 = 16$, $C_2 = 8$, $\lambda_1 = 20$, $\lambda_2 = 6$, $w_1 = w_2 = 1$ and $v_1 = v_2 = 2$. The policy that maximizes the expected return per unit time is given in Figure 4.

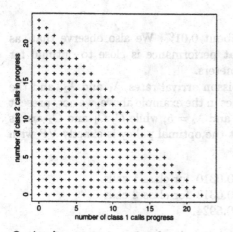

Optimal acceptance region for class 1.

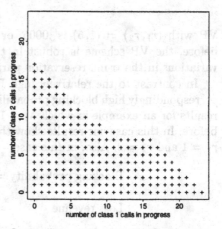

Optimal acceptance region for class 2.

Fig. 4. Policy that maximizes the expected return per unit time. $K = 2, C = 24$, $C_1 = 16$, $C_2 = 8$, $\lambda_1 = 20$, $\lambda_2 = 6$, $w_1 = w_2 = 1$ and $v_1 = v_2 = 2$

We note that the optimal policy, even though it is not exactly virtual partitioning, is nevertheless of a very similar form. The policy is monotone, in that if a call of class i is rejected in state (i, j), (where $i + j < C$), it is also rejected in states $(i + 1, j)$ and $(i, j + 1)$. Furthermore, if a call of class 1 is rejected in state (i, j), then it is rejected in state $(i + 1, j - 1)$; and similarly, if a call of class 2 is rejected in state (i, j) it is also rejected in state $(i - 1, j + 1)$. All the examples that were investigated had monotone optimal policies.

In Table 4 we present numerical results on the performances of the optimal policy and the VP scheme in the example considered above (see Fig. 4). We observe that there is little to be gained over VP schemes by using a strictly optimal policy. The difference in lost revenue between the optimal policy and

Table 4. Blocking probabilities and lost revenue in the virtual partitioning scheme and the optimal policy for the example in Fig. 4.

		Blocking probability class 1	class 2	Lost revenue
VP	$(r_1, r_2) = (1,4)$	.2311	.1335	5.7992
	$= (2,4)$	.2559	.1171	5.9164
	$= (1,5)$	.2310	.1357	5.7989
	$= (2,5)$	.2558	.1221	5.9163
	$= (1,6)$	.2309	.1369	5.8011
	$= (2,6)$	.2557	.1234	5.9189
Optimal policy		.2311	.1340	5.7984

VP with $(r_1, r_2) = (1,5)$ is .0005, or about 0.01%. We also observe that, as before, the VP scheme is robust in that performance is close to optimal for variations in the trunk reservation parameters.

In contrast to the relatively high Poisson arrival rates, λ_1 and λ_2, and the correspondingly high blocking probabilities in the example above, we now present results for an example in which $\lambda_1 = 12$ and $\lambda_2 = 5$, while C, C_1 and C_2 are as before. In this case our results show that the optimal policy is exactly VP with $r_1 = 1$ and $r_2 = 2$. Also, in this case,

$$\text{Blocking probability} = 0.0310 \text{ for class 1 calls}$$
$$= 0.0318 \text{ for class 2 calls}$$
$$\text{Lost revenue} = 0.5924.$$

In fact, in all the examples we have studied in which the arrival rates and the blocking probabilities are low, the optimal policy is VP.

The above investigations on robustness have been for fixed traffic patterns. To investigate further the adequacy of the VP scheme when traffic patterns vary dynamically we also considered the model described above, but with Markov modulated Poisson arrival streams, rather than fixed arrival rates. We found that a virtual partitioning policy was very close to optimal (or even optimal) in many situations, particularly in the two extreme cases where arrival rates modulated either very slowly or very rapidly. This corresponds to the modulation varying on a different time-scale than the arrivals and departures for the system. We do not give details of this here.

4 Conclusions

We have proposed a robust, fair and efficient scheme for sharing an unbuffered resource, such as bandwidth or capacity, by various services. The scheme is a form of virtual partitioning which is implemented by state-dependent priorities. The numerical results show that the scheme is robust, fair and efficient.

Future work will address services with multiple rates. The motivation for the extension comes from multiple services offered on ATM networks. The notion of effective bandwidth translates to a specific rate or bandwidth requirement for each traffic source type. The central ideas presented here are amenable to generalizations in the multi-rate framework.

Acknowledgements

Ilze Ziedins would like to thank A.T.& T. Bell Labs for support while working on this project.

References

Ash, G.R., Chen, J.S., Frey, A.E., Huang, B.D.: Real-time network routing in an integrated network. Proceedings of the International Teletraffic Congress (ITC-13), Copenhagen (1991).

Foschini, G.J., Gopinath, B., Hayes, J.F.: Optimum allocation of servers to two types of competing customers. IEEE Trans. Comm. **COM-29** (1981) 1051–1055.

Gopal, I.S., Stern, T.E.: Optimal call blocking policies in an integrated services environment. Proceedings of the Conference on Information Sciences and Systems, Johns Hopkins University (1983) 383–388.

Kelly, F.P.: Loss networks. Ann. Appl. Prob. **1** (1991) 319–378.

Kraimeche, B., Schwartz, M.: Traffic access control strategies in integrated digital networks. Proceedings of Infocom '84 (1984) 230–235.

Mason, L., Dziong, Z., Tetreault, N.: Fair-efficient call admission control policies for broadband networks. Proceedings of Globecom '92 (1992).

Ross, K.W.: Multirate Loss Models for Broadband Telecommunications Networks. Springer (1995).

Ross, S.M.: Introduction to Stochastic Dynamic Programming. Academic Press (1983).

Tijms, H.C.: Stochastic Modelling and Analysis: A Computational Approach. Wiley (1986).

A Performance Study of the Local Fairness Algorithm for the MetaRing MAC Protocol

G. Anastasi, M. La Porta, L. Lenzini

University of Pisa, Department of Information Engineering
Via Diotisalvi 2 - 56126 Pisa, Italy
anastasi@iet.unipi.it, lenzini@iet.unipi.it

Abstract
The MetaRing is a Medium Access Control (MAC) protocol for Gigabit LANs and MANs with cells removed by the destination stations (*slot reuse*). Slot reuse increases the aggregate throughput beyond the capacity of single links but can cause starvation. In order to prevent this the MetaRing MAC protocol includes a fairness algorithm. Two types of fairness algorithms have been proposed: "global" and "local". The paper analyses the local fairness algorithm specified in [CHEN93] by focusing on a worst-case network scenario which is particularly critical with respect to slot reuse. This scenario consists of a group of N equally spaced stations located in one half of the ring and which all send data traffic to a gateway only. The analysis can be divided into two parts. In the first part stations are assumed to operate in *asymptotic conditions* (never empty queues) and closed formulae for the aggregate and station throughputs are derived. In the second part of the paper the performances of the local fairness algorithm are studied in normal conditions, that is when the offered load is lower (*underload conditions*) or slightly higher (*overload conditions*) than the aggregate throughput achievable in asymptotic conditions. The analysis is performed (by simulation) on the same worst-case network scenario. The results obtained show that in asymptotic conditions the local fairness algorithm always allows a complete utilization of the medium capacity. Although it is not perfectly fair, the unfairness can be reduced by choosing of the algorithm parameters appropriately. This choice guarantees fairness in overload conditions as well, but causes unfairness in underload conditions.

1. Introduction
The MetaRing is a MAC protocol for high speed LANs and MANs with spatial reuse which can achieve an aggregate throughput much higher than the network capacity ([OFEK94]). However, spatial reuse can cause starvation at stations covered by upstream highly loaded stations. A fairness mechanism is thus required to prevent this from occurring. Fairness protocols have already been proposed and they can be classified into "global" ([CIDO93]) and "local" ([CHEN93], [MAYE95]). A "global" fairness algorithm regulates the access to the shared media by considering the network as a single communication resource. By contrast, a "local" fairness algorithm views the media as a distributed collection of communication resources. Hence, a local fairness algorithm regulates the transmissions of the interfering stations without affecting the others so that local fairness algorithms can exploit the advantages of spatial reuse better than global algorithms.

A simulative analysis of a MetaRing MAC protocol with a local fairness algorithm was performed in [CHEN93] where the network was analyzed in asymptotic conditions (i.e., when stations always have traffic to send). Station and aggregate throughputs were measured for various network scenarios. From the results reported in [CHEN93] it follows that the throughput achieved by a station is not influenced by the behaviour of stations which do not interfere with it. Thus, stations in the networks can be divided into disjoint groups. Within a single group the throughput achieved by each station, as well as the aggregate throughput (i.e., the sum of the throughputs of all the stations in the group), strictly depends on the interactions among stations belonging to the group.

In this paper we further advance the analysis reported in [CHEN93] by focusing on a single group of N equally spaced stations located in one half of the ring and which all send traffic to a gateway only (Figure 1).

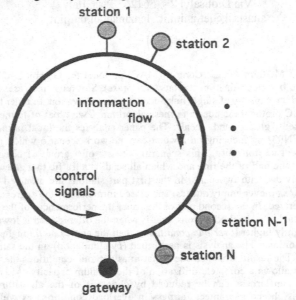

Fig. 1. Network scenario.

It can be easily verified that this scenario is particularly critical with respect to slot reuse. In fact, since slots are emptied by the gateway, a slot can be utilized at most once during its journey from station $\{1\}$ to the gateway. Therefore the spatial reuse capability is used to the minimum extent possible. Also, stations observe different patterns of empty/free slots. Specifically, station $\{1\}$ always observes empty slots, while station $\{N\}$ is covered by traffic transmitted by its upstream stations and therefore it can send a cell only if a slot has been left empty by stations $\{1,2,...,N-1\}$.

In the first part of the paper we derive closed formulae for the aggregate and station throughputs when all the stations operate in *asymptotic conditions*; i.e., their queues are never empty.

In the second part of the paper we study the performances of the local fairness algorithm in normal conditions, that is when the offered load is lower (*underload conditions*) or higher (*overload conditions*) than the aggregate throughput achievable in asymptotic conditions. The analysis is performed by simulation, and the average access delay in underload conditions and the packet loss in overload conditions are the performance measures of interest.

The paper is organized as follows. Section 2 describes the MetaRing MAC protocol with local fairness. Section 3 reports the analysis in asymptotic condition. Section 4 is devoted to the analysis in underload conditions, while Section 5 reports the results related to the analysis in overload conditions. Summary and conclusions are drawn in Section 6.

2. The MetaRing MAC Protocol with Local Fairness

The MetaRing is a Medium Access Control (MAC) protocol for LANs and MANs operating at a speed of 1 Gbit/sec and above ([OFEK94]). It connects a set of stations by means of a bidirectional ring made up of full duplex point-to-point serial links. The protocol provides two types of services: asynchronous and synchronous ([WU92]). This paper only deals with the asynchronous type. A MetaRing network can operate under two basic access control modes: *buffer insertion* for variable size packets, and *slotted* for fixed length cells. In this paper only the slotted access mode is considered.

According to the slotted access mode, information is segmented into cells and each cell transmitted in one slot. Slots are structured into a header and an information field. The header includes a busy bit which indicates whether the slot is empty (busy bit=0) or busy. A cell can be accommodated on an empty slot. Since the ring is bidirectional the MetaRing MAC protocol uses the *shortest path* criterion to choose one of the possible directions. Cells are removed by the destination station which frees the slot by resetting the busy bit to zero. After cell removal the slot can be reused by the same station or by its downstream stations. Slot reuse increases the aggregate throughput beyond the capacity of single links but can cause starvation.

In order to prevent this from occurring a fairness mechanism is required. The fairness algorithm analyzed in this paper is *local* and is triggered by any station whenever it foresees a potential starvation. The algorithm is based on two control signals: *REQ* (REQuest) and *GNT* (GraNT) which propagate in the opposite direction with respect to the information flow they regulate. According to this algorithm a station can be in a finite set of states and, correspondingly, in two different operation modes: *non-restricted* and *restricted*. In the former there are no limitations on the number of cells a station can transmit, while in the latter a station can only send a predefined quota, Q_r, of cells.

Note that in [CHEN93] the condition under which a station *establishes* that it will become starved is not specified. Therefore, to carry out the performance evaluation analysis reported in the paper we had to integrate the local fairness algorithm specified in [CHEN93] with the following definition: a station observing more than Q_s contiguous busy slots becomes starved. Clearly, other definitions are possible and each one may lead to different performances.

With our definition of starved station, the local fairness algorithm works as follows. If no conflict occurs (i.e., if no transmitting station is covered by the traffic originated by upstream stations) all the stations are in the *Free Access* (FA) state. In this state a station operates in non-restricted mode and therefore it can transmit a cell every time it observes an empty slot.

Once a station (say station $\{i\}$) becomes starved, it triggers the local fairness mechanism by sending the *REQ* signal to the upstream station $\{i-1\}$ and then enters the *Tail* (T) state. Upon reception of a *REQ*, station $\{i-1\}$ enters the restricted mode of operation and if its upstream station ($\{i-2\}$) is idle, it will enter the *Head* (H) state. On the other hand, if station $\{i-1\}$ observes busy slots from upstream stations it will forward the *REQ* to station $\{i-2\}$ and enter the *Body* (B) state. While in states T, B and H a station can send at most a quota Q_r of cells (restricted mode).

Upon satisfaction (i.e., transmission of the predefined quota) or as soon as the buffer becomes empty, the Tail station sends a *GNT* signal to its upstream station and switches back to the non-restricted FA state. Upon receiving this *GNT*, the upstream station follows similar rules: if it is in the state B, it transits to state T and will forward the *GNT* upon satisfaction or when its buffer becomes empty (whichever occurs first). If a station is in state H, it switches to state FA and the local fairness cycle involving a set of stations located on that ring segment is terminated. The local fairness mechanism thus creates a *Request Path* which contains unique and distinct head and tail stations for each segment in the ring where there are interfering stations. Furthermore, since there might be multiple initiators of the fairness algorithm (and, therefore, of a *Request Path*), two or more distinct *Request Paths* may overlap. When this occurs the overlapping *Request Paths* are merged into a unique *Request Path*. Further details on the local fairness algorithm can be found in [CHEN93].

3. Analysis in Asymptotic Conditions

In this section we analyze the local fairness algorithm behaviour when every station tries to seize all the medium capacity (*never-empty station queues*). This is the situation which causes the maximum interference between stations, since all compete in grabbing as much medium bandwidth as possible. Although this situation should never occur in a real network, its investigation is very useful as it provides insight into the MAC protocol's capability to solve conflicts between stations fairly. Performance measures of interest in asymptotic conditions are the aggregate throughput (ρ) that the MetaRing can manage, and the throughput that each station can achieve on the ring (γ_i, $i=1,2,...,N$). The ratio between the most and the least station throughputs achieved has been adopted as a metric for the MetaRing fairness in asymptotic conditions.

Hereafter, the throughput achieved by a station is expressed as the average number of slots made busy by that station in a ring latency interval, normalized to the number (S) of slots circulating in the ring (ring latency expressed in slots). Therefore, the aggregate throughput gives a measure of the number of times a slot is utilized by stations during a ring round. Thus, in our scenario, the aggregate throughput can never exceed 1.

In asymptotic conditions the MetaRing behaviour implementing the local fairness algorithm is deterministic. It can be shown that ([LAPO95]) at any time instant there is at most one Request Path on the ring and that at most two control signals, *REQ* and *GNT*, are circulating on the ring. When the Request Path is complete station $\{N\}$ is the Tail while station $\{1\}$ is the Head.

Before deriving closed formulae for the station and aggregate throughputs we need to introduce some terminology and concepts.

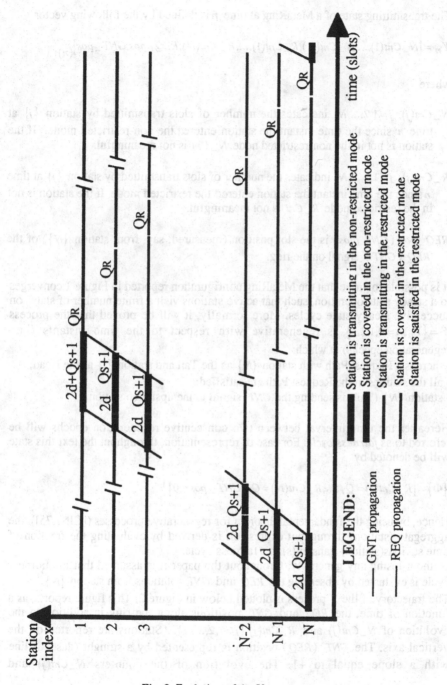

Fig. 2. Evolution of the Y process.

The transmitting state of a MetaRing at time n is defined by the following vector

$$Y_n = \left[N\_Cnt(1),...,N\_Cnt(N),R\_Cnt(1),...,R\_Cnt(N),REQ\_pos,GNT\_pos\right]_{(n)}$$

where

$N\_Cnt(j)$, $j = 1,2,...,N$, indicates the number of slots transmitted by station $\{j\}$ at time n since the time instant the station entered the non-restricted mode. If the station is not in the non-restricted node N_Cnt is not meaningful.

$R\_Cnt(j)$, $j = 1,2,...,N$, indicates the number of slots transmitted by station $\{j\}$ at time n since the time instant the station entered the restricted mode. If the station is not in the restricted node R_Cnt is not meaningful.

REQ_pos (GNT_pos) is the slot position (measured, say, from station $\{N\}$) of the REQ (GNT) signal on the ring.

It is possible to prove that the MetaRing configuration reported in Figure 1 converges to a steady state operation, such that active stations visit a finite number of states on successive steady state cycles. More formally, it will be proved that the process $Y = \{Y_n, n = 0,1,2,..\}$ is regenerative with respect to the time instants (i.e., regeneration epochs) at which:
- there is a Request Path with station $\{N\}$ as the Tail and station $\{1\}$ as the Head;
- all the stations in the Request Path are satisfied;
- station $\{N\}$ (Tail) is sending the GNT signal to the upstream station.

Hereafter the time interval between two consecutive regeneration epochs will be referred to as *fairness cycle*. For ease of representation, throughout the text this state will be denoted by

$$\{\Phi\} = \left[R\_Cnt(1) = Q_r,...,R\_Cnt(N) = Q_r, GNT\_pos = 0\right]^{1}$$

Hence, by using the fundamental theorem for regenerative processes ([CINL75]), the aggregate (station) throughput expression is derived by evaluating the fraction of time stations(station) transmit(s) in a fairness cycle.
Without losing any generality, throughout the paper it is assumed that the fairness cycle is evaluated by observing the REQ and GNT rotations from station $\{N\}$.
The trajectory of the Y process is plotted below in Figure 2. This figure reports, as a function of time, the REQ and GNT positions along the ring in addition to the evolution of $N\_Cnt(j)$ and $R\_Cnt(j)$ ($j = 1,2,...,N$). Stations are reported on the vertical axis. The GNT (REQ) position is represented by a straight (dashed) line with a slope equal to +1. The evolution of the counters $N\_Cnt(j)$ and

---

[1] Only meaningful components are reported.

$R\_Cnt(j)$ $(j=1,2,...,N)$ are represented by horizontal segments. Since a station can alternatively be in the non-restricted mode or in the restricted mode, at a given time instant only one of the two counters is meaningful. In Figure 2 segments representing the N_Cnt counter are drawn by a continuos double size line, while a continuos single size line is used to represent the evolution of R_Cnt counter of a specific station. In the non-restricted mode of operation a station is not transmitting when it is covered by the upstream ones. This is represented in Figure 2 by dashed double size segments.

On the other hand, in the restricted mode a station can refrain from transmission either because it observes busy slots from upstream (dashed single size segments) or because it is already satisfied (dotted single size segments).

With this type of representation fairness cycles can be identified straightforwardly, as it is easy to realize when the process Y restarts visiting the same set of states already visited in the previous cycle.

In the following two sections we report closed formulae for the aggregate and station throughputs as a function of the number of active stations (N), the distance between stations (d) which is assumed constant, and the local fairness parameters Q_r and Q_s. The distance between stations d is expressed in slots and is assumed to be an integer greater than or equal to one.

3.1 Aggregate Throughput

The aggregate throughput is always equal to 1 and is not influenced by the values of Q_r and Q_s. Specifically, if we consider a particular slot it is easy to verify that this slot will always be used during its journey around the ring, and just once. The second part of the previous statement immediately follows from the network scenario (Figure 1) analyzed in the paper. In order to prove the first part we assume that the slot under consideration arrives empty at station $\{N\}$. Station $\{N\}$ can be in non-restricted mode or, alternatively, in restricted mode. In the former case it certainly makes the slot busy. In the latter case, station $\{N\}$ is the Tail station of the Request Path. Therefore, it uses the slot because (by definition of Tail station) it cannot yet be satisfied. Hence, in any case the slot is used by station $\{N\}$.

3.2 Station Throughputs

To calculate the station throughputs we need to analyze the evolution of the Y process starting from the state $\{\Phi\}$ at time (for example) $t_0 = 0$. At this time instant station $\{N\}$ forwards the GNT to the upstream station and enters the FA state. When station $\{j\}$ $(j=N-1,N-2,..,2)$ receives the GNT from a downstream station, since it is already satisfied, it switches from state B to T and, immediately after, sends the GNT, then transits to state FA and enters the non-restricted mode. The behaviour of station $\{1\}$ upon GNT arrival is slightly different from the other stations. Since $\{1\}$ is the Head of the Request Path it transits immediately to state FA. The above consideration implies that, after GNT release, station $\{j\}$ $(j=N,N-1,...,2)$ observes a train of $2d$ empty slots. Since $\{j\}$ operates in non-restricted mode it uses all of them before being covered by the upstream stations.

Now, let us analyze the *REQ* propagation. When station $\{N\}$ is covered by the upstream traffic it counts for Q_s+1 busy slots and then performs the following operations: it sends a *REQ*, switches to state T and enters the restricted mode of operation. Stations numbered from $\{N-1\}$ down to $\{2\}$ execute the same operations. Each of them receives the *REQ* from the downstream station at the same time that it has counted Q_s+1 consecutive busy slots (and hence it would be ready to send a *REQ* upstream). The simultaneous occurrence of the two events is managed by a station as follows: it first moves from state FA to H and, immediately after, forwards the *REQ* upstream entering the state B.

The trajectory of the *REQ* is thus perfectly parallel to the trajectory of the *GNT* as shown in Figure 2. Since station $\{N\}$ issues a *REQ* $2d+Q_s+1$ time slots after the *GNT* release, the *REQ* itself reaches station $\{1\}$ exactly $2d+Q_s+1$ slots after the *GNT* arrival at station $\{1\}$ itself. This implies that station $\{1\}$ transmits $2d+Q_s+1$ cells in the FA state before receiving a *REQ* from the downstream station. Upon reception of the *REQ*, station $\{1\}$ changes its state from FA to H, enters the restricted mode, and sends Q_r more cells before stopping transmitting because it has been satisfied. After a time interval equal to d slots station $\{2\}$ observes empty slots and thus it sends its quota (Q_r) of cells before stopping transmitting. The same sequence of events occurs for stations numbered from $\{3\}$ to $\{N\}$. In particular, when station $\{N\}$ is satisfied, i.e., after the transmission of its quota of slots in the T state, it sends the *GNT* upstream, transits to the FA state and enters the non-restricted mode. At this point the *Y* process is in the same state as it was at time $t_0=0$ and from this time onward the process regenerates itself cyclically. Hence, the *Y* process is regenerative with respect to the time instants at which station $\{N\}$ sends the *GNT* to the upstream station.

To calculate the duration of the regenerative cycle T_Y we proceed as follows. If $t_0=0$ is the time at which station $\{N\}$ sends the *GNT* because it is satisfied, then the *GNT* reaches station $\{1\}$ at time $t_1=(N-1)d$. In fact, the quantity $(N-1)d$ is the propagation delay from station $\{N\}$ to station $\{1\}$. On the other hand, the *REQ* reaches station $\{1\}$ at time $t_2=t_1+Q_s+2d+1$ since station $\{1\}$ transmits Q_s+2d+1 slots in non-restricted mode. Starting from time t_2, station $\{1\}$ sends Q_r cells and then stops transmitting. After d slots station $\{2\}$ observes empty slots, and therefore, can transmit its quota before stopping. All the other stations, from $\{3\}$ to $\{N\}$ follow the same transmitting procedure. Specifically, station $\{N\}$ starts observing empty slots at time $t_3=t_2+(Q_r+d)(N-1)$ and becomes satisfied (hence, it sends the *GNT*) at time $t_4=t_3+Q_r$. Thus, the following equality holds

$$T_Y=t_4-t_0=NQ_r+2Nd+Q_s+1 \tag{3.1}$$

Therefore, the station throughputs can easily be calculated by evaluating the fraction of time a station transmits during a fairness cycle. From an analysis of Figure 2 and from (3.1) it follows

$$\gamma_i = \begin{cases} \dfrac{Q_r + 2d + Q_s + 1}{NQ_r + 2Nd + Q_s + 1} & if \ i = 1 \\[3mm] \dfrac{Q_r + 2d}{NQ_r + 2Nd + Q_s + 1} & if \ i = 2,3,...N \end{cases} \qquad (3.2)$$

As stated in Section 3.1, (3.2) implies that the aggregate throughput $\rho = \sum_{i=1}^{N} \gamma_i$ is exactly equal to 1 for whatever Q_r and Q_s values.

Furthermore, from (3.2) it clearly appears that stations can be partitioned into two distinct classes. The first class includes station $\{1\}$ only and is characterized by the higher throughput value. The second class includes all the remaining stations which achieve the same throughput. If we assume, as a fairness metric, the ratio between the throughputs of the most and least favoured classes, i.e., $\eta = \gamma_1/\gamma_2$, the following relation holds

$$\eta = \frac{Q_r + 2d + Q_s + 1}{Q_r + 2d} = 1 + \frac{Q_s + 1}{Q_r + 2d} \qquad (3.3)$$

From (3.3) two considerations follow in order. First, η is always greater than 1 and this means that the protocol is not perfectly fair. Second, the unfairness decreases as the ratio between Q_r and Q_s increases. This can be observed from Figures 3 and 4 which both relate to a MetaRing configuration with $N = 10$ stations, $d = 7$, and where we considered different values for the ratio between Q_r and Q_s. Furthermore, Table 1 reports the value of η for each curve reported in Figures 3 and 4 respectively.

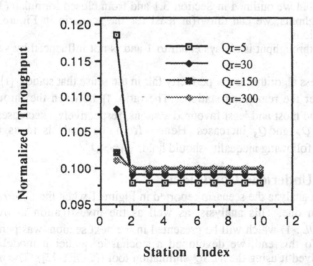

Fig. 3. Station throughputs for a network configuration with $N = 10$ stations, $d = 7$, $Q_s = 3$.

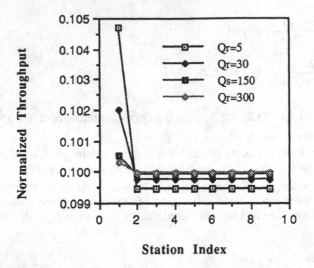

Fig. 4. Station throughputs for a network configuration with $N = 10$ stations, $d = 7$, $Q_s = 0$.

	$Q_r = 5$	$Q_r = 30$	$Q_r = 150$	$Q_r = 300$
$Q_s = 3$	1.211	1.090	1.024	1.013
$Q_s = 0$	1.053	1.022	1.006	1.003

Tab. 1. η values for a network configuration with $N = 10$ stations and $d = 7$.

On the basis of what we outlined in Section 3.1 and from closed formulae (3.2) for the station throughputs, we can draw (at least for the scenario in Figure 1) the following conclusions.

1. The aggregate throughput is always equal to 1 and is not influenced by values of Q_r and Q_s.

2. The local fairness algorithm is not perfectly fair in the sense that station {1} has an advantage over the remaining stations. The ratio η between the throughputs achieved by the most and least favoured stations, respectively, decreases as the ratio between Q_r and Q_s increases. Hence, from the analysis in asymptotic conditions the following inequality should hold: $Q_r >> Q_s$.

4. Analysis in Underload Conditions

In this section we analyze the scenario reported in Figure 1 when the *Offered Load* (*OL*) is less than one. This analysis, as well as the investigation in overload conditions (i.e., $OL > 1$) which will be presented in the next section, was performed by simulation. To this end, we developed a stochastic queueing model of the MetaRing and solved it using the *RESQ* simulation tool ([GORD92]). The network scenario analyzed is the same as the one considered in the previous section. The parameter values which completely define our model are reported in Table 2.

- Stations were spaced equally along half of the ring
- Capacity of each ring = 150 Mbps
- Slot size = 53 octets
- Slot duration *(ts)* = 2.82 msec
- Length of each ring (S)= 150 slots (85 km)
- Ring latency = 423 msec

Tab. 2. Network parameter values.

To have acceptable simulation times we decided, for most of the experiments, to choose the channel bandwidth of each ring equal to 150 Mbps, which is in the range of the channel bandwidth for the current MANs.

We performed a simulation analysis with several workloads, each characterized, besides the offered load, by the distributions of the message interarrival times and message lengths. In our experiments we considered exponential (coefficient of variation *C* equal to one), hypoexponential (*C* < 1), and hyperexponential (*C* > 1) interarrival time distributions. Furthermore, the message length *Msg* was assumed constant in each experiment and equal to 1, 4, and 8 cells.

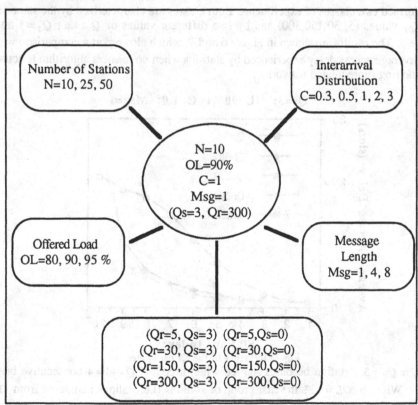

Fig. 5. Organization of underload experiments.

In underload conditions the performance measure of interest for us is the *station average access delay*, i.e. the average time interval between a cell arrival at the station queue and its transmission onto the media. Obviously, in a fair network with equally loaded stations, each station should experience the same average access delay. Any deviation from this behaviour has been assumed as being a measure of the *MAC* protocol unfairness in underload conditions.

The station average access delay depends upon several parameters which characterize the workload, the number of active stations (N), the *MAC* protocol parameters $(Q_r$ and $Q_s)$, and the MetaRing network configuration (i.e., ring length, slot size, etc.). It was thus decided to partition the experiments into the five classes shown in Figure 5.

Beginning with the following set of parameters $(N = 10, OL = 90\%, Msg = 1, C = 1, Q_s = 3, Q_r = 300)$ all the other experiments were performed by varying the parameters one at a time (except for Q_s and Q_r which were varied together) between simulation runs. The rest of this section discusses the results.

Influence of Qs and Qr

First we considered various couples of values for Q_r and Q_s. Specifically, we performed two classes of experiments. Both classes are characterized by the same set of Q_r values (5, 30,150,300) and by two different values of Q_s: i.e., $Q_s = 3$ and $Q_s = 0$. The results are shown in Figure 6 and 7, which also report a comparison with the average access delay experienced by stations when no fairness algorithm is active on the ring (unregulated network).

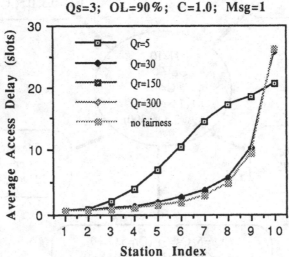

Fig. 6. Influence of the Qr parameter.

When $Q_s = 3$ a station becomes starved when it counts $Q_s + 1 = 4$ consecutive busy slots. With an $OL = 90\%$ the first group of stations (i.e. stations numbered from {1} to {3}) observes a large number of empty slots and thus the probability that a station in this group becomes starved (i.e. counts 4 consecutive busy slots) is very small. This implies that stations in this group are rarely involved in Request Paths triggered by stations close to the gateway which see longer trains of consecutive busy slots. Thus, the average access delays experienced by stations from {1} to {3} are not significantly affected by the Q_r values because for most of the time these stations are in state FA. This is clearly visible from Figure 6 from which we can observe that the average access delays experienced by stations {1}, {2} and {3}, with different values of Q_r, are very close to the delay these stations would experience if no fairness algorithm were activated.

On the other hand, variations in Q_r affect the performances of stations {4} to {10}, since if Q_r increases the network unfairness increases too.

In addition, Figure 6 shows that the average access delay curves with $Q_r = 30$, 150 and 300 are very close to the curve of an unregulated network (in particular, the curves related to $Q_r = 150$ and $Q_r = 300$ are completely overlapped with the curve obtained without fairness). This can be justified by observing that in underload conditions, stations globally use a fraction (which may be small) of the channel capacity with the result that the average queue length of each station could be smaller than quota Q_r. Thus, when Q_r is sufficiently greater than the mean queue length a station will never be forced to wait because it will receive the GNT signal before it has transmitted the predefined quota Q_r of cells. Clearly, if a station is never forced to wait to transmit, its average access delay approaches the average access delay that the station would experience in an unregulated network. On the other hand, when $Q_r = 5$ a station in the restricted mode will reach satisfaction quite often by transmitting Q_r slots. Hence, it will be forced to wait for the GNT signal before transmitting other cells. Therefore, with respect to the previous case the average access delay of stations close to station {4} increases while the average access delay of stations close to station {10} decreases. Figure 6 shows that fairness decreases as Q_r increases.

In the second series of experiments ($Q_s = 0$) a station needs to count only one busy slot to become starved, thus the Request Path propagation speed is increased with respect to the case with $Q_s = 3$. Some Request Path is thus very likely to reach stations close to station {1} despite the fact that such stations observe long trains of consecutive empty slots. Therefore, when $Q_s = 0$, a variation in the value of Q_r also affects the average access delay of stations close to station {1} (see Figure 7).

The influence of Q_r on the average access delay is very similar to what we have already described for the first series of experiments: if we increase the value of Q_r ($Q_r = 30$, 150 and 300) the network tends to behave as an unregulated network. The difference from the previous case is that when $Q_r = 5$, the highest average access delays are experienced by stations {7} and {8} rather than station {10}.

This can be explained by observing that with $Q_r = 5$ and $Q_s = 0$ we have two opposite effects. On the one hand, the average access delay tends to increase with the station index, since stations located close to the gateway observe more busy slots than stations with lower indices. On the other hand, since $Q_s = 0$ it is possible that more stations, located close to the gateway, simultaneously start a Request Path by sending a *REQ* upstream. The various Request Paths will then be merged into a unique one with station {10} as the Tail station. Since the Tail station is the first one to leave the restricted mode, stations located close to the gateway remain in the restricted mode longer and longer as the station index decreases. Thus, stations {10} and {9} have more access opportunities than stations {8} and {7}. Obviously, this second effect becomes more and more meaningful as the distance between stations increases.

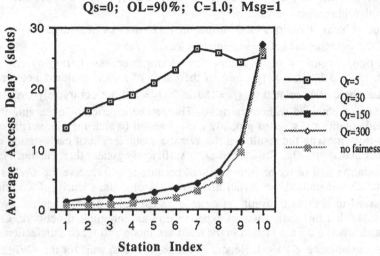

Fig. 7. Influence of the Q_r parameter.

The following conclusions can be drawn from the above considerations.
- The Q_s parameter influences the *REQ* propagation "speed" and "depth". If Q_s decreases the *REQ* signals propagate faster and can "penetrate" more upstream.
- The Q_r parameter influences the possibility to force a station in the restricted mode to wait for the next cell transmission.
- The fairness decreases as the value of Q_r increases.
- For the values of Q_r and Q_s which guarantee a good degree of fairness in asymptotic conditions, the local fairness mechanism is hardly ever active and thus, the network operates closely to an unregulated network. However, since in underload conditions the difference between the greatest and the smallest average access delay is not very large, in the next experiments we will assume $Q_s = 3$ and $Q_r = 300$.

Influence of Offered Load

Figure 8 shows the influence of the offered load (*OL*) on the station average access delay. The initial shape of the curve can be explained by noting that the first group of stations (stations {1} – {6}) observes a large number of empty slots and thus their station average delays are not significantly affected by the *OL* considered. On the other hand, the last group of stations (stations {7} – {10}) observes trains of

201

contiguous busy slots with increasing length as the *OL* increases. This explains the sharp increase in the average access delay in the curve for these stations.

Qs=3; Qr=300; C=1.0; Msg=1

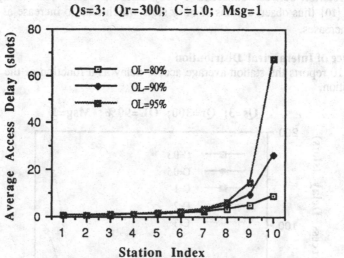

Fig. 8. Average access delay vs station index (*N* = 10).

Influence of Message Length
Figure 9, which shows the station average access delay as a function of message length, highlights that the MetaRing unfairness increases significantly with message length.

Qs=3; Qr=300; OL=90%; C=1.0

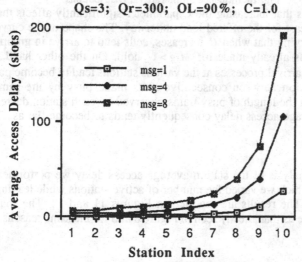

Fig. 9. Influence of Message length.

This can be intuitively justified by observing that one message arrival contributes to the aggregate *OL* with a number of cells equal to the message length. Therefore each station will transmit a batch of cells rather than a cell at a time, with the result that the correlation between consecutive busy slots increases. Stations approaching station {10} thus observe trains of busy slots which tend to increase as the station index increases.

Influence of Interarrival Distribution

Figure 10 reports the station average access delay as a function of the interarrival distribution.

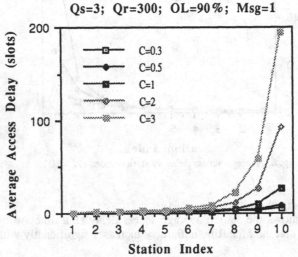

Fig. 10. Influence of Interarrival Distributions.

This figure shows that increasing the C parameter significantly affects the average access delay at least for the offered load underway. The shape of the curves can be justified by observing that when *C* increases, cells tend to arrive in groups. Hence, the considerations already made for $Msg > 1$ hold. On the other hand, when *C* approaches zero, arrival processes at the various stations tend to become periodic. In fact, the correlation between consecutive slots made busy by the same station decreases, so that the length of busy trains observed by each station decreases. The curve of the average access delay consequently tends to become flat as *C* tends to zero.

Influence of *N*

To further the analysis of the station average access delay we performed a set of experiments in which we varied the number of active stations while leaving the ring length constant. The results are reported in Figures 11 and 12. The shape of the curves is the same as that already seen in Figure 8 and for the same reasons.

Qs=3; Qr=300; OL=90%; C=1.0; Msg=1

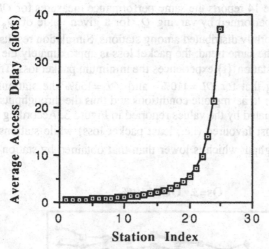

Fig. 11. Average access delay vs station index ($N = 25$).

Qs=3; Qr=300; OL=90%; C=1.0; Msg=1

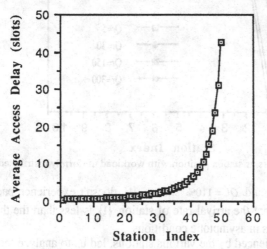

Fig. 12. Average access delay vs station index ($N = 50$).

5. Analysis in Overload Conditions

The network scenario is still the same (Figure 1) but the offered load is greater than the maximum achievable aggregate throughput. The performance measure that we considered in overload conditions is the packet loss. In this section we investigate how the packet loss is distributed among the stations as a function of the Q_r and Q_s parameter values, of the OL, and of the distribution of the workload among stations. The buffer size of each station was chosen in order to have a very low packet loss in underload conditions, even with values for the OL very close to 100%. By simulation we found that a buffer size of 400 can satisfy this requirement. Figure 13 shows the packet loss vs station index for five different values of Q_r and for $OL = 110\%$, while Figure 14 reports the same performance measures for $OL = 150\%$. The experiments were performed by varying Q_r for a given value of Q_s ($Q_s = 3$) and for a workload uniformly distributed among stations. Simulation results reported in both figures exhibit the same trend: the packet loss is approximately the same for stations $\{2 + 10\}$, while station $\{1\}$ experiences the minimum packet loss. This can be explained by observing that for $OL = 110\%$ and $OL = 150\%$ the stations tend to operate closer and closer to asymptotic conditions and thus the throughputs they can achieve can be approximated by the values reported in Figure 3. According to Figure 3, station $\{1\}$ is the most favoured (i.e., least packet loss) while stations $\{2 + 10\}$ achieve the same throughput which is lower than that obtained by station $\{1\}$ (i.e., greatest packet loss).

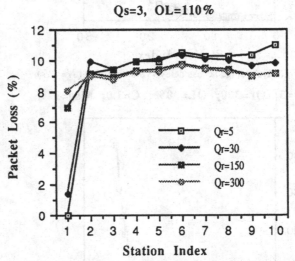

Qs=3, OL=110%

Fig. 13. Packet loss vs station position with workload uniformly distributed.

Notice that with $Q_r = 5$ and $OL = 110\%$ station $\{1\}$ doesn't experience packet loss. This is due to the fact that the arrival rate of station $\{1\}$ is less than the throughput the same station achieves in asymptotic conditions.

The packet losses experienced by the various stations, led us to analyze the possible influence of a group of stations operating in overload conditions over another group of stations operating in underload conditions. We performed two sets of simulation experiments. In the first set, stations operating in overload were located upstream with respect to underload stations, while in the second set overload stations were downstream of the underload stations. For both sets of experiments we assumed $Q_s = 3$ and considered different values for Q_r. The results obtained, as well as the workload distribution among stations, are reported in Figures 15 and 16 respectively.

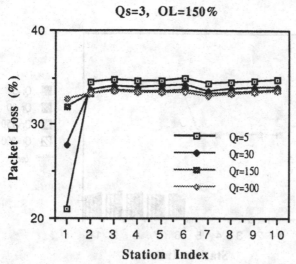

Fig. 14. Packet loss vs station position with workload uniformly distributed.

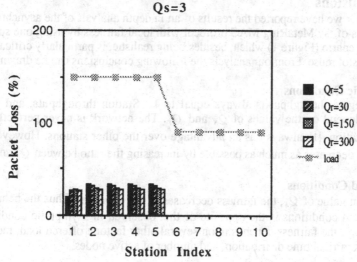

Fig. 15. Packet loss vs station position with unbalanced workload.

By analysing of Figures 15 and 16 we can conclude that in all the experiments performed no packet loss was recorded in the underload stations. Furthermore, the fairness in terms of packet loss decreases as Q_r increases.

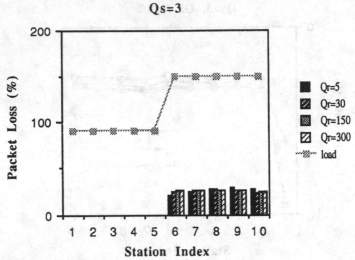

Fig. 16. Packet loss vs station position with unbalanced workload.

6. Conclusions

In this paper we have reported the results of an in depth analysis of the asynchronous capabilities of the MetaRing MAC protocol with local fairness by using one specific network scenario (Figure 1) which, besides being realistic, is particularly critical with respect to slot reuse. From our analysis the following conclusions can be drawn.

Asymptotic Conditions
The aggregate throughput is always equal to 1. Station throughputs, and hence fairness, depend on the values of Q_s and Q_r. The network is never perfectly fair because station {1} always has an advantage over the other stations. However, the unfairness decreases as much as possible by increasing the ratio between Q_r and Q_s.

Underload Conditions
For a given value of Q_s, the fairness decreases as Q_r increases. Thus the behaviour in underload conditions is the opposite of that observed in asymptotic conditions. Furthermore, the fairness depends upon several other factors: offered load, message length, interarrival time distribution, and number of active nodes.

Overload Conditions
When the workload is uniformly distributed among stations, station {1} experiences the least packet loss while all the other stations have the same packet loss. Furthermore, the unfairness decreases as the ratio between Q_r and Q_s increases. When the workload is not uniformly distributed among stations (Figures 15 and 16), the underload stations never experience packet loss.

References

[CHEN93]   J. Chen, I. Cidon, Y. Ofek, "A Local Fairness Algorithm for Gigabit LAN's/MAN's with Spatial Reuse", IEEE JSAC, Vol. 11, No. 8, October 1993, pp. 1183, 1192.
[CIDO93]   I. Cidon, Y. Ofek, "MetaRing, a Full Duplex Ring with Fairness and Spatial Reuse", IEEE Transaction on Communications, Vol. 41, No. 1, January 1993.
[CINL75]   E. Çinlar, "Introduction to Stochastic Processes", Englewood Cliffs, NJ: Prentice-Hall, 1975.
[GORD92]   R. F. Gordon, P. G. Lowener, E. A. MacNair "The Research Queueing Package Version 3. Language Reference Manual", IBM Research Report, Yorktown 1992.
[LAPO95] M. La Porta, "Valutazione delle Prestazioni del Meccanismo di Fairness Locale REQ/GNT applicato alla MAN al Gigabit/sec MetaRing", Laurea Thesis (in Italian).
[MAYE95]   A. Mayer, Y. Ofek, M. Yung, "Approximating Max-Min Rates via Distributed Local Scheduling with Partial Information", private communication.
[OFEK94]   Y. Ofek, "Overview of the MetaRing Architecture", Computer Networks and ISDN Systems, Vol. 26, Nos. 6-8, March 1994, pp. 817-830.
[WU92]   H. T. Wu, Y. Ofek, K. Sorhaby, "Integration of Synchronous and Asynchronous Traffic on the MetaRing Architecture and its Analysis", Proceedings of ICC '92.

Experiences with Multimedia Teleshopping Applications over a Broadband Network - The Project ESSAI

Daniel Felix, ETHZ (CH)
Ettore Paolillo, COSI srl (I)
Franco Mercalli, Centro di Cultura Scientifica A. Volta (I)
Yung-Shain Wu, Ascom Tech AG (CH)

Abstract

In the framework of the RACE Programme of the European Union on advanced communications, the project ESSAI - Experimental Service Sale Automation on an IBC Network, has investigated the issues related to the provision of Teleshopping Services for services and goods based on a broadband communication infrastructure.

The Project ESSAI has focused on a very broad spectrum of issues spanning the identification of the users' requirements, the assessment of the different architectural and technological options for the provision of the teleshopping service, the analysis of costs and benefits of teleshopping in different market segments and the definition of a methodology for the assessment of the users' acceptance.

Most importantly, ESSAI has designed and developed two demonstrators of the proposed teleshopping system and has operated them during three extended service usage trials. In these trials experimental, but "real", teleshopping services were offered to real users who could buy services and goods in the areas of entertainment and travel, tourism and transportation through the system. The assessment of users' acceptance of the ESSAI system and service during these different trials has enabled the evaluation of the market potential for network-based teleshopping services and the definition of a suitable market introduction strategy in view of the forthcoming availability of the first commercial IBC (Integrated Broadband Communication) services throughout Europe.

This paper presents some of the most relevant technical results of ESSAI and reports the experience gained with real users in the public service usage trials. Many of the issues addressed here are considered crucial for a successful deployment of teleshopping services as soon as commercial IBC services become available.

1.   Introduction

The Project ESSAI is an Advanced Communication Experiment (ACE) which, in the framework of the European Union RACE II Programme, has investigated the issues related to the provision of Teleshopping Services of services and goods over a broadband communication infrastructure.

There is more and more evidence, also supported by the results of ESSAI, suggesting that automated service [1] sales are needed and will be common by the end of the century. The project ESSAI has demonstrated that there are feasible technical solutions. With the emergence of broadband communication networks, new possibilities in the field of automated sale of services have in become available. In particular:

A system can be devised in which a Control Centre, responsible for the provision of teleshopping services, acts as the collector of offerings from a number of goods/service suppliers and relays these to multiple geographically scattered multimedia terminals;

The multimedia catalogues, (which describe with the help of video, audio, still pictures and text the offerings from different suppliers), can be implemented and easily maintained in the Control Centre; the catalogues are then made available for consultation to the remote vending terminals in a shared mode through broadband network facilities;

Sophisticated user interfaces can be implemented in the terminals by using high quality pictures, video and sound to guide users through all phases of the purchase transaction, i.e. catalogue browsing, availability checking, reservation, payment and ticket/voucher printing;

An effective security policy can be implemented allowing centralised checking of users' identity and authorisation to use services and make purchases;

The operational costs of service provisioning can be minimised by means of full control of the system operation and maintenance directly from the teleshopping service centre;

Advanced help functionalities are also possible: for all cases which cannot be foreseen by the application software a video/sound connection with an help desk operator at the Control Centre can be set up, thus giving human answers to human problems.

2.   ESSAI in General

2.1   Overall Project Objectives and Assumptions

The main goals of project ESSAI have been to:

---

[1]   Goods selling can also be considered a service: vending terminals can deliver vouchers allowing customers to get their purchase at special (possibly automated) stores or via mail delivery

- determine the technical and economical viability of the sale of services and goods by means of teleshopping applications based on an IBC network infrastructure;

- define the architecture of a teleshopping system aligned with the social, legal and cultural European environment and matching the users' requirements for automated sales applications with the possibilities offered by IBC;

- develop sufficient know-how about the technical, social and marketing implications of teleshopping systems in order to be able to offer teleshopping services as soon as IBC becomes commercially available.

In order to achieve these major goals and validate the results obtained, the project has adopted an experimental approach based on a significant involvement of users (both consumers and goods/service suppliers) in the service definition phase and, most importantly, in real service usage trials. This approach has led the project to:

- design and develop two experimental teleshopping systems based on the ESSAI architecture (terminal and a Control Centre) for the sale of services and goods based on a broadband infrastructure;

- conduct three controlled service usage trials with real customers, buying real goods or services, for real money in order to determine acceptance of teleshopping systems by the general public.

These general goals were pursued on the basis of a number of basic assumptions which have guided all technical and architectural choices made by the project. The assumptions are as follows:

that the teleshopping system architecture is generic;
ESSAI has defined a general purpose teleshopping system for the sale of services and goods which aims at satisfying the very clear user requirement of having access, through a multimedia vending terminal, to a broad and diversified spectrum of offerings by different suppliers and in different market segments;

that the system is designed to satisfy the requirements of generic end-users in public locations;
this means that the intended user of the system is an untrained end-user passing by the vending terminal in public, well-frequented locations and wanting to make a purchase independent of any assistance from professional operators;

that the system is designed with operations of the vending terminals in public environments (stores, streets, public buildings, railway stations, airports, etc.) in mind;
possible household applications of teleshopping have been considered to be outside the scope and perspective of ESSAI;

that the proposed system and service architecture are intended to fully exploit the capabilities of IBC;

this means that physical centralisation of all system resources in the teleshopping Control Centre has been adopted and maximum exploitation of the available communication throughput is done in order to connect remote vending terminals. [2]

2.2 Project's Achievements

The Project ESSAI began on 1st January , 1992 and ended on 30th April 1995. During this period a large number of achievements were accomplished. These are summarised below:

Basic teleshopping system aspects were defined: i.e. identification of market demand and requirements of teleshopping system users (both end-users and service operators). Also, a definition of the functional architecture of future systems, system modelling and verification, identification of basic threats in a teleshopping system and a definition of a suitable security policy were established.

A functional design of the two ESSAI demonstrators was also created. It identified the technological options available for the demonstrators in Milan (I) and Basel (CH), and defined the resources available in the two demonstrators.

The first demonstator in Milan was specified (4), implemented, integrated and the experimental set and the teleshopping applications were tested.

A methodology to measure the requirements of users of the proposed Teleshopping System and its level of acceptance was defined.

The first demonstrator in Milan was publicly operated in a real teleshopping environment over the periods December '93-February '94 (first usage trial), and June-July '94 (second usage trial).

---

[2] The issue of centralisation vs. de-centralisation of system resources is clearly dependent upon a number of factors such as communication costs, hardware/software complexity and cost of centralised system resources, expected number of remote terminals which çan be supported from a central location, application specific issues (e.g. variability in time of the multimedia catalogues content). In ESSAI this issue has been solved by designing the teleshopping applications as real distributed applications (without regard to the actual physical location of the involved resources) and adopting two different approaches for the demonstrators. In the first demonstrator in Milan (I), the video database was loaded directly in the vending terminal; on the contrary, for the second demonstrator in Basel (CH) video storage was provided in the Control Centre and made accessible via the ATM network from the terminal.

The first ESSAI demonstrator was operated in Milan according to the following architecture.

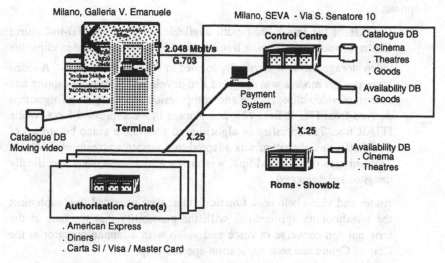

Fig. 2.1   Architecture of the First ESSAI Demonstrator in Milan

The terminal was located in Galleria V. Emanuele, a very central zone of the city while the Control Centre was operated in Via S. Senatore, 10 (roughly 3 Km away). The Terminal and the Control Centre were connected through a 2,048 Mbit/s G.703 line. During the first trial period the terminal was situated in the premises of the Information Office of the Municipality of Milan which is open to the public from 8:00 a.m. to 8:00 p.m.. The teleshopping applications offered the sales of tickets for cinema events in five movie-halls, and the sales of tickets for theatrical events in Verona (at ÑL'Arena di Verona" and ÑTeatro Romano"). During the trial period extensive measurements and observations of system users' acceptance were carried out.

The second public trial was carried out in the public telephone area of SIP (now TELECOM Italia) in Galleria V. Emanuele as a means to derive additional results from system operations and experience with the public. This second trial, which was carried out with only one teleshopping application (the sale of theatrical events for the summer season of L'Arena di Verona) turned out to be much more successful in terms of users' number and interest than the previous trial. The reason for this is the better location, which allowed a higher exposure to the public. This fact shows that the selection of a suitable terminal location is an essential factor.

(The assessment of users' acceptance during the first trial in Milan will be analysed in further detail in Section 5 of this paper).

2.3   Design, Implementation and Testing of all Hardware and Software Modules

The design, implementation and testing of both hardware and software modules of the second demonstrator were carried out in Basel on the EXPLOIT

ATM test-bed. With respect to the first demonstrator in Milan, the experimental set-up of the Basel demonstrator was upgraded with several new technological options.

Because of the larger bandwidth available for the Terminal-to-Control Centre connection, allowing transmission of the moving video clips, the video storage was now physically located in the Control Centre. [3] A video mass-storage module was designed and developed: it allowed capture and editing of video clips, coding and compression according to an algorithm derived from ITU-R Rec. 723, and storage in a compressed form. While ITU-R Rec. 723 specifies an algorithm to generate a video bit stream of 34 Mbit/s, this algorithm was adapted to generate a stream of 7 Mbit/s. Transmission on the ATM link, was via an AAL1 sub-system specifically designed and developed.

Audio and video help desk functionalities were provided to complement the teleshopping applications with the possibility that the user at the terminal can converse in voice and video with a human operator at the Control Centre and seek application-specific help.

Signalling functions according to Q.2931 were implemented to manage connections on the EXPLOIT ATM test-bed. In particular, dynamic management of moving video connections was performed. [4]

2.4 Development of the Teleshopping Applications Deployed in Basel.

Two suppliers have provided their services through the ESSAI System during the public trial in Basel: DANZAS TRAVEL AG and SBB, the Swiss Federal Railways. DANZAS, which is one of the major tour operators in Europe and a large "ticket broker" in a position to sell flight tickets at very low prices, offered the following services:

sale of "last-minute" flight tickets, i.e. discounted price tickets for flights available in the same week when the ticket is purchased; the offers change from hour to hour and the use of a fully computerised system which included the sales channel, such as ESSAI is provided offering major advantages with regard to conventional sales channels.

sale of travel packages (i.e. flight + hotel + additional services).

---

[3] In the demonstrator in Milan, where a 2 Mbit/s G.703 connection was available, the video clips were stored in the Terminal on a laser disk.

[4] This allows a conservative approach to bandwidth usage which is allocated only when needed.

The Swiss Federal Railways sold "flat rate travel" packages consisting of a train or bus ticket combined with the entrance ticket to a show or a venue.

2.5 Public Operation of the 2nd Demonstrator in Basel in a Real Teleshopping Environment During the Period March - May '95.

The second ESSAI demonstrator was operated in the Basel (CH) area on the EXPLOIT ATM test bed infrastructure operated by ASPA, Association of Swiss PTT and ASCOM. The teleshopping terminal was located in the "Gartenstadt" Shopping Centre in Münchenstein while the Control Centre was hosted in the Swiss PTT Building "Grosspeter". The physical configuration of the Basel demonstrator is examined later on in this paper in Section 3.3.

2.6 Assessment of the ESSAI System Market Potential in the TTT Sector.

Several market segments have been analysed in the TTT sector with a view to assessing the economic viability of an automated system, such as ESSAI, in comparison to traditional sales channels. Results are encouraging. A segment of the TTT sector is concerned with the way a large section of the younger generation travel: comfort, security, time, travel agent advice and, sometimes, even destination are not really important. It is the price which is the major factor when deciding. In this respect, good market chances exist for teleshopping, as the sales cost of a journey can be reduced by approximately 50% using automation.

3. Configuration of ESSAI Demonstrators

3.1 Introduction

The future of teleshopping definitely falls into the category of interactive communication with the use of moving pictures. Interactive communication requires a user-friendly interface and a broadband communication network because transmission of moving pictures is only conceivable where this is available. Bearing these two key issues in mind, we can roughly sketch a possible layout of functional units for a teleshopping system (see Figure 3.1). In this scenario, customers can get the information about the supplies and order goods by using teleshopping terminals (TS-TE) distributed throughout a particular region (town or country). The goods ordered in this way would be eventually dispatched to the customer's address from the stock locations indicated. It is envisaged that the system could have a Control Centre which may also contain the operator positions for the "on-line help" facility and the master copies of the audio and video sequences.

According to this concept, the future teleshopping system should therefore contain the following four functional blocks:
(i)   Teleshopping Terminal (TS-TE)
(ii)   Payment System
(iii)   IBC-Network
(iv)   Service Centre

The functional block "Service Centre" is connected with the IBC-Network at the reference point NNI (Network-Node Interface), while the functional block "Teleshopping Terminal" is connected with the network at the reference point UNI (User-Network Interface). The "Payment System" can be connected either with the "Teleshopping Terminal" or the "Control Centre" of the "Service Centre".

Fig. 3.1 represents these key functional blocks in a Functional Architecture Model.

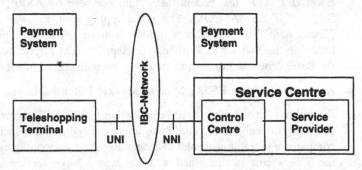

Fig. 3.1 General Functional Architecture Model of a Teleshopping System

Two demonstrators have been built in the project ESSAI. While the first ESSAI Demonstrator showed only the functionality of a future teleshopping system in Milan, the second one demonstrated the full capabilities of a future teleshopping system, in particular, the ATM-signalling and the transmission of real-time audio and video over the ATM-based EXPLOIT testbed in Basel were demonstrated.

3.2 The Physical Configuration of the 1st ESSAI Demonstrator

The first ESSAI Demonstrator is composed of a Control Centre with a Service Provider and a Teleshopping Terminal (TS-TE). This physical configuration is based on the following assumptions:
(i)   The *analog* multimedia database was located *locally* in the TS-TE.
(ii)   The data communication was based on TCP/IP over AAL 3/4 and ATM.
(iii)   The network used was a PDH-based one G.703 with 2Mb/s, transparent.
(iv)   The teleshopping system featured no signalling capabilities.

Fig. 3.2 shows the physical configuration of the first ESSAI Demonstrator, in which the SPARC 10 in the Control Centre is in charge of the application development, both the SPARC 2 are mainly dedicated to the data communication. For a detailed specification of this physical configuration, refer to [1].

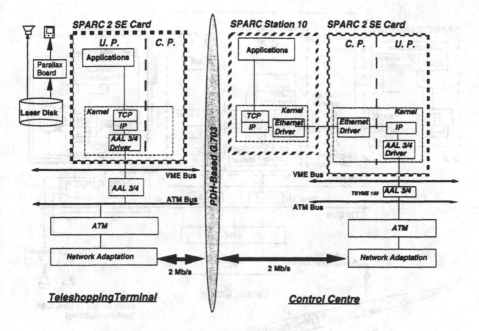

Fig. 3.2 Physical Configuration of the 1st ESSAI Demonstrator

3.3 The Physical Configuration of the 2nd ESSAI Demonstrator

The second ESSAI Demonstrator is composed of a Control Centre with two Service Providers and a Teleshopping Terminal (TS-TE). This physical configuration is based on the following assumptions:

(i)   The *digital* multimedia database was located *centrally* in the Control Centre for easier management of the database.

(ii)   The data communication is based on TCP/IP over AAL 3/4 and ATM.

(iii)   The network used was an ATM-based EXPLOIT testbed of the RACE Project R2061 [2].

(iv)   The audio/video communication is based on dedicated software and hardware.

(v)   The teleshopping system features with signalling capabilities.

Fig. 3.3 shows the physical configuration of the second ESSAI Demonstrator, which is composed of a Control Centre and a Teleshopping Terminal too. In the ESSAI Control Centre, the SPARC 10 is in charge of the application development, SPARC 2 is dedicated to the data communication as well as the connection set-up & release and the Motorola 68000 is served as the audio/video- multimedia database.
A SARC 2 is used in the ESSAI Teleshopping Terminal for playing the multimedia sequences.
For a detailed specification of this physical configuration, refer to [3].

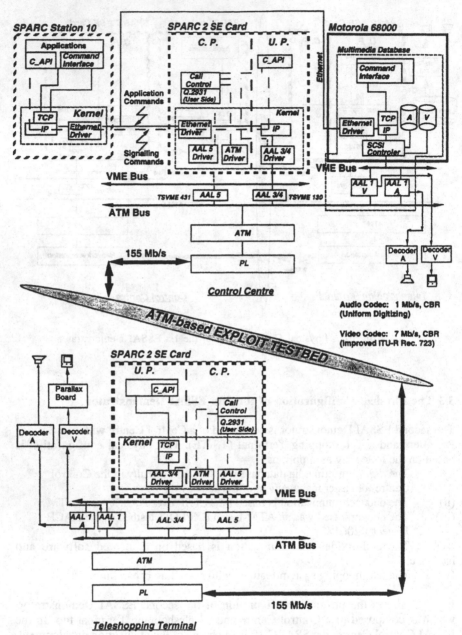

Audio Codec: 1 Mb/s, CBR
(Uniform Digitizing)

Video Codec: 7 Mb/s, CBR
(Improved ITU-R Rec. 723)

Fig. 3.3 Physical Configuration of the 2nd ESSAI Demonstrator

4. General Structure of a Teleshopping Application

The general structure of the teleshopping application is shown in Fig. 4.1.

This structure is the result of the implementation of two demonstrators in the frame of the ESSAI project and it has been designed taking into account a number of conflicting requirements for which convenient trade-offs have been devised.

The application is built around the *event driven* programming metaphor. It consists of a set of panels shown on the touchscreen of the vending terminal, the navigation among which is driven by the occurrence of *events*, caused either by the end-user (e.g.: pressure of a button on the touchscreen surface) or by the system (e.g.: arrival of an authorisation message from the payment system). This allows to make use of the *hypermedia* paradigm for developing multimedia applications, whose potential has been recently demonstrated through the surge in World-Wide Web (WWW) activity [4].

In the following we will give some details about each one of the phases illustrated in Fig. 4.1.

Idle demo

It is a sequence of continuously rolling panels which has the goal to attract customers to the terminal. The sequence can be interrupted at any time by the customer presence. A suitable detector tracks the event *customer presence*, which, in turn, stops the idle sequence and starts the real application.

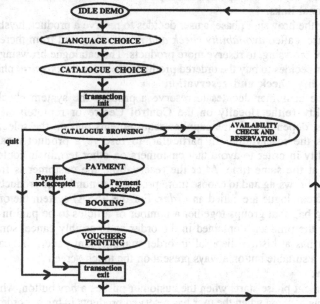

Fig. 4.1   General application structure

Language choice

The ideal locations for an ESSAI terminal are crowed and public places like stations, airports, shopping centres and so on, attended by many people which are likely to come from different countries of European Union and outside. The ESSAI application is consequently designed to cope with multilinguism. For this reason it is necessary to allow the customer to choose the language of preference: after this choice all the texts and sounds in the application will be in the language selected by the user.

Catalogue choice

In the ESSAI system suppliers are in general more than one. The ESSAI experience shows that each supplier strongly prefers that its own sales transactions are managed separately from the ones of other suppliers. This suggested to use the *catalogue metaphor*, for which the end-user is forced to choose a given supplier by choosing the corresponding *catalogue*. After that, sales transactions are possible only for that supplier. To change supplier it is necessary to exit the current catalogue, terminating the corresponding sales transaction, and then to choose a new one.

Catalogue browsing

These panels present the supplier's goods and services offer to the customer. These panels depend strictly on the products to sell and on the style desired for the MMI. For this reason its detailed design is to be worked out in strict co-operation with the supplier, accordingly to its specific requirements.

The design of this part of the teleshopping application involves strict MMI requirements, deeply analyzed in [5]. However, useful guidelines are provided by recent research results in the area of hypertext and hypermedia technology. In particular, in the ESSAI project, the Hypertext Design Model (HDM, see [6]) has been used during the specification phase, in order to guide the structuring of navigational links.

If during the browsing phase, a user decides to reserve a product, he/she can go to the next phase, called *availability check and reservation*, and from there return back to catalogue browsing, to reserve more products. The catalogue browsing ends when the customer decides to buy the ordered products, going to the *payment* phase.

Availability check and reservation

When the customer decides to reserve a product the system checks the product availability (either locally on the Control Centre or remotely at the supplier's premises, depending on the supplier requirements and on the selected goods) and manages the reservation. In particular, to reserve a product means to limit its availability in order to avoid that customers on other terminals could buy the same product at the same time. After the reservation, the customer is free to return to catalogue browsing and to choose more products. A number of products reserved from the same catalogue are called an *order*. For the ESSAI system the order is a sort of shopping bag that groups together a number of articles to be paid in a single batch. To view the products contained in the order and possibly cancel some of them, the end-user has at his/her disposal an order management panel, that can be called by pressing a suitable button always present on the touch-screen.

Payment

The payment phase starts when the customer pushes a *pay* button, which is activated on the touchscreen when the user has reserved products in his/her order bag. He/she is invited to insert a payment card, either credit or debit card, into the card reader. A payment authorisation is then requested on-line to the Authorisation Centre of the relevant credit or debit card company. The payment phase ends with either the confirmation message or the rejection message reception from the Authorisation Centre. If the payment authorisation is rejected the following two phases are skipped.

Booking
In this phase, which is performed only after a successful payment authorisation request, the products are sold to this customer, making them no more available to other users.

Vouchers printing
As a receipt of the sales transaction the ESSAI terminal provides the customer with a set of printed vouchers. A number of these vouchers are the proof of the articles' purchase (e.g.: tickets, if the products sold are cinema movies). Besides them a payment voucher is always present, with the information concerning the payment authorisation, coming from the payment card company.

5. Experiences Gained from the ESSAI Demonstrators

Two field experiments were conducted within the project ESSAI:
A first test in Milan (I) in the Winter of 1993/94, where tickets to the Arena di Verona and tickets for several local cinemas were sold, as well as information on the municipality of the city of Milan was given.
A second test in Basel (CH) in the Spring of 1995, where tickets for day trips of the Swiss Federal Railway and last minute travel tickets from a tour operator were sold.
The objective of the test was to assess user acceptance of a system like the ESSAI teleshopping terminal. The methods used to assess the acceptance were a questionnaire, video recording of user behaviour and the registration of log files of all user actions at the terminal [7].
From the first test, questionnaires of 45 users were analysed. Most users were in the age group of 20 to 29 years old (n = 16), however 7 users were older than 50 years, which indicates that the system is attractive to all age groups. Approximately 80% of users reported that they have used a bank automat before and other automatic services were reported to be used as well. These various systems are used mostly once per month by about 50% of the users and 40% use them more frequently. This means that the automat experience level is high with the users of the ESSAI system. However, the users were not experienced computer users. In the second test, 23 questionnaires were analysed. The age profile shifted slightly towards the younger ages. However all age groups were represented, but it was noted that the users of the second test were much more experienced computer users.
The ergonomic aspects such as screen height, positioning of the system and so on were in both tests judged to be adequate, confirming research previously carried out [8]. The sound level was found to be a little too low, which was due to a problem with the adjustment of sound from the video laser drive and from the hard disk. The video clips, used to show more about a specific offer, were judged to be helpful and many users viewed them. They were highly attractive and in the second experiment the impression was gained that users specifically searched for these clips.
Navigating in the system was not an easy task for the users. The range of goods and information offered made it necessary to have a clear model of the whole system, a task which was not understood by all users. More than a third of the users had no clear image of what they were looking for in the system (pure browsing). Therefore, navigation aids must lead the users to the most important parts of a program where they can learn what choices there are. In the first experiment, many users reported that they had to touch the screen too often to reach their set goal. For users who know what they are looking for, a quicker path must be provided. This was improved

in the second test, and the results showed less frustration. The overall orientation was rated much better in the second test.

The overall acceptance was very high. Most users would use the system again because they believe that such a system is a good idea and that the goods and information offered are of interest to them. Many users even claimed that the ESSAI system offered a quicker method to purchasing goods than traditional shops.

The best time for users seemed to be around lunch time and in the evening. Most users were observed during these time periods. The average number of users per day is about 25, a number which must be much higher for real commercial operation. However, most users only briefly used the system and mainly looked at the introductory screens, so the percentage of serious users is even lower. The mean time spent at the system is 4.7 minutes. This time includes the introductory screens and the termination procedure (most interactions were terminated by the system idle time-out).

In general, the system seems to be accepted by the users who actually used the system. However, for a commercial operation the number of users and transactions are too low. It was observed that the longer the system had been installed, the more the system was used, and transactions (sales) could be registered. This means that the following must be implemented:

- better communication about the offers of the system
- better communication about the advantages of the system
- more marketing for the system
- more financial benefits for the user.

The findings are congruent to other projects in the same area, such as a bank-information system in Switzerland [9]. The initial barrier for the users is high and only if they know what to expect, what their benefit will be and that the system is trustworthy and reliable, will they willingly buy their goods at a system like the ESSAI terminal. A quicker path to a goal for users who know exactly what they are looking for must also be implemented besides the navigational aids for users who only wish to browse. Goods of real interest to the users, preferably not available at any other source, must be offered. On the whole, however, it seems that systems like this have a future and if they are designed correctly, can be a profitable investment.

Acknowledgements

The authors are grateful to the **European Union** and the **Bundesamt für Bildung und Wissenschaft** (BBW) of Switzerland for the financial support of this project during the period 1.1.92-30.4.95. This work would not have been successful without the following ESSAI partners: Ascom Autelca (CH), SEVA (I), Thomson Broadband Systems (F), Syseca (F), OST (F), STZ (D), Delta (Dk), proXima (D), SNV(D), ShowBiz (I), SBB (CH), Danzas (CH) and COOP (CH).

222 223

Key Words

AAL	ATM Adaptation Layer
ASPA	Association of Swiss PTT and Ascom
ATM	Asynchronous Transfer Mode
ESSAI	Experimental Service Sales Automation on an IBC network
EXPLOIT	Exploitation of an ATM Technology Testbed for Broadband Experiments and Applications
IBC	Integrated Broadband Communication
MMI	Man-Machine Interface
NNI	Network-Node Interface
PDH	Plesiochronous Digital Hierarchy
RACE	Research and development on Advanced Communications in Europe
TCP/IP	Transmission Control Protocol/Internet Protocol
TS-TE	Teleshopping Terminal
TTT	Travel, Tourism and Transportation
UNI	User-Network Interface

References

[1] Deliverable 6 of the RACE Project R2029-ESSAI: First Demonstrator Specification, 31st December 1992
[2] Deliverable 27 of the RACE Project R2061-EXPLOIT: Enhanced ATM Platform, 31st May 1995
[3] Deliverable 9 of the RACE Project R2029-ESSAI: Second Demonstrator Specification, 28th February 1994
[4] Bieber M., Isakowitz T., Designing Hypermedia Applications, *Communication of the ACM 38*, 8th August 1995;
[5] Maino D., Mercalli F., Negrini R., A Multimedia Database for an Advanced Teleshopping Application, *3rd International Symposium on Database Systems for Advanced Applications*, Taejong (Korea), April 1993;
[6] Garzotto F., Paolini P., Schwabe D., HDM - A Model for the Design of Hypertext Applications, *ACM Transactions on Information Systems 11*, 1st January 1993.
[7] Felix, D. and Krueger, H.: ESSAI - Interactive Sales System on an IBC-Network. Proceedings of HCI'93 (5th International Conference on Human-Computer-Interaction), Orlando. Elsevier, Amsterdam 1993
[8] Stulzer, W. and Felix, D.: Gestalterische und ergonomische Anforderungen an das Terminal. Unpublished, ETH Zurich, 1992
[9] Steiger, P. and Ansell Suter, B.: Experiences with an Interactive Information Kiosk for Casual Users. In: Bischofberger, W. R. and Frei, H.P. (Eds.): Computer Science Research at UBILAB, UVK Informatik, Konstanz 1994

Multimedia Multipoint Teleteaching over the European ATM Pilot[‡]

Simon ZNATY[1], Thomas WALTER[2], Marcus BRUNNER[2],

Jean-Pierre HUBAUX[1], Bernhard PLATTNER[2]

[1]TCOM Laboratory
Telecommunications Services Group
Swiss Federal Institute of Technology Lausanne
CH-1015 Lausanne
E-Mail: {znaty, hubaux}@tcom.epfl.ch

[2] Computer Engineering and Networks Laboratory
Swiss Federal Institute of Technology Zürich (ETH)
ETH Zentrum, Gloriastr. 35, CH-8092 Zürich
E-Mail: {walter, brunner, plattner}@tik.ee.ethz.ch

Abstract

In this paper, we present the design and implementation of the BETEUS (Broadband Exchange for Trans-European Usage) communication and application platform. BETEUS is a European project aiming at developing generic, stable, flexible and scalable communication and application platforms which provide support for a collaborative work environments. In terms of collaborative work, BETEUS concentrates on distributed classroom, tele-seminar, multimedia document archival and retrieval and tele-tutoring. The problems found during the realization of the BETEUS platform are outlined and the proposed solutions explained. The follow-up of BETEUS in Switzerland through the TELEPOLY project which concerns operational teleteaching between the two Swiss Federal Institutes of Technology (EPFL and ETHZ) is also presented.

1 Introduction

Support for Multimedia is an important issue that drives the request for ATM (Asynchronous Transfer Mode) communication. The usage of multimedia features in a *collaborative work environment* is the core of BETEUS - Broadband Exchange for Trans-European USage [1]. The functions that have been developed in the project are best captured by the so-called *virtual community paradigm*. A community is a group of people with common interests participating in a set of operations towards reaching a common goal. In the case of BETEUS the operations are *working, teaching, learning, project management* and *technical design* [2]. The virtuality of the community is due to the fact that people who participate in the common operations are located in geographically distant locations. The BETEUS platform provides the services required for

---

‡   The material is based upon work support by the CEC under grant Project Number
     M1010. EPF Lausanne and ETH Zürich have been supported by the "Bundesamt für
     Bildung und Wissenschaft" under contract number 94.0069.

interconnecting distant locations to establish a virtual community [3]. BETEUS concentrates on distributed classroom, tele-seminar, multimedia document archival and retrieval and tele-tutoring, as well as collaborative work environments.

The BETEUS communication platform, i.e. the part of the BETEUS platform providing services for interconnecting distant locations, is built on top of the European ATM pilot network. As services, both end-to-end native ATM and IP are provided. In its current implementation, the BETEUS application platform, i.e. the part of the BETEUS platform implementing services for collaborative work support, can communicate via UNIX-sockets [4]. This decision has been taken for practical reasons. Firstly, the development of the BETEUS application platform is thus independent from the actual support of the underlying communication infrastructure. Although all BETEUS user sites now run FORE ATM switches [5], this was not obvious when the design of the communication platform was started. However, at that time, it was apparent that IP connectivity was available for all partners either directly over ATM [6] [7] or SMDS (Switched Multimegabit Data Service). Secondly, the communication platform can be extended so that IP multicast facilities [8] can be used for point-to-multipoint communication as is requested by teleteaching applications.

The BETEUS application platform is not a stand-alone application like most of the today's collaborative teleconferencing systems, which implement all components, from low-level transmission and processing up to the user interface, from scratch. It is a platform for implementing different scenarios like distributed classroom (i.e. teleteaching) or tele-seminar (something like a virtual face-to-face meeting). Therefore, we could evolve the final scenarios out of a series of prototypes that could be tested on the BETEUS network. This led to a significantly reduced effort as compared to stand-alone prototype systems for every scenario.

The paper is structured as follows: Section 2 discusses the services offered by the European ATM pilot. BETEUS user sites are connected by a fully meshed network, based on the semi-permanent virtual path service (Section 2.2). Such a network is considered appropriate with respect to the design considerations of the BETEUS communication platform (Section 3.1). The realization of the BETEUS communication platform is discussed in Section 3.2. We give details on the overall network topology, the local site configurations and multicast. This section also introduces the BETEUS management platform responsible for the management of the communication and application platform. In Section 3.1 we briefly discuss the BETEUS application platform. Section 4 explains the use that has been made of the infrastructure and summarizes some pragmatic conclusions that have been drawn; it sketches future plans. We conclude with an assessment of our achievements.

2 European ATM Pilot - Topology and Services

The introduction of the *asynchronous transfer mode (ATM)* is currently being driven by the need for fast and hopefully cheap data communication in public and private networks.

In the context of the BETEUS project (a follow up to Betel [9]), ATM technology is used to provide end-to-end pure ATM communication between BETEUS sites: CERN in Geneva (Switzerland), EPFL (Switzerland), ETHZ (Switzerland), EURECOM in Sophia Antipolis (France), KTH in Stockholm (Sweden) and TUB in Berlin (Germany). The goal of the communication infrastructure is to run multipoint multimedia teleteaching applications such as *distributed classroom*, *tele-seminar* and *multimedia document archival and retrieval*.

2.1 European ATM Pilot Services

The European ATM pilot provides its users with two main services, a *semi-permanent virtual path service* (Section 2.2) and *adaptation services*. Within the latter category, we find LAN bridging (Ethernet), circuit emulation (E1 emulation) and point-to-point SMDS (Switched Multimegabit Data Service).

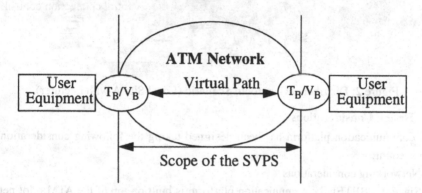

Figure 1: Semi-permanent Virtual Path Services

2.2 Semi-permanent Virtual Path Service (SVPS)

This service supports communication in both directions between reference points T_B/V_B (Figure 1). The establishment modes are either reserved (occasional or periodic) or *permanent*. Only the physical and the ATM layer are defined. Higher layers have end-to-end significance and can be chosen according to the needs of an application. In BETEUS we use IP over ATM adaptation layer 5. The provision of SVPS is based on virtual path (VP) connections in the ATM pilot network. The physical bandwidth at the access points is defined by the existing interfaces. For the ATM pilot, two interfaces, at 34 Mbit/s (E3 interface) and 155 Mbit/s (STM-1 interface) are used. A parameter "peak cell rate" is associated with each VP. The peak cell rate corresponds to the usable transfer bit rate, from the user's point of view, divided by the payload of an ATM cell in bits (48 * 8). A user can subscribe to any value of the usable information transfer peak cell rate that is available at the user interfaces.

3 The BETEUS Platform

The BETEUS platform consists of two parts: communication and application platform. The BETEUS communication platform interconnects BETEUS sites using semi-permanent virtual path services (SVPS). It provides the application platform with native ATM functions as well as a socket interface.

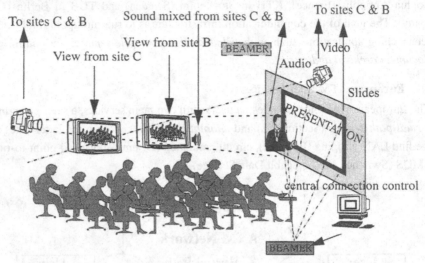

To sites C & B

Sound mixed from sites C & B

To sites C & B

View from site B

BEAMER

Video

View from site C

Audio

Slides

PRESENTATION

central connection control

BEAMER

Figure 2: Distributed Classroom Scenario (Local Site A with Lecturer)

3.1 Design Considerations

The communication platform has been designed taking the following considerations into account:

- Networking considerations
  Since the BETEUS communication platform is built on top of the ATM pilot network, the design decisions must take into account the services offered by the ATM pilot (as described in Section 2) and the variations of the network configurations in the local sites. For instance, the maximum bandwidth is 34 Mbit/s for most of the BETEUS partners (except ETHZ which is operating two STM-1 links). ATM signalling and point-to-multipoint services at the ATM level are not provided by the ATM pilot, neither the switched virtual connections (SVC). These restrictions have consequences for the provision of multicast in the BETEUS communication platform (see Section 3.2).

- BETEUS application requirements
  The BETEUS applications are *interactive multimedia applications*. This class of applications imposes stringent network requirements as summarized as follows: guaranteed throughput, bounded end-to-end delay, low packet loss, connection oriented transport, multicast and support for real-time data services.

- Bandwidth requirements
  The different BETEUS supported scenarios have different bandwidth require-

ments. In the distributed classroom scenario (Figure 2), for instance, every distributed classroom sends its classroom view to all other classes and receives classroom views from them. Images of the speaker are also distributed to other classrooms where he/she is not physically present. Distributed classroom has predictable bandwidth requirements since the scenario and the maximum number of sites involved are generally fixed during the session. Tele-seminar, on the other hand, is quite more demanding with respect to bandwidth requirements, since users may freely join and leave sessions.

However, the bandwidth required for connecting *two* sites can (more or less) accurately be estimated. For running a distributed classroom session the bandwidth of a connection has been estimated to 3 Mbit/s, composed of low quality video streams at about 1 Mbit/s of a quarter of PAL resolution and with 12 frames per second (fps) for classroom views, a speaker image at the same resolution but with a higher quality video stream at 25 fps (~ 2 Mbit/s). Assuming use of medium quality audio at 64 Kbit/s and shared workspace application at 60 Kbit/s data exchange rate, the total bandwidth is about 3 Mbit/s per connection.

Experiences have shown that the described application requirements are met by the BETEUS communication platform.

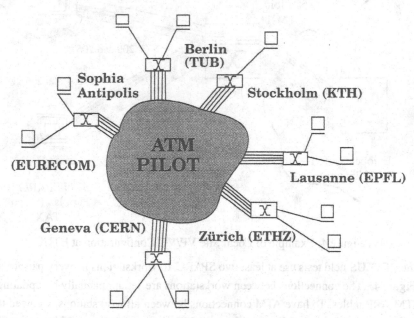

Figure 3: BETEUS Communication Platform Topology

3.2 Realization of the BETEUS Communication Platform

The BETEUS communication platform [3][10] is built on top of the ATM Pilot with a fully meshed VP topology network (Figure 3). Since access to the ATM pilot was not possible for all partners at the same time, the fully meshed network was the optimum solution for launching tests as early as possible.

Each site is connected with the other five sites through the ATM pilot by means of bi-directional VPs of 3Mbit/s peak bandwidth. On top of the VPs the Internet Protocol (IP) over ATM adaptation layer 5 is running [6]. The use of IP over ATM has been straightforward as the ATM hardware interfaces used with the workstations are supporting the necessary mapping.

Local Site Configuration

The BETEUS partners are using ASX-200 ATM switches from FORE Systems [5] and SUN workstations (Figure 4). The switch software is ForeThought version 3.2.0. The ATM switches are connected to the ATM pilot by E3 or STM-1 interfaces and to workstations by 100 Mbit/s TAXI interfaces. The workstations are equipped with a PAR-ALLAX video board (with JPEG compression done on the video board) and with a SBA-200 ATM adapter card (also from FORE Systems [11]). The workstations are running SunOS 4.1.3, ForeThought 3.0.2 and SNMP agents.

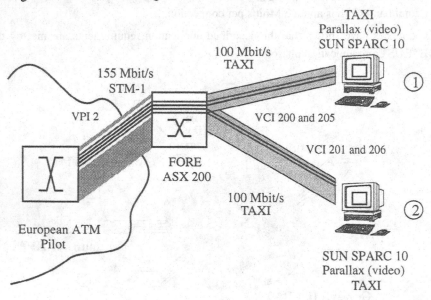

Figure 4: Example of Local Site VP/VC Configuration at ETHZ

The BETEUS field tests use at least two SPARC 10 workstations in every partner's site (Figure 4). The connections between workstations are set up manually by updating the ATM ARP tables. To have ATM connections between all workstations, we need to set up four virtual channels (VC) at each site (2 VCs per workstations). This configuration gives us the possibility to access every workstation in the BETEUS network from every other workstation.

For example, the ATM switch at ETHZ (Figure 4) is configured to send all traffic to EPFL on virtual path identifier (VPI) 2. Workstation 1 at ETHZ reaches workstation 1 at EPFL using virtual channel identifier (VCI) 200 and workstation 2 at EPFL using VCI 205. Similarly for workstation 2 at ETHZ which reaches workstation 1 and 2 at EPFL using VCIs 201 and 206, respectively.

Multicast in BETEUS

Since the core of BETEUS applications is to provide multimedia features in a collaborative work environment, the BETEUS communication platform should support *multicast services* for point-to-multipoint connections.

Multicasting means that a group of recipients can be addressed in a single data transfer. In principle, ATM can support applications with multicast services. This is usually done by using signalling protocols, e.g. SPANS (Simple Protocol for ATM Network Signalling) [12] or UNI (ATM User-Network Interface) [13].

Currently no signalling is used in the BETEUS communication platform, mainly due to some ForeThought limitations, e.g. in the version we are using, signalling is only possible on VP 0. With it IP multicast couldn't be employed too.

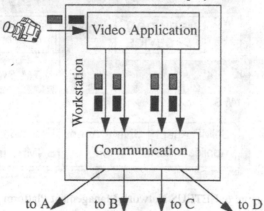

Figure 5: BETEUS Multicast Support

Within BETEUS we therefore have implemented multicast functions in the application as a short term solution. For instance, a video stream is transmitted to several remote sites by duplicating all video packets in the application (Figure 5). Quite obviously, this introduces an unnecessary load on an end system but, on the other hand, it has been easy in its realization.

BETEUS Network Management Platform

The BETEUS platform is managed by the BETEUS management platform. The BETEUS management architecture is compliant with the TMN principles [14]. A *telecommunications management network (TMN)* is a network to provide surveillance and control of another network. Two TMN management layers are considered within BETEUS: the *element management layer (EML)* which manages BETEUS sites on an individual basis and the *network management layer (NML)* which is responsible for

232

the management of all BETEUS sites. The EML level makes use of the *simple network management protocol (*SNMP) [15] while the NML level operates with *common management information protocol* (CMIP) [16] to exploit the capabilities of OSI management, such as event reporting. In each BETEUS site, a CMIP/SNMP gateway (EML level) is used to query the local SNMP agents (ATM switch, ATM adapter cards, host applications) on behalf of the network management centre (NMC) (Figure 6). This gateway is directly managed by the NMC using CMIP.

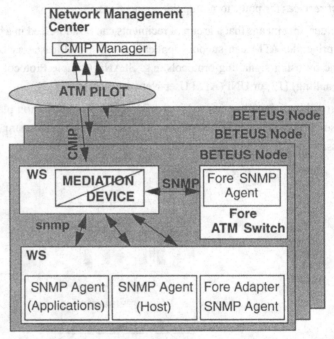

Figure 6: BETEUS Network Management Platform

The BETEUS network management platform focuses particularly on aspects like faults, performance and accounting management. Statistics are collected during multimedia sessions and analysed off line. This analysis enables the evaluation of the *grade of service* given by the European ATM pilot and the local ATM switches, correlated to the QoS measured and estimated by users.

In the following we present a sample of performance statistics [17] that have been collected during a BETEUS session at the ATM adapter card of the workstation dedicated to video applications. It should be noted that applications use JPEG video compression and performs a kind of traffic shaping by maintaining as much as possible a near constant throughput.

During three hours running a video application, we observe a stable behaviour generating AAL5 PDUs. An average of 73 ATM cells per AAL5 PDU with a standard deviation of 16.46 cells/AAL5 PDU were emitted. The average AAL PDU size then equals to 3504 bytes, including AAL5 PDU trailer and its payload data. Application data represent 98.74% of the payload of the 73 ATM cells (only 1.26% is due to protocol headers and tailers).

During three hours, 37 AAL5 PDUs were discarded out of 394'413 received PDUs, which leads to a AAL5 layer loss rate of 93.8E-6. If we assume that the cell loss occurrence is uniformly distributed over the time (pessimistic), then this results in an ATM cell loss rate of 1.28E-6.

3.3 The BETEUS Application Platform

The application platform [2][18] provides all the elements needed for running collaborative work sessions between several end points of the BETEUS virtual community. The application platform includes software for application set up, collaborative work and control.

Architectural Environment

The BETEUS applications exhibit the notion of a *site* (Figure 7). A site is a collection of workstations, media I/O devices and switching devices that are, in terms of control, tightly correlated.

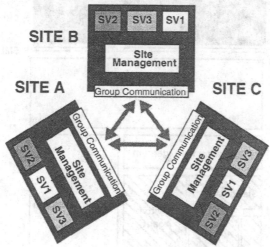

Figure 7: Inter-site Communication

Site control is centralized within the *site management*. The control aspects of an application within BETEUS are caught by the abstraction of a *session*. A session can be regarded as being typed with a certain application. The abstraction for an instantiation of an application within a site is a *session vertex (SV)*. A BETEUS application consists of a set of session vertices that are distributed over various sites (e.g. all *SV1* session vertices in Figure 7). Session vertices communicate with human users through user interfaces, with the local site management through the site management interface (SMI) and with each other by means of some specific functions that the site management interface provides (Figure 8). Connection management within a session is a func-

234

tion that is distributed to site management entities. Site management entities interact with each other by means of a group communication protocol.

Component Description

The site architecture is depicted in Figure 8. Plain blocks stand for entities that have only a single instantiation within a site. Hatched blocks can have many instantiations. The activities within a site are coordinated by the site management. The site management establishes local connection endpoints via the *connection management* on an abstract level. The connection management maps abstract device names to physical addresses and communicates with *station agents* for the establishment of audio, video and application sharing connections, and with *switch agents* for the establishment of *analog* connections within a site (if, for instance, a peripheral audio/video switch is used). A station agent is found in every workstation that can be source or sink of a audio, video or application sharing connection.

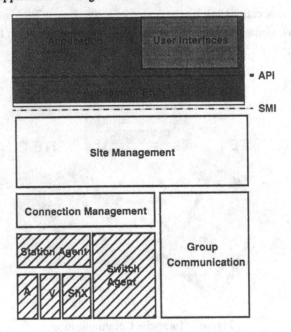

Figure 8: Site Architecture

The BETEUS audio component implements silence detection at the sending site with an adjustable threshold value and is built on top of the realtime transport protocol (RTP) [19], which in turn uses UDP for transmission. The receiving site supports audio mixing. Both sender and receiver generate activity events that can be graphically displayed on the user interface. The two audio encodings that are supported are 8kHz sampling rate with 8bit resolution and 16kHz sampling rate with 16bit resolution.

Video transmission is built around the XVideo board from Parallax. The compression of the Parallax board follows the JPEG standard for the compression of still images

[20]. On connection set up the video sender allows to specify a maximum data rate and a frame rate. The window size is adjustable during the lifetime of a connection. To enforce the maximum data rate, a control loop compares the maximum and the measured data rate and modifies the JPEG compression factor according to the result of the comparison. This allows to have a constant frame rate, which results in excellent video quality, because the human eye is extremely sensitive to frame rate irregularities.

The application sharing component (Xwedge [21]) allows BETEUS session members to share any X11 application running at his node with all other session members. Xwedge is a distributed shared window system that has agents running at all involved client and server sites. A distributed approach has been taken in order to improve performance. Another design goal of Xwedge has been to keep it policy-free, i.e., not to prescribe a default admission and floor control. The system that integrates Xwedge has thus the possibility to employ its own sharing policies.

A BETEUS application is built on top of the site management. It consists of a collection of (identical) processes that are distributed over various sites.

The site management exports a message based control interface (SMI) to the session vertices that use its services. To relieve the application programmer from message exchange semantics, a skeleton for the construction of applications is provided. This is the *application stub* in Figure 8. The actual application programming interface (API) is provided by the functions of the application stub rather than the site management itself. Asynchronous events of which the site management wants to notify a session vertex are treated by callback functions for which the application stub provides a framework.

4 Utilization and Future Work

The previously described infrastructure has been used between the different sites in order to support tele-seminar, teleteaching as well as multimedia document storage and retrieval. A major event has been the usage of the BETEUS platform during the first International Distributed Conference on High Performance Networking for Teleteaching November 1995, in order to distribute a panel session.

Tele-seminar is best understood as a "virtual face-to-face meeting". Every participant has its own workstation. Every workstation acts as a communication unit that transmits, receives and processes multiple video, audio and data streams. A tele-seminar does not have a fixed structure but participants may join and leave a tele-seminar freely so the size (in terms of participants) of a tele-seminar varies dynamically.

As soon as a first prototype of the BETEUS platform was available we used this prototype for our own purposes, mainly for organizing weekly meeting. Every partner site operated its own workstations that were connected by the BETEUS communication platform. Video and audio connections and shared workspace connections were established. We were able to perform technical discussion and to prepare presentations, which includes the editing of slides. From our experiences we can conclude that the functionality of the BETEUS platform is sufficient in order to support collaborative work sessions.

As for teleteaching, a few courses have been broadcast from EPFL to the other sites. Each of the five sites involved received a video stream from all of the other sites. The audio produced by a given site was broadcast to all other sites; on each site, the four incoming audio streams were mixed in order to let everybody be able to discuss with everybody at any time.

The courses were given by Professor Rossi from EPFL and covered the area of acoustics in multimedia communication; a few acoustic (music, voice) samples were provided to illustrate the concepts introduced during the course. The following conclusions have been drawn from this experiment:

- The application and communication platforms of BETEUS does allow to support this kind of course. With appropriate precautions, it is indeed possible to send complex audio samples to remote sites.
- Conventional transparencies are acceptable and can be read by people at the remote sites, if they are properly designed (larger fonts than usual, simple graphics); electronic slides shared by means of a distributed application are obviously better, but a pointing device is mandatory.
- From the point of view of the lecturer, it is highly desirable to be able to monitor the cameras (rotation and zooming functions) located in the other sites. This allows him or her to glance in the direction of the pair of remote participants whose concentration is reducing and whose (noisy) chats may disturb the other attendees of that class.

The third function, multimedia document storage and retrieval, was also tested in the framework of the project. At CERN several audio/video clips were stored on a server equipped with Uniflix from Paradise software. JPEG is used with this solution. Both hardware and software compression techniques were experimented, demonstrating a clear (and expected) superiority of the former. An application developed by the consortium allows to access a Web page displaying available video clips. When selecting a given clip, the full potential of ATM is exploited: instead of transferring the whole file from the server to the requesting end system, as it is usually done on the Web, the video is immediately read on the server, transmitted and played at the same pace as recorded on the end system. As an example, a session was played with 20 frames per second and SuperCIF format. It required a bit rate of 5 Mbit/s.

Based on the experience gained within the BETEUS project, EPFL and ETHZ have decided to start a follow up project called TELEPOLY which aims at providing an operational teleteaching system between the two Institutes.

Three phases are foreseen within this project. The first phase (early 1996) consists in weekly seminars between both sites, on the topic of networking and multimedia systems. During the second phase (Spring 1996), a regular course of the 6th semester on Computer Networks (in English) will be transmitted from Lausanne to Zurich. Finally, in phase 3 (Fall 1996), a first year course of mathematics or physics will be transmitted from Zurich to Lausanne. The intention here is to have the course given in German, in order to favour the integration at EPFL of students whose mother language is German. In this way, teleteaching can foster mobility within a multilingual country such as Switzerland.

5 Conclusions

In this paper we have discussed the realization of the BETEUS communication platform which is used in the BETEUS project to support distributed multimedia applications like distributed classroom, tele-seminar and multimedia document archival and retrieval. We have started out from a discussion of ATM network services offered by the public network operators. It has turned out that some essential networking features which would have been advantageous for the realization of the BETEUS communication platform have not been available. Essential features would have been the signalling over the European ATM pilot and with this the possibility of multicasting in the ATM network. Therefore, we have decided to build up an alternative. This has mainly affected the multicast functionality of the BETEUS application platform. Providing multicast services by the BETEUS communication platform rather than by the BETEUS application is still an unresolved issue. Although a new ATM switch software release has been provided by FORE, no optimal solution is available. Nonetheless, we have been able to implement a communication platform which is stable, flexible and scalable. Furthermore, the important application requirements (as discussed in Section 3) are fulfilled. Five sites (i.e. CERN, EPFL, ETHZ, EURECOM and TUB) have been connected. All BETEUS scenarios are supported and tests have been conducted to interconnect to another European project IBER, which has similar goals, but explores a different technical approach.

We expect that a fully operational ATM infrastructure providing all ATM services, particularly signalling and multicast, will improve utilization of resources in the BETEUS community. For instance, we expect that the error prone manual installation of VP and VC connections will be improved by ATM signalling functions.

To go further in teleteaching over broadband networks, EPFL and ETHZ have launched the TELEPOLY project, which should be operational from beginning on January 96.

Acknowledgement: We are grateful to all members of the BETEUS consortium for participating in the field trials: C. Blum, P. Dubois, B. Dufresne, M. Goud, C. Isnard, J. Kawalek, X. Logean, R. Molva, O. Schaller and K. Traore.

6 References

[1] BETEUS, "Broadband Exchange For Trans-European Usage", Technical Annex, Project Number: M1010, 1994.

[2] BETEUS Consortium, "BETEUS Application Platform Detailed Specification", Deliverable D6, November 1994.

[3] BETEUS Consortium, "BETEUS Communication Platform Specification", Deliverable D5, October 1994.

[4] Stevens, W. Richard, "UNIX Network Programming", Prentice Hall, 1990.

[5] FORE Systems Inc., "FORE Runner ASX 200 ATM Switch User's Manual", Software Version 3.2.x, May 1995.

[6] Heinanen, J., "Multiprotocol Encapsulation over ATM Adaptation Layer 5", RFC 1483, July 1993.

[7] Laubach, M., "Classical IP and ARP over ATM", RFC 1577, January 1994.

[8] Deering, S., "Multicast Routing in a Datagram Internetwork", PhD Thesis, Stanford, 1991.

[9] Pusztaszeri, Y.-H., Biersack, E., Dubois, P., Gaspoz, J.-P., Goud, M., Gros, P., Hubaux, J.-P., "Multimedia Teletutoring over a Trans-European ATM Network", 2nd IWACA Conference, Heidelberg, September 1994.

[10] Walter, T., Brunner, M., Loisel, D., Znaty, S., Dufresne, B., "The BETEUS Communication Platform", Proceedings of the first International Distributed Conference IDC'95, Madeira, November 1995.

[11] FORE Systems Inc., "FORE Runner SBA100/200 ATM SBus Adapter User's Manual", FORE Systems, 1994.

[12] FORE Systems Inc., "SPANS: Simple Protocol for ATM Network Signaling, Release 2.3", FORE Systems, 1994.

[13] ATM Forum, "ATM User-Network Interface Specification Version 3.1", ATM Forum Specification, September 1994.

[14] ITU-T Rec. M.3010, "Principles for a telecommunications management network", ITU 1993.

[15] Schoffstall, M., Fedor, M., Davin, J., Case, J., "A Simple Network Management Protocol (SNMP)", RFC1161, October 1990.

[16] Stallings, W., "SNMP, SNMPv2 and CMIP", Addison-Wesley, 1993.

[17] Besson, M., Traore, K., Dubois, P., "Control and Performance Monitoring of a Multimedia Platform over the ATM Pilot", Proceedings of the first International Distributed Conference IDC'95, Madeira, November 1995.

[18] Blum, C., Dubois, P., Molva, R., Schaller, O., "A Semi-Distributed Platform for the Support of CSCW Applications", Proceedings of the first International Distributed Conference, Madeira, November 1995.

[19] IETF Internet Draft: "RTP: A Transport Protocol for Real-Time Applications", Audio-Video Transport WG, ftp://ds.internic.net/internet-drafts/draft-ietf-avt-rtp-07.txt, March 1995.

[20] Wallace, G.K., "The JPEG Still Picture Compression Standard", Communications of the ACM, April 1991.

[21] Gutekunst, T., Bauer, D., Caronni, G., Hasan and Plattner, B., "A Distributed and Policy-Free General-Purpose Shared Window System", IEEE/ACM Transactions on Networking, February 1995.

A Universal Scaling Principle for ATM Based Connectionless Servers

Christian M. Winkler

Lehrstuhl für Kommunikationsnetze
Technische Universität München, D-80290 München, Germany

chris@lkn.e-technik.tu-muenchen.de

Abstract. The Direct Connectionless Service in B-ISDN makes use of servers to supply the datagram routing function. An economical service provision requires servers which are extendable. Therefore, a generic and universal scaling principle, that can be applied to the existing, non scalable servers is essential. This paper first determines the relevant properties of such a scaling principle. Following, a new modular approach is presented. The key component of the architecture is a Packet Distribution Unit, that forwards the incoming packets from the network to the appropriate server modules.

1. Introduction

One of the first services to be supported by the Broadband Integrated Services Digital Network (B-ISDN[1]), that is currently developed by the public carriers to provide wide area integrated services, will be the transport of connectionless computer data traffic. B-ISDN is based on the ATM transport technology and is connection oriented. However, two methods for a connectionless service provision in B-ISDN have been identified by ITU: The Indirect and the Direct Connectionless Data Service [3].

The Indirect Service simply uses B-ISDN as a connection oriented transmission system. Each pair of communicating end-systems is linked by a unique connection. There are several proposals for an efficient Indirect Service [6]-[10]. They all have specific disadvantages like high setup delay, low bandwidth utilization or the need for complex add-on functions to the ATM switches, etc. But in certain cases, for example to interconnect high volume data sources, e.g. MANs or LANs, they can be a reasonable alternative to the Direct Connectionless Service.

The Direct Connectionless Service offers a connectionless routing functionality embedded in the ATM network. Connectionless servers within the public network receive the datagrams from the sources and route them to their destination based on addressing information contained within the message headers. The connectionless servers CLS (in ITU terms: CLSF, connectionless service function) are linked to each other by ATM-VPCs. The logical network comprised of servers as nodes and ATM connections as edges is called connectionless overlay network. An economical direct service provision calls for a scalable and flexible server architecture. But the customary servers are not or only inadequately scalable. The solution is a universal

---

[1]Refer to the appendix for a list of abbreviations

scaling method that can be applied to all ATM based servers regardless of their individual structure.

The remainder of this paper presents a new universal scaling method for connectionless servers. Section 2 briefly recapitulates the CL-Service and QoS terminology. Section 3 defines the requirements that have to be met by a universal scaling principle and the resulting architectural constraints. Section 4 describes the new scaling principle.

2. Direct Connectionless Service

The ITU recommendations I.364, F.812 and I.211 describe the "Support of Broadband Connectionless Data Service on B-ISDN" [1]-[3]. They introduce the CLNAP (Connectionless Network Access Protocol) datagram layer protocol and outline some supplementary services and QoS aspects. The service definition is very similar to the Switched Multi-Megabit Data Service (SMDS) defined by BELLCORE [4][5], which complements the ITU recommendations.

CLNAP offers a service similar to the MAC sub-layer service. The addresses are based on the geographical international ISDN numbering plan E.164, which includes publicly administered individual and group addresses. The overlay network ensures, that the sender of a data unit cannot pretend a fraudulent source address (Source Address Validation). It includes network capabilities for charging and interworking to other data services.

In addition to the connectionless bearer service, CLNAP/SMDS defines some supplementary services: The Source/Destination Address Screening allows restrictions to be enforced on the delivery of data units from particular sources and the transmission of data units to particular destinations. This mechanism enables the construction of virtual private networks. The Egress Rate Control is a traffic smoothing function provided by the network, that can assist terminals in handling traffic bursts.

Besides the service aspects, there are QoS parameters concerning the maximum end-to-end delay, lost, errored, duplicated and misdelivered packet ratio and the allowable packet lifetime. An important QoS requirement with respect to the scaling issue is the observance of the datagram sequence by the overlay network.

3. Requirements for Scaling of Connectionless Servers

This section describes the requirements to be met by an universal scaling principle for ATM based connectionless servers. The resulting design constraints are derived and illustrated.

3.1. Properties of a Universal Scaling Principle

The objectives of a universal scaling principle are [12]:

- It is applicable to customary connectionless servers.
- It has a modular design.
- It offers at least the same services and QoS as customary servers.

□ It allows adequate scalability.

□ It is independent of the AAL.

□ It fits seamlessly into the ATM environment.

□ It is transparent to the overlay network.

3.2. Module Capacity and Scalability

The requirement of adequate scalability can be met only if the limits of the individual modules do not impose an upper bound on the server size. For example, in Fig. 1, each server module can send packets to all subscribers and therefore has to store and manage the complete subscriber data. Here, the limits of the individual modules determine the maximum server size. Consequently, a module should have to manage only a sub-set of the subscribers connected to the server.

3.3. The Supplementary Service "Egress Rate Control"

The Egress Rate Control is a traffic smoothing function provided by the network, that can assist terminals in handling traffic bursts. If several modules can send packets to the same terminal simultaneously (see Fig. 1), a coordination of their sending rates is required to maintain the subscribed egress rate. This enormously complex control on burst level would generate an uneconomical volume of coordination traffic. So the individual subscriber's egress rate should be controlled by exactly one module, which is therefore the only packet source.

3.4. QoS- Characteristic "Datagram Sequence"

A modular approach implies some kind of load sharing. A load sharing method that randomly distributes packets among the different modules will not maintain the datagram sequence. So either a resequencing unit will have to rearrange the datagrams, or the load sharing mechanism by itself will have to preserve the sequence. Besides the fact, that the current CLNAP-PDU definition does not include sequence information, the known crucial disadvantages of resequencing make it an unfavorable solution. Hence, a suitable load balancing must be based on the end-to-end communication relation as the unit of sharing.

3.5. Bandwidth Control on the Network VPCs

As stated in the introduction, the servers are linked to each other by VPCs. For two reasons, these VPCs and the VCs inside should be shared between the modules: First to increase traffic concentration and thus resource utilization. Second, to avoid the otherwise necessary reconfiguration of the overlay network when adding new modules. Hence, a distributed shaping mechanism is required, that coordinates the sending rates of the multiple modules. This task is entirely different from the egress rate control, because the shaping operates only on the aggregate traffic of many subscribers and furthermore there are just a few VPCs to be controlled.

3.6. Related Work

Our objective is the development of a generic and universal approach to scalability, that is applicable to customary connectionless servers. Research is presently more concerned with the technological problems of server realization [13]-[16]. The question of scalability, in particular of generic scaling methods, is still unanswered.

To illustrate the above-mentioned design constraints, the next paragraph comments on a rudimentary scaling approach, which was presented in conjunction with a specific server architecture [13]. The proposed connectionless server comprises of identical server modules, which all have connections toward each subscriber (Fig. 1). Each module has to store and manage the routing, security, etc. information of all subscribers. Consequently, the maximum server size is limited by the amount of memory and the processing power of the individual modules. The decentralized datagram forwarding prevents the implementation of an egress rate control.

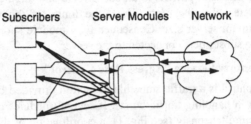

Fig. 1. Scaling approach [13]

Each module uses its own separate connections to the other servers in the network. This has an unfavorable effect on the bandwidth utilization. Moreover, the internal structure of the server is not hidden from the overlay network. So the addition of a new module requires new connections to all subscribers and to the neighboring servers. And the load splitting between the modules cannot be adjusted locally but is determined by the neighboring servers.

4. Universal Scaling Principle

4.1. Overview

The universal scaling principle (Fig. 2) uses customary connectionless servers as components of a modular approach.

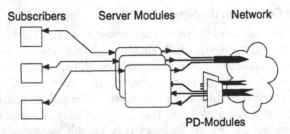

PD-Modules

Fig. 2. Universal scaling approach

To comply with the datagram sequence QoS objective, all datagrams of an end-to-end relation pass through the server on the same path. The data concerning a specific subscriber is centralized on a single CLS-module. This module is responsible for sending and receiving packets to/from the subscriber and for performing the security and rate control functions. To achieve balanced utilization, the set of subscribers a module is responsible for has to be chosen appropriately.

The connections from the network carry packets addressed to any subscriber in the domain of the CLS. Because the internal architecture of the server, including the number of installed modules, should not be visible from the outside, a packet distribution function is provided. The design supports multiple distribution modules (PD-Modules). Every subscriber is connected (bidirectionally) to exactly one CLS-Module. The incoming network connections terminate at the distribution modules. The outgoing network connections originate at the CLS-Modules and are concentrated to aggregate traffic flows. The scaling principle is independent of the AAL. However, the PD-Module of course has to be designed according to the the type of AAL used in the CLS modules.

4.2. Realization in the ATM Environment

The seamless integration of the scaling concept into the ATM environment calls for ATM based module interconnection. Therefore all modules, the CLS-Modules which offer an ATM interface anyway, and the PD-Modules, are connected to an ATM switching network. The inter-module traffic, as well as the necessary control and coordination traffic, is transported through the ATM switching network.

The traffic from a subscriber is screened and checked by the corresponding CLS-Module and forwarded to the network (Fig. 3, example: bold lines). In case the destination is located in the same domain, the packets are either sent directly to the destination terminal or via a PD-Module to the responsible CLS-Module.

Packets from the network are forwarded through a PD-Module to the particular CLS-Modules (Fig. 3, example: normal lines).

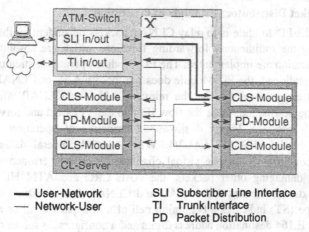

— User-Network   SLI   Subscriber Line Interface
— Network-User   TI   Trunk Interface
                             PD   Packet Distribution

Fig. 3. Block diagram

4.3. Assignment of Subscribers to CLS-Modules

Subscribers are identified by their E.164 numbers, which relate to geographical areas. In general, dependent on its location in the overlay network, a server has to manage a few uncontiguous sub-sets of E.164 numbers. The PD-Module forwards packets from the incoming network connections to the CLS-Module, that is responsible for the destination subscriber. A free allocation of individual subscribers (i.e. E.164 numbers) to specific CLS-Modules allows an ideal load balancing between the modules. For performance reasons, the forwarding table lookup in the PD-Module is done hardware based e.g. using CA-Memory [11]. For large servers with powerful CLS-Modules the granularity of the address resolution can be decreased by configurable selection of the E.164 digits to be taken into account. This measure limits the size of the forwarding tables.

4.4. Packet Stream Identification

The packet streams from the PD-Modules terminating at one CLS-Module have to be distinguishable. For performance reasons a PD-Module does not alter AAL or CLNAP information fields. Hence, the packet streams have to be distinguished by the ATM connection identifier. To ease and speed up the switch routing table operation, a solely VPI-based lookup is preferred.

The following identification scheme allows a maximum number of concurrent packet streams. The PD-Modules map unique (with respect to their interface) packet stream IDs into the VPI fields of the cells and leave the VCI unchanged. In the switch lookup table, the physical routing information is determined based on these packet stream IDs (VPIs) and the VPI is changed to a Module-ID unique with respect to the destination CLS-Module. Although this is just the normal lookup table operation, the table entries in this case have to be preset by switch management commands instead of signaling procedures.

4.5. The Packet Distributor: Principle of Operation

The task of the PD-Module is to relay CLNAP packets dependent on their destination address. Only the rudimentary forwarding functions which are absolutely necessary for a safe operation are implemented. The PD-Module operates in streaming mode.

As already mentioned, the PD-Module does not modify AAL or CLNAP fields in the cell body. It extracts, on the fly, the relevant AAL and CLNAP information bits, determines the destination, maps the flow ID into the VPI field and forwards the cells to the switch interface. Figure 4 shows the principle of operation of the Packet Distribution Module PD for AAL3/4. With AAL3/4, several datagrams can be transmitted concurrently on one virtual channel. To prevent erroneous cells of one packet from damaging other packets, the AAL CRC and ATM HEC have to be checked. Next, the relevant ATM, AAL and CLNAP data is extracted. If the AAL Segment Type (ST) indicates a beginning cell of a new packet, all or a configurable subset of the E.164 destination address digits and a configurable subset of the VPI bits are used as address into the module selection table (CAM) to determine the flow ID. In addition, the destination information is cached in the active connection table. In the case of continuation cells, the output information is recalled from the active connection table addressed by the AAL MID and the ATM VPI/VCI. Cells from packets directed to a congested output port and erroneous cells are discarded.

The ATM data interface of the PD-Module is designed according to the UTOPIA specification (ATM-FORUM). The control interface is operated via an ATM connection through the switch. It allows e.g. programming the selection table or in case of congested CLS-Modules, setting up a selective packet based cell discard.

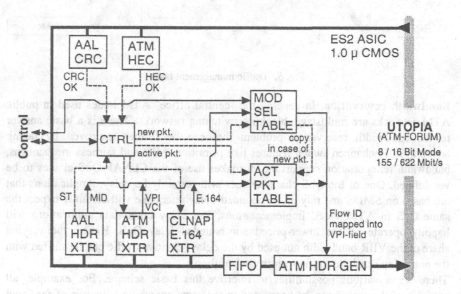

Fig. 4. The packet distributor for AAL 3/4

Our VHDL test implementation indicates that the PD-Module can be realized for 622 Mbit/s using an 1.0µ CMOS ES2 process. Current innovations on the ATM chip market include generic ATM/AAL processors that integrate an additional RISC kernel (e.g. LSI L64360). Hence, future realizations of the PD-Module can be based on standard VLSI.

The AAL5 based datagram distributor operates similarly but it has a simpler design. AAL5 does not support packet level multiplex. Because an erroneous last cell of a packet inevitably destroys the following packet on the same VC, it is useless to check the AAL5 CRC for that purpose. Besides, the number of concurrent packets is reduced by a ratio of 1024 (MID) in comparison to AAL3/4. Hence the lookup tables can be smaller.

4.6. Traffic Management Problems

There are two principal traffic management problems linked with the modular approach (Fig. 5). First, on both the sending and the receiving side, the bandwidth to be reserved in the ATM switch for the individual traffic flows has to be determined. Second, on the sending side, the aggregate rate of the multiple CLS-Modules concurrently sending datagrams on the same VPC has to be controled.

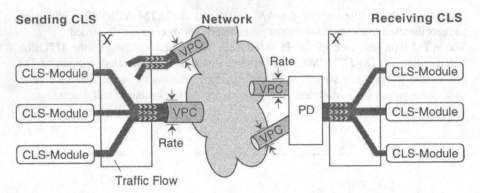

Fig. 5. Traffic management tasks

Bandwidth reservation. In general, the "central office" ATM nodes used in public ATM networks are multistage, blocking switching networks. There is a basic answer to the bandwidth reservation problem that can be further improved by taking advantage of enhanced switch features like priorities, ABR and fairness mechanisms, bandwidth renegotiation support etc. It makes use of the CLNAP QoS classes to be yet defined. One of the QoS characteristics will certainly be delay. Applications that are based on SMDS and rely on its bounded-delay guarantee will certainly expect the same QoS in ATM based implementations. But many other data applications will happily operate on the lower-priced non-bounded-delay QoS. Hence, the varying share of the VBR bandwidth not used by the delay sensitive traffic can be stuffed with the non delay sensitive traffic (priority scheduling).

There are a various possibilities to improve this basic scheme. For example, all modules of the server can be connected to the same switching networks of the input and the output stages. The intra-switch flows are routed such that they travel the same inner stages and the same links and are not splitted before the last stage. The rate of these flows is definite and known. Only in the last stage switching network, managing exclusively server modules, additional bandwidth has to be reserved to cope with the statistical traffic variations. If, further, the switching networks themselves are non-blocking, the full capacity can be used. The prerequisite for such improvements is, that the switch provides the necessary enhanced features and management operations. For example, the predetermination of the intra-switch routing is not supported by the regular user signaling.

Rate control. Customary servers, in our case each CLS-Module, shape their outgoing traffic to a characteristic, that lends itself to statistical multiplexing (VBR, CBR). To improve the medium term bandwidth utilization of the outgoing VPCs, that carry the aggregate traffic of all CLS-Modules, the allocation of the VPC bandwidth to the individual modules is constantly recalculated based on traffic measurements, which are taken anyway for tariffing and management purposes (cp. [7][8]). The shaping units in the server modules are adjusted accordingly. Again, the QoS-delay based priority scheduling assists the individual modules in maximizing the utilization of their traffic flows.

4.7. The Process of Upgrading

The process of upgrading should be transparent to the overlay network. The time of degraded QoS must be kept as short as possible. But the different QoS characteristics delay, datagram sequence and loss are contradictory with respect to the reconfiguration. For example, buffering and flush mechanisms are necessary to maintain the datagram sequence, even if they inadmissibly increase the delay. So compromises have to be made according to the order of precedence of the QoS objectives. The QoS issues are still for further study [1]. Future QoS definitions (per packet, per end-to-end flow, per subscriber) might allow improved reconfiguration methods.

The procedure for adding of a new CLS-Module is shown in Fig. 6. This example keeps the sequence but temporarily increases the delay.

Determine an appropriate allocation of subscribers to CLS-Modules (load balancing).
Select a module **M** (e.g. the new module) that shall henceforth handle other/new sets of subscribers and has enough free capacity to instantly take over a sub-set *N* of them.
P is the module that controls *N* at the moment.
Load the subscriber data for *N* to **M**.
Reprogram the Module Selection Tables of all PD units such that **M** receives the packets from the network addressed to *N*.
Reprogram the VPI/VCI tables of the SLI units such that **M** receives the packets from the subscribers in *N*.
Let **M** collect and hold the packets until **P** signals that the last packet to/of *N* has been sent out (separately for send and receive direction).
Let **M** take over and start sending the buffered packets.
Delete the entries for *N* in **P**.
Until the new allocation is fully operational

Fig. 6. Example reconfiguration procedure

The same method is applied to adapt the load distribution between the CLS-Modules. With similar procedures, PD-Modules can be added.

4.8. Overall Operation

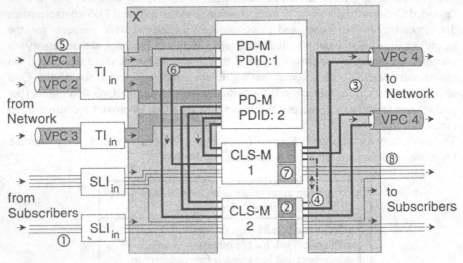

Fig. 7. The overall operation

The configuration shown in Fig. 7 illustrates the operation of the complete system. Bold lines represent aggregate packet streams.

Cells from the subscribers ① are switched to the appropriate CLS-Module. The CLS-Module processes the packets, carries out the security and tariffing operations and sends out the cells through the shaping unit ② to the outgoing VPC ③. The CLS-Modules communicate via a control connection ④ to calculate the fair share of the allocated VPC bandwidth and adjust the shaping units accordingly. The VPCs from the network ⑤, are routed through the switch to a PD-Module. The PD-Module forwards the individual packets on a cell by cell base to the appropriate CLS-Module. The streams ⑥ are identified by their Module ID which is unique with respect to the terminating CLS-Module. All cells from a particular incoming VPC, which are forwarded to the same CLS-Modul, carry the same Module-ID in the VPI field and their original VCI and MID. The CLS-Modules process the packets and route them through the egress rate control unit ⑦ to the destination subscriber ⑧.

5. Conclusion

An economical connectionless service provision calls for a modular server concept, that allows the adaptation of the servers to growing demands and offers a construction set for servers of different sizes. But research in this area is dominated by the technological problems of fast datagram routing. There is still too little attention paid to the question of scalability, which will become decisive as soon as the public connectionless service starts operation.

In this paper, a new universal scaling approach has been presented, with is applicable to customary connectionless servers. It has a modular design using only two types of modules: Server modules, which are customary servers, and simple packet distribution modules. The scaling principle fits seamlessly into the ATM environment and is independent of the AAL used. The internal structure of a server, e.g. the number of installed modules, is hidden from the overlay network. This allows a server extension transparent to the network.

The key element of the architecture is the packet distribution module that routes incoming packets from the network to the appropriate server modules based on the E.164 destination address. The server modules and the distribution modules are interconnected via an ATM switch. Several measures are taken to reduce switching overhead and maximize the utilization of the overlay network.

The proposed new universal scaling principle helps to provide an economical public connectionless service.

6. Appendix

Abbreviations

AAL	ATM Adaptation Layer
ABR	Available Bitrate
ATM	Asynchronous Transfer Mode
B-ISDN	Broadband Integrated Services Digital Network
CAM	Content Addressable Memory
CLNAP	Connectionless Network Access Protocol
CLS	Connectionless Server
CLSF	Connectionless Function
CRC	Cyclic Redundancy Check
HEC	Header Error Control
ITU	International Telecommunication Union
LAN	Local Area Network
MAC	Media Access Control
MAN	Metropolitan Area Network
PD	Packet Distributor
PDU	Protocol Data Unit
QoS	Quality of Service
RAM	Random Access Memory
SLI	Subscriber Line Interface
SMDS	Switched Multi-Megabit Data Service
TI	Trunk Interface
UNI	User Network Interface
VBR	Variable Bitrate
VC(I)(C)	Virtual Channel (Identifier)(Connection)
VP(I)(C)	Virtual Path (Identifier)(Connection)

References

[1] ITU-T Study Group XVIII: Support of Broadband Connectionless Data Service on B-ISDN, Recommendation I.364, Helsinki, March 1993

[2] ITU-T Study Group I: Broadband Connectionless Data Bearer Service, Recommendation F.812, Geneva, August 1992

[3] ITU-T Study Group XVIII: B-ISDN Service Aspects, Recommendation I.211, Helsinki, March 1993

[4] Bellcore: Generic System Requirements in Support of Switched Multi-Megabit Data Service, Technical Reference, TR-TSV-000772, Issue 1, May 1991

[5] Bellcore: Switched Multi-Megabit Data Service Generic Requirements for Exchange Access and Intercompany Serving Arrangements, Technical Reference, TR-TSV-001060, Issue 1, December 1991

[6] ATM Forum SWG Drafting Group: LAN Emulation Over ATM: Draft Specification, LAN Emulation, ATM Forum 94-0035R2, Revision 2, 1994

[7] ATM Forum TMgmt Working Group: Traffic Management Specification Version 4.0, ABR Service, ATM Forum 95-0013-R3, 1995

[8] Pierre E. Boyer et al.: A reservation principle with applications to the ATM traffic control, Computer Networks and ISDN Systems 24, 1992, pp. 321-334

[9] Jean-Yves Le Boudec et al.: Connectionless data service in an ATM based customer premises network, Computer Networks and ISDN Systems 26, 1994, pp. 1409-1424

[10] M. Gerla et al.: Internetting LAN's and MAN's to B-ISDN for Connectionless Traffic Support, J. Sel. Areas of Comm., Vol. 11, No. 8, Oct. 1993, pp. 1145-1159

[11] L. Chisvin et al.: Content-addressable and associative memory: alternatives to the ubiquitous RAM, IEEE Comp. Mag., July 1989, pp.51-64

[12] Christian M. Winkler: Connectionless Service in public ATM networks: scaling issues, Papers on ATM, Networks and LANs, EFOC&N'95, June 1995, pp. 106-109

[13] Daniel S. Omundsen et al.: A Pipelined, Multiprocessor Architecture for a Connectionless Server for Broadband ISDN, IEEE/ACM Transactions on Networking, Vol. 2, No. 2, April 1994, pp. 181-192

[14] Brett J. Vickers et al.: Connectionless Service for Public ATM Networks, IEEE Communications Magazine, August 1994, pp.34-42

[15] Masafumi Katoh et al.: A Network Architecture for ATM-Based Connectionless Data Services, IEICE Trans. Commun., Vol. E.76-B, No. 3, March 1993, pp. 237-248

[16] SIEMENS: EWSXpress Connectionless Server, Product Description, PROD REL1-1E, A2207-X9111-E-1-18

The UMTS Mobility Server: a Solution to Support Third Generation Mobility in ATM

Johan De Vriendt*, Leo Vercauteren*,
Konstantinos Georgokitsos**, Abdelkrime Saïdi*

*Alcatel CRC Antwerp, F. Wellisplein 1, B-2018 Antwerpen, Belgium
tel/(fax): + 32 2 240 7521 (9932), Email: vriendtj@btmaa.bel.alcatel.be
**Alcatel CRC Stuttgart, Holderäckerstr. 35, D-70499 Stuttgart, Germany

Keywords: UMTS, Mobility, B-ISDN, Integration

Abstract

For the third generation mobile systems (UMTS, FPLMTS) an integration with B-ISDN is foreseen. This integration is not limited to service integration (common services for mobile and fixed networks), but also aims at reusing B-ISDN switching and transport infrastructure. In this paper the UMTS Mobility Server is presented as a solution to support third generation mobility in ATM. The Mobility Server is a dedicated network element that handles the specific mobility functions. As such, it has a very limited impact on the existing B-ISDN switches. The mobility functions that are performed by the Mobility Server are: transport interworking/integration, bridging, macrodiversity and combining, and seamless handover.

The Mobility Server can be located at CSS and/or LE level. All mobility functions can be implemented at both levels. Some of these are radio interface dependent. The choice of the allocation of the Mobility Server depends on the specific environment (e.g. urban city, rural low density, highway) and on the required network functionalities (e.g. handover, macrodiversity). The Mobility Server can be used for the GSM, DECT or new radio interfaces such as ATDMA and CDMA.

The Mobility Server does not only limit the impact on the standard broadband switches or B-ISDN transport infrastructure, but also provides a modular introduction of mobile communications in a fixed B-ISDN network and allows for a gradual network evolution. Furthermore, the infrastructure can be extended in a flexible way as the mobile communication segment increases.

1. Introduction

Mobile communications is today one of the largest growing telecommunication segments. As a result of this market pull, there is a rapid evolution of the second generation mobile systems such as GSM (Global System for Mobile communications) and DECT (Digital European Cordless Telecommunications) towards DCS-1800 (Digital Cellular System at 1800 kHz), CTM (Cordless Terminal Mobility) and PCS (Personal Communication System). In addition, progress in radio techniques ([9,10]), the support of new services on the air and the trend to integrate fixed, mobile and satellite networks ([8]) result in the need for third generation mobile service concepts (e.g. UMTS: Universal Mobile Telecommunication System, [7]).

At the same time, fixed telecommunication networks are moving towards broadband, characterised by a high speed transport network infrastructure based on the ATM transfer mode ([2]). Moreover, a flexible control infrastructure based on Intelligent Network (IN) concepts is developed for service definition and creation, and to a certain extent also network control. A major aim of third generation mobile systems will therefore be the use of B-ISDN transport functions and IN concepts for the service and control aspects. The purpose of this integration is cost saving by means of maximum and efficient reuse of resources. Furthermore, future B-ISDN services would benefit from a transparent service provision over a mobile access network.

Third generation mobile systems will evolve from second generation systems, by a gradual introduction of third generation mobile features (cfr. [4]). Therefore, UMTS will be a federation of mobile systems. This paper concentrates on one of the third generation mobile features, namely the use of ATM in the fixed part of the mobile access system. The requirements relevant for this are based on the two overall directives user demand (maximum QoS, services) and minimal system cost (installation, operation, evolution, maintenance) ([6]):

- integration of UMTS into B-ISDN (common services)
- reuse of fixed B-ISDN switching and transport infrastructure (maximum sharing of functions and protocols)
- minimise impact on existing broadband switches and transport network
- adding mobility in a modular way
- low complexity and optimised allocation of complexity
- minimise the number of 'sites' (nodes)
- light, small and simple BTSs (Base Transceiver Station)
- seamless handover support.

The basic transport functions required for the support of mobility are:
1/ *Bridging* is the function that changes a part of the connection from an old connection element to a new one, between two bridging points, one in the Mobile Terminal (MT), and one in the network.
2/ For some radio interfaces (e.g. CDMA, i.e. Code Division Multiple Access), macro-diversity (MD) will be performed. In uplink macro-diversity a number of copies of a specific piece of information are transmitted through uncorrelated paths between the mobile terminal and a specific point in the network, while in downlink MD, the information is multicasted in a point of the network, and transmitted to the MT over a set of base stations. There are two sub-processes to be considered: *multicasting and combining*. Multicasting is the copying of the incoming information stream and the distribution of the information. Combining is covering the incoming information streams and converting these to a single outgoing stream by applying quality information associated to the incoming streams. A synchronisation of the different streams is required for combining (*MD synchronisation*). Note that the use of macrodiversity in a CDMA system results in an increase of the capacity of up to a factor of 2, whilst maintaining the same quality of service level.

3/ Seamless handover (with and without MD) means that the QoS is not affected by a change of the connection elements. This means no data or call loss, nor QoS affecting delays, and maintaining the data flow sequence. *Buffering* could for instance be required to support seamless handover, handling the difference in path delay between the old and the new path, and the possible break in the communication on the radio interface during handover (MT has to synchronise to another base station). In addition, in-band signalling between the bridging nodes can be required for exchanging the necessary signalling messages (e.g. start and stop buffering).

4/ As the capacity of the radio interface is limited, bandwidth reducing source coding (*transcoding*) is used for some services (e.g. voice). The transcoder will also adapt the transport format used in the fixed network to a radio interface compatible format. The transcoding operation is thus radio access technique dependent.

In [1], it has been shown that on the one hand ATM offers a number of inherent functions supporting mobility (e.g. VP/VC routing), and that on the other hand mobile procedures set additional requirements on the lower layers of the B-ISDN transport network. These requirements are the MD synchronisation, the cell sequence sustaining and the definition of connection elements for the change of active connections.

In section 2, the possible radio access system architectures and the transport functions are discussed. In section 3, the general architecture of the access network including the Mobility Server, which handles the mobility transport functions, is described. Furthermore, the advantages and disadvantages of the Mobility Server are discussed. As an example, in the next section the influence of the radio access technique on the network architecture is discussed for the GSM and the ATDMA (Advanced Time Division Multiple Access) radio interface. In section 5, a brief discussion of the standardisation efforts for the third generation mobile systems is given. Also the relation of this paper with the ongoing standardisation is identified.

2. Radio Access System Architecture and Transport Functions

2.1. Radio Access System Architecture

The UMTS RAS (Radio Access System) architecture is influenced by the services provided in the specific environments (e.g. urban city, rural low density, highway), and the required network functionalities (e.g. handover, macrodiversity) (see [6]).

From the transport perspective, following nodes in the network are foreseen (see figure 1):

1/ The MT (Mobile Terminal), which is the assumed end-point of the communication connections (from the mobile network point of view).

2/ The BTS (Base Transceiver Station), containing an antenna for radio coverage, and responsible for the transport interworking (physical and data link layer interworking) between the fixed part of the network (in which the transport is ATM based) and the radio interface (with a radio access specific transport mechanism, e.g. ATDMA, CDMA or GSM TDMA).

3/ The CSS (Cell Site Switch), is an optional switch in the access network, which has depending on the environment following functions: concentration of traffic between the radio cells and the LE (Local Exchange), and gateway to an environment specific transport architecture (e.g. a ring or bus architecture).

4/ The LE (Local Exchange), is assumed to be a standard LE as used for fixed B-ISDN communications (ATM switch). A possible shared use of the LE for mobile and fixed communications has to be considered.

Independent of the physical implementation, logical link interconnections are defined between these nodes for the routing of the mobile traffic.

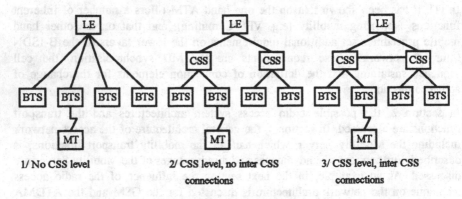

Figure 1: Nodes and logical links in the fixed part of the UMTS access network.

Three basic architectures or a mixture (see figure 1) can be identified: (1) no CSS level, (2) CSS level without inter CSS connections, and (3) CSS level with inter CSS connections. The first architecture minimises the number of 'sites' (nodes) which has a positive impact on the system cost. The advantages of an additional switching level in the access network are: the CSS acts as a traffic concentrator/multiplexer, some mobile specific functions (e.g. multicasting and combining) can be performed closer to the radio interface (e.g. resulting in a smaller signalling load in the network due to the exchange of radio signal quality measurements) and a smaller impact of mobility on the fixed network local exchange (FN LE). Logical inter CSS links allow fast inter CSS handovers, macro-diversity with BTSs belonging to different CSSs, and minimise the impact of handover and macro-diversity on the LE. They however require the definition of re-routing in the fixed part of the mobile access network, for route optimisation after the fast handover.

2.2. Transport Interworking/Integration

In the fixed part of the access network, (at least) two points in the network can be identified where the transport mechanism of the user traffic has to be changed: the BTS and the TRA (Transcoder and Rate Adaptor).

In the ATM context, the TRA represents the point in the network where the information format is adapted between an access network format and the core network format. The main functions of the TRA are source coding and decoding (e.g. in GSM between the 64 kbit/s A-law and the 13 kbit/s in the GSM network for voice), and the adaptation of the transport format. As a result, ATM is used in the core network up to the TRA in a B-ISDN 'standard' way, while between the TRA and the BTS, ATM is used in a radio interface compatible way, taking into account the bearer connection types and the radio interface dependent burst/frame formats. E.g. for GSM, one speech frame (320 bit between BTS and TRA, i.e. on the Abis interface) should be put in one ATM cell (partially filled). Filling the ATM cells completely would introduce extra cell filling delay and an increase of the complexity because both in the TRA and the BTS, the speech frame structure has to be rearranged.

Several options can be identified for the allocation of the transcoding point (TRA) in the network (see figure 2): collocated with either the BTS, CSS or LE (options 1, 3, 5 and 7) ; between these nodes (options 2 and 6) or attached to the CSS or LE (options 4 and 8). The higher the TRA is located in the access network, the more efficient it is because of the sharing (centralising) of the complexity. The options where the TRA is attached to the CSS or LE (i.e. options 4 and 8) are also advantageous because this means an easier sharing of switching and transmission infrastructure between fixed and mobile communication. Furthermore, if the mobility functions are located in the same node as the TRA (i.e. in the Mobility Server, see further), the impact of mobility on the existing switching infrastructure is limited. Therefore, the options 4 and 8 will be further analysed in the remaining of the paper.

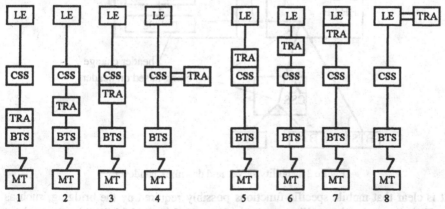

Figure 2: Options for allocating the transcoding point in the network

2.3. Allocation of Bridging, Multicasting and Combining

2.3.1. Bridging

The bridging in the network can be performed on the three switching levels of the network, the BTS, the CSS and the LE level. On the CSS and LE level, the switching functions are required to alter the connection from the old to the new connection element, i.e. to change the route. For the LE, the implementation of the basic bridging functions to support seamless handovers is not trivial, due to the possible buffering, in-band signalling, the temporal use of multicast during handover execution, the data sequence sustaining functions, etc... As a result, the impact on the LE is possibly high.

An alternative approach is to use the switches for what they are made for, switching, and to allocate the additional mobility functions in the transcoding entities as defined in options 4 and 8 of figure 2. The resulting entities can have standard ATM interfaces with the switches, and operate as 'Mobility Servers' in the B-ISDN network, or they can be integrated in the switch architecture. The connection elements in the network are defined between the Mobility Server and the BTSs, and are routed over the ATM switch(es) and the ATM based logical interconnections.

As an example, figure 3 gives a schematic overview of the handover execution in the downlink direction. After the transcoding (i.e. source coding and rate adaptation), a header with a VPI/VCI (Virtual Path/Virtual Channel Identifier) is added to the outgoing ATM cells. This header is used in the switches to address the right BTS. During handover, the bridging is performed by providing a new VP/VC identifier to the ATM cells, resulting in a switching of the cell towards a new BTS. Bridging is thus split into a 'routing' operation in the Mobility Server, and a 'switching' performed in the BB LE or CSS.

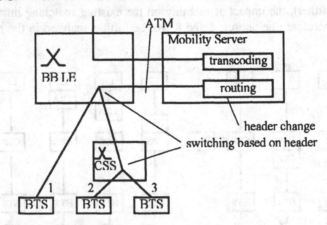

Figure 3: Mobility server and downlink handover.

It is clear that mobile specific functions possibly required by the bridging, such as buffering or in-band signalling can be implemented in the Mobility Server, without affecting the switches, nor the rest of the ATM transmission access network. The same reasoning can be made for the uplink direction. As a result, the radio access technique specific transport functions are allocated in the Mobility Server and the BTS. The other network elements, i.e. the BB-LE, the CSSs and the ATM access network can be shared with fixed network connections.

Although the overall connection in the access network 'consists' of two connection elements during handover, it (normally) only uses one of them for information transfer. However, during the handover process, the necessary bandwidth on both connection elements has to be reserved. A dedicated VP/VC based resource management (e.g. the activation of a connection element has to be performed faster for handover than for call set-up as handover is a time critical process) is required in the fixed part of the mobile access network. The resource management can be integrated with the radio resource management on the radio interface. This is however out of scope of this paper and will not be discussed further.

2.3.2. Multicasting

Similar as for bridging, the multicasting (to support macro-diversity, or used during some handover execution algorithms) can be performed either in the ATM switch or in a dedicated network element, the Mobility Server. For multicasting in the latter option, a copy function is required, while in-band signalling can be used for the synchronisation of the different streams.

2.3.3. Combining

The combining of the different incoming information streams in the network can be performed on layer 1 or layer 2. On layer 2, the combiner selects on block (cell), burst or even bit level, i.e. on the basis of ATM cells or on smaller entities. In the latter case a specific combine function (an interworking function) is required, changing the data contents of the cells and combining them into a new cell. The impact on the switch can be reduced if this combine function is put in the Mobility Server (figure 4).

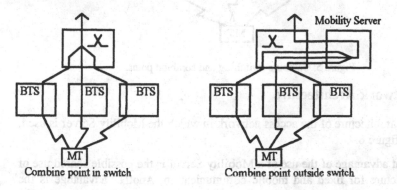

Figure 4: Specific combiner.

If selective combining is done on cell basis, selecting a connection element of the MD-group is based on a radio signal quality measurement in the BTSs. While the combine control decision has to be performed centrally (e.g. at the CSS or LE level), the combine switching can either be distributed over the BTSs or centralised at the CSS or LE level (see figure 5). In the first case, only one BTS transmits its ATM cell at each point in time. Both the combine switching and the combine point can be performed in the switch or outside the switch. Combine switching and combine point in the same Mobility Server outside the switch (see figure 5) has the least impact on the switches, though increases the traffic in the fixed part of the access network.

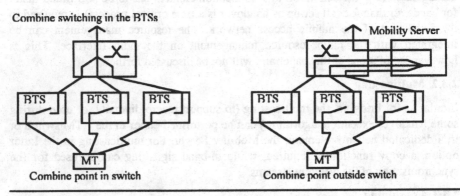

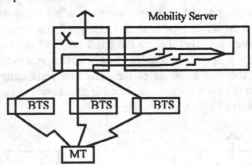

Figure 5: Combine switching and combine point.

3. The Network Architecture

The general architecture of the access network in which the Mobility Server is used, is shown in figure 6.

An important advantage of the use of a Mobility Server is the possible shared use of the infrastructure for fixed and mobile communication. Another advantage is the modular introduction of mobile communications in a 'fixed' B-ISDN network. The infrastructure can be extended in a modular way as the mobile communication segment increases. Also the different radio access techniques can be easily merged by the use of parallel Mobility Servers (each adapted to a certain radio interface) or by the use of 'dual mode' Mobility Servers.

The Mobility Server based network architecture shows also some disadvantages. As the traffic has to pass twice through the LE and/or the CSS, this puts an extra load on the switches. The increase of the load will be higher if there is a Mobility Server on CSS level as well as on the LE level. Because of the centralised solution of the Mobility Server, the signalling load (e.g. radio signal strength reports,...) in the access network will also be higher as these messages now have to be transported to this centralised point in the network. This increase will be even more significant if the Mobility Server is located only on the LE level.

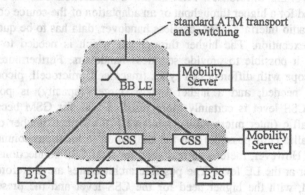

- BB LE: BroadBand Local Exchange
- BTS: Base Transceiver Stations (GSM, DECT, ATDMA, CDMA air interfaces)

Figure 6: General architecture of the UMTS access network.

If the Mobility Server is used for handover (bridging) in the LE, this LE will still be involved after an inter-LE handover. This is similar to the anchor-MSC (Mobile services Switching centre) concept in GSM. Rerouting is possible, though also involves the tranfer of the control point (e.g. for charging). The issue of control point transfer (CPT) is out of the scope of this paper and will not be discussed further.

4. Influence of Radio Access Techniques on Network Architecture

For each radio access technique some specific characteristics (use of macro-diversity, cell types) influence the network architecture. One option in the deployment of the network is to install a CSS level or not, and when a CSS level is chosen, if CSS interconnections are considered. Another option is the allocation of a Mobility Server at CSS level, and if the TRA is part of this Mobility Server. It is important to remark that the radio access technique dependent functions are allocated in the Mobility Server and the BTSs. As examples, the GSM and ATDMA radio interfaces are considered.

4.1. UMTS with GSM or ATDMA Radio Interface

For GSM, there is a low requirement on handover, as the break in the radio interface is in the order of 100-200 ms (Mouly and Pautet [5]). In this case no seamless handover can be performed, as the bitrate on the radio interface is fixed for a given connection. As a consequence, no inter CSS connections for fast handover are needed. The need for a CSS level depends on cost and performance requirements. The possible architectures are combined in figure 7.

The ATDMA radio interface is of the same type (TDMA) as the GSM radio interface, and can even be based on the GSM radio access. However, there are a number of important differences. First of all, a connection can have a variable bandwidth (used for a higher throughput or an adaptation of the source coding to the quality of the radio interface). For seamless handover, data has to be queued during the handover execution. The higher throughput, which is needed to empty the queues, makes it possible to provide seamless handovers. Furthermore, the radio interface can cope with different cell types (macrocell, microcell, picocell), faster handovers are needed, and a higher traffic (larger capacity) is possible. For ATDMA, the CSS level is certainly more justified than for GSM because of the higher local traffic (inter microcellular handover), the higher number of handover initiations (smaller cells possible) and the adaptation to local environment specific characteristics. However, there is no real need for inter CSS connections due to the Mobility Server at the LE level. The possible architectures are therefore similar to those for GSM, with the higher need for the CSS level and the presence of the Mobility Server (bridging) at CSS level.

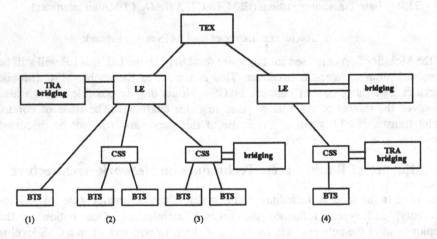

Figure 7: Possible network architectures for the GSM and ATDMA radio interface.

The ultimate choice between the different network architectures will be decided considering the following characteristics and their relative importance:

- The complexity of the network, i.e. the number (volume) of the network elements.

- The complexity of the Radio Resource Management (RRM).
- The complexity of the handover process.
- The required network resources (processing and transmission capacity).
- The choice between a centralised and a decentralised solution.
- The signalling load.
- The synchronisation requirements.
- The performance aspects (QoS).
- The environment (i.e. urban city, rural low density, highway,...).
- The size of the system.

Comparing for example solution 2 and 3 (see figure 7), it can be noticed that solution 2 is a centralised solution resulting in a less complex network (no bridging on CSS level), a complete sharing of the bridging resources (smaller number of Mobility Servers required) and a smaller delay (no delay in the Mobility Server at CSS level). On the other hand, solution 2 can have a larger impact on RRM (possibly more complex due to the centralisation), a possibly more complex handover process (controlled from the LE level, even for intra-CSS handovers). Solution 4 (TRA at CSS level) has the advantage that the synchronisation requirements between TRA and BTS (used e.g. in GSM) can be more easily fulfilled as there are less nodes between the TRA and the BTS. The thorough evaluation of the above mentioned characteristics, needed for an objective choice of the appropriate network architecture, is for further study.

5. Standardisation Efforts

The European Telecommunications Standards Institute (ETSI) as well as the International Telecommunications Union (ITU) define a framework for the future mobile networks so called third generation mobile networks. In ETSI this third generation mobile networks are called UMTS. UMTS is jointly standardised within several STCs of which ETSI Sub Technical Committee (STC) SMG5 has taken a co-ordinating role. In ITU, SG 11 Question 8 defines the network architecture and IN support for FPLMTS (Future Public Land Mobile Telecommunication System) which is also called IMT 2000 (International Mobile Telecommunications). Radio aspects for FPLMTS are treated in ITU R T 8-1 which has together with the Inter Sector Co-ordination Group a co-ordinating role on FPLMTS within the ITU. These are the main fora where the overall standards for UMTS/FPLMTS are defined. However since this next generation of mobile networks is expected to have impact on the broadband network other standardisation groups have been involved.

The CEC sponsored within the RACE II program several projects of which RACE 2066 is called MONET. MONET carried out research on UMTS functionality, - protocols and the required architectures. Currently the MONET project contributes to STC NA5 via partner companies concentrating on the impact of mobile networks on the ATM transport and adaptation layer level.

For NA5 a work item entitled "Requirements of Mobile Networks on B-ISDN" was created, which will result in an ETSI Technical Recommendation (ETR). This document will compile the requirements of UMTS upon B-ISDN, as well as the requirements from other mobile networks (e.g. GSM). Requirements are derived from the basic functions required for mobile network operation: e.g. bridging, multicasting and combining, buffering, and transcoding, which impose new capabilities on networks with ATM transport. This paper described the impact of these basic functions on architecture of BB networks with mobile capabilities.

Aside from NA5 further contact has been established to the STCs within ETSI TC SPS in order to incorporate requirements of mobile networks in signalling related standards.

6. Conclusions

In this paper a solution to support third generation mobility in B-ISDN is presented. The radio access technique specific transport functions are allocated in the BTS and in the UMTS Mobility Server. The UMTS mobility transport functions which are performed in this Mobility Server are: bridging, multicasting and combining, transcoding and rate adaptation, buffering, etc... The server is radio dependent as it also handles the radio related functions. By the use of a Mobility Server no significant changes are required to the broadband switches or B-ISDN transport infrastructure. It also provides a modular introduction of mobile communications in a fixed B-ISDN network. Furthermore, the infrastructure can be extended in a modular way as the mobile communication segment increases.

7. References

[1] L. Van Hauwermeiren et al, "Requirements for mobility support in ATM", IEEE Globecom '94, pp. 1691-1695, 1994.
[2] M. De Prycker, "Asynchronous transfer mode: solution for B-ISDN", Ellis Horwood Limited, 1993.
[3] M. Mc Tiffin et al, "Mobile access to an ATM network using a CDMA air interface", IEEE JSAC, vol. 12, pp. 900-906, 1994.
[4] J. Bursztejn, "Alcatel's viewpoint on evolution towards UMTS", Workshop "Towards 3rd generation mobile communication systems", Brussels, january 1995.
[5] M. Mouly and M.-B. Pautet, "The GSM system for mobile".
[6] CEC deliverable, "UMTS system structure document".
[7] J. Rapeli, "UMTS: targets, system concept, and standardization in a global framework, IEEE Personal Communications, Vol. 2, pp. 20-28.
[8] E. Buitenwerf et al., "UMTS fixed network issues and design options", IEEE Personal Communications, Vol. 2, pp. 30-37.
[9] A. Urie et al, "An advanced TDMA mobile access system for UMTS", IEEE Personal Communications, Vol. 2, pp. 38-47.
[10] P.-G. Andermo and L.-M. Ewerberg, "A CDMA-based radio access design for UMTS", IEEE Personal Communications, Vol. 2, pp. 48-53.

VSAT Satellite Networks Providing ATM Service

Michael H. Hadjitheodosiou

Communications Research Centre (CRC)
3701 Carling Avenue, PO Box 11490, Station H
Ottawa, Ontario, CANADA K2H 8S2

Abstract. We discuss the role of Very Small Aperture Terminal (VSAT) satellite networks in the provision of access for remote users to the Integrated Broadband Communications Network. The need for optimisation of the channel capacity allocation scheme is discussed, using a preliminary performance analysis for a Dynamic Reservation TDMA system which takes advantage of the flexibility and the statistical multiplexing capabilities of ATM and supports various VBR and CBR services. Investigation on the effect of the satellite propagation delay shows that improvements in performance are possible if "next generation" satellite technology is used. Some bottlenecks, introduced by the satellite link, in the performance of protocols such as TCP/IP are highlighted. The limitations of the satellite link capacity and the earth station buffer size and their effect on the system's performance are also discussed.

Keywords: VSAT, ATM, Multiple Access, Satellite Communications, TDMA, Network Performance, Teletraffic.

1. Introduction

Asynchronous Transfer Mode (ATM) has been adopted as the main technology for the implementation of the Integrated Broadband Communications Network (IBCN). However, the deployment of an ubiquitous terrestrial infrastructure to support this technology could probably take many years [CHIT94], and the traffic demands on this network are as yet unknown. Satellite networks, offering broad geographical coverage and fast deployment, appear to be an attractive option for the early deployment of the IBCN and could play a major role in its development, provided a number of difficulties arising from the nature of satellite systems can be overcome.

The main advantages that communication satellites offer are:

• B-ISDN services can be provided over a large area, without the need of excessive investment in the early phase.

• Satellite communication systems can be complementary to terrestrial networks, especially for widely dispersed users.

• Common alternative channels can be provided for routes where demand and traffic characteristics are uncertain, so that resources are used at maximum efficiency.

• The broadcast nature of satellites can be used for cases where the same message has to be transmitted to a large number of stations (point-to-multipoint transmissions).

• A wide range of customer bit rates and circuit provision modes can be supported.

• Satellite networks offer transparency to the type of services carried.

• New users can be accommodated swiftly with simply installing new earth stations at customer premises. Thus possible network enlargement is not a significant planning problem.

• By using satellites the full range of services could be offered, rapidly and without the need of major investment, to areas where the terrestrial network infrastructure is not very well developed, as is the case of the Eastern European countries.

For these reasons a number of research projects, such as the UNIVERSE project [BURR86], the SATINE project and the COST 226 projects [HARR91] investigated high speed networking via satellite, and more recently under RACE II, the CATALYST project [HADJ94], was the first to look at the use of satellites for broadband island interconnection.

Satellite networks can play two main roles:

• the interconnection of a few geographically distributed broadband networks, usually called "broadband islands".

• the provision of a network interface to a large number of thin-route users that want access to the IBCN.

For the first scenario large earth station gateways would be required to accommodate the high and usually continuous traffic rates that would be expected. Some aspects of these were investigated by CATALYST, a RACE II research project that , using a series of demonstrators investigated the applicability and compatibility of satellite networks with the terrestrial IBCN [HADJ94], [POLE95].

For the second scenario, Very Small Aperture Terminal (VSAT) satellite networks could be used. VSATs have been very successful over the last decade as private networks offering a variety of data communication services. In these networks, a large number of medium/small size (below 3.4 m in diameter) stations, carrying traffic of bursty nature, are interconnected by a satellite (Fig. 1). In this paper we discuss how these systems could be adapted for this new service and investigate some of the problems that need to be resolved.

The 155 and 622 Mbits/s transmission rates conventionally associated with ATM are well above the maximum rates possible with today's VSAT technology. However, in practice, most individual users will usually require significantly lower traffic rates, especially if there are only a few data or voice terminals located at a remote location. This large number of users with bursty traffic will need a cost-efficient way to occasionally access the B-ISDN/ATM network. There is therefore a need for a scheme that allows them to share a channel, takes advantage of the "bandwidth on demand" property of ATM while at the same time provides a compatible interface to the ATM technology.

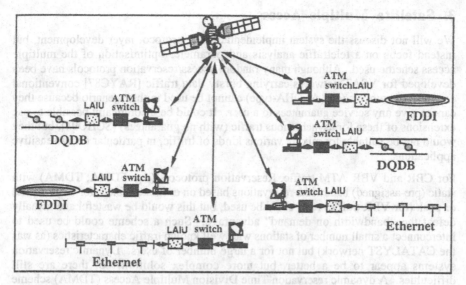

Fig. 1 ATM Services via Satellite

2. CBR/VBR Traffic

Detailed description of models of Constant Bit Rate (CBR) and Variable Bit Rate (VBR) Traffic Sources is beyond our scope here. A large amount of work has recently been devoted to this topic, and [ONVU94] contains a detailed description of the state-of-the-art and a comprehensive reference list. For the purpose of our analysis we focus on the basic characteristics of CBR and VBR services.

CBR services generate traffic at a constant rate, and can be simply described by fixed bit rates. A CBR source is considered active during the duration of the connection, therefore a constant bandwidth is reserved for the whole connection. Typical examples of CBR services include voice and video transmission, document retrieval and remote file transfer.

Most typical traffic sources alternate between active and silent states and have a peak-to-average bit rate much greater than one. If we try to present this to a network as CBR traffic by means of buffering or controlling the bit generation rate, this would result in the under-utilisation of the bandwidth resources and degrade the quality of service. Although this would simplify the network management task, it would be more natural to provide VBR service to VBR sources. This is especially advantageous in a VSAT satellite network where the nature of the traffic is bursty, bandwidth is a very expensive commodity and hence a "bandwidth-on-demand" approach is usually the best option. The flexibility of ATM allows the statistical multiplexing of VBR traffic from a number of traffic sources at each node (VSAT), and by using an efficient allocation of the satellite channel capacity we can achieve a multiplexing of traffic from different nodes.

3. Satellite Multiple Access

We will not discuss the system implementation or protocol layer development, but instead focus on a teletraffic analysis and parameter optimisation of the multiple access scheme used. Although many random access/reservation protocols have been developed for VSAT networks carrying bursty data traffic [RAYC87] conventional contention protocols (e.g. ALOHA-type) cannot be used in this scenario because they cannot give any service guarantees to a user. It could be theoretically possible to use extensions of these for asynchronous traffic (with no guarantees) [SCHO94], but they would not be suitable for mixing various kinds of traffic, in particular delay-sensitive applications.

For CBR and VBR ATM traffic, Reservation protocols (e.g. FDMA, TDMA) with static (pre-assigned) bandwidth reservations based on estimated fixed (for CBR traffic) or peak (for VBR traffic) rates could be used, but this would be wasteful and actually defeat the "bandwidth on demand" advantage. Such a scheme could be used to interconnect a small number of stations with well known traffic characteristics (as was the CATALYST network) but not for a large number of users. Dynamic reservation systems appear to be a better, but more complex solution, but there are still difficulties. A dynamic reservation Time Division Multiple Access (TDMA) scheme appears to be the best choice, since it allows the satellite transponder to operate at maximum efficiency. Also, the flexibility of TDMA allows ATM cells to be mapped into the variable length TDMA sub-bursts as the basic unit (slot) of transmission. Fig. 2 shows how ATM cells could be packed into TDMA frames. The example is based on the CATALYST project [POLE95] and assumes the use of a EUTELSAT-II SMS transponder.

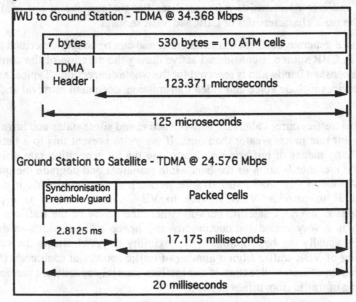

Fig. 2 ATM Cells into TDMA Frame Format

Satellite systems accommodating both synchronous and asynchronous traffic using TDMA-based techniques have been recently proposed, e.g. [ZEIN91], [CELA91]. An efficient use of the capacity of the TDMA system can be achieved if we provide a combination of pre-assigned and on-demand capacity in each link, matching the broad mix of VBR and CBR traffic we expect to have.

4. System Performance

We next focus on a TDMA Reservation System consisting of a number of VSATs, a central reservation/control unit (Hub), and a single channel, shared by all active VSATs. The Hub co-ordinates the channel allocation. In a typical VSAT system there could be a number of sources (data terminals, telephones, fax, video) connected to a VSAT via an ATM switch, or the VSAT could act as a gateway for a LAN, in which case a specially developed LAN-ATM Interface Unit (LAIU) is needed to perform the required protocol layer translation (Fig. 1).

There are two possible architectures:

• A *Star* network, in which the VSATs always communicate with each other via the Hub; this would be ideal for an application involving point-to-multipoint transmission, e.g. a Hub located at the company headquarters sending information to a number of branches.

• A *Mesh* network, in which the Hub acts as a Network Management Centre for channel allocation and policing, but VSATs having established a connection talk directly to each other via the satellite. In this case the propagation delay is a single satellite hop, while it is twice that for the Star network. A typical example of a Mesh application would be telephone service provision.

A VSAT with a message to transmit, sends its request with the message size to the Hub, which will then let the terminal know when to start the message transmission. The TDMA frame (Fig. 3) is divided into Reservation slots and Data slots. Reservation slot is that period of time in which terminals report their requests to the reservation unit. There are only a few Reservation slots available and a VSAT selects one at random without knowing whether another station is using the same slot. There is a possibility of collision, and if, after a time-out, the VSAT does not receive an acknowledgment of its request, it waits a random time and re-transmits it. The acknowledgment from the Hub to VSATs is transmitted in TDM mode over a separate, faster channel.

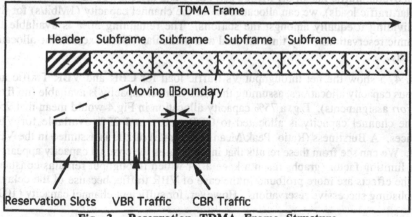

Fig. 3   Reservation TDMA Frame Structure

Data slots represent the part of the frame in which a terminal can transmit its message. In every frame there are many data slots and the Hub will assign enough sequential data slots to a particular successful request. A data slot is assigned to at most one terminal and therefore there is no possibility of collision. The available data slots can be reserved for the duration of a CBR connection or allocated on a burst-by-burst basis to VBR services. We can see that the frame partition (i.e. number of slots allocated to reservation, CBR and VBR traffic) needs to be adjusted to the traffic load and mix we have, and clearly an adaptive allocation using a moving boundary would represent a more efficient (but much more complex to implement!) use of the capacity, if the traffic load changes frequently with time. Unless the number of reservation slots/frame is carefully adjusted the result would be either low capacity utilisation and long delays (too many reservation slots, less capacity available for information transmission) or network backlog (too few reservation slots resulting in successive collisions and high delay).

A possible implementation scenario would be to allocate a portion of the capacity to each VSAT according to its mean traffic rate (derived for the load of all the sources connected to that station) and its QoS requirements, and allow it to request additional capacity using the reservation slots or by piggy-backing the request with the data sent in the data slots. In this way, a service with strict low delay/jitter requirements (e.g. voice or video) could use the allocated capacity, while a less demanding service (e.g. data transmission) could take advantage of the capacity reservation, thus maximising the efficient use of the channel. The actual procedure might be very delicate and details need to be investigated in greater length for a realistic implementation. Each VSAT must monitor its traffic arrival rate and the network load (e.g. from the time it takes to establish a reservation, number of successive collisions) and be able to block new users if there is a danger of exceeding a certain threshold.

4.1. Traffic Analysis

For the analysis we consider a network consisting of 10 VSATs, with a total channel capacity of 2 Mbits/s. Unless otherwise stated we assume an error-free channel and a mesh configuration, where each satellite hop takes approx. 0.27 sec. Part of the capacity can be assigned to each VSAT based on expected load, while the remaining capacity could be left for dynamic reservations. For example, if we expect to have a mean overall traffic rate of 0.1 MBit/s per VSAT and there are 10 active VSATs (with similar traffic loads), we can allocate 50% of the channel capacity (1Mbit/s) for this by dividing it equally amongst the stations. The remaining 50% is available for dynamic reservations every time the load at a particular VSAT exceeds the allocated capacity.

Figs. 4, 5 show the net throughput Vs traffic load for CBR and VBR Traffic and various capacity allocations, assuming the total channel capacity is available (no fixed *a priori* assignments). E.g. a 75% capacity allocation in Fig.4 would mean that 75% of the channel capacity is allocated to CBR traffic and 25% available for VBR services. A Burstiness (Ratio Peak/Mean traffic rate) of 200 is assumed in the VBR case. We can see from these results that in both cases the channel capacity appears to be a limiting factor (graphs reach a knee after which throughput remains constant), but the effects are more profound in the case of VBR traffic, because of the delay in establishing successive reservations. However, for sufficient channel capacity (100%

allocation for either CBR or VBR), the maximum achieved throughput is higher in the VBR case, because statistical multiplexing enables a better channel utilisation.

Clearly, a capacity allocation based on the expected mix of VBR/CBR services would offer the best performance. Combining Figs. 4,5 with a delay performance analysis for specific services could determine the capacity partition required to guarantee a particular Quality of Service, based on the traffic load.

4.2. Effect of Propagation Delay

If a satellite system is used to provide ATM services, the long channel propagation delay imposes important limitations on the system's capabilities. However, future developments in satellite communications could reduce the delay and thus improve the system's performance. The introduction of on-board processing satellites [COAK95] would half the reservation delay, as only one hop will be required. The introduction of Medium Earth Orbit (MEO) and Low Earth Orbit (LEO) systems would decrease the propagation delay from 270 ms (for GEO) to around 80-100ms (MEO) and 16-20ms (LEO). Fig. 6 represents a simple indication of the possible improvement in performance for the system studied here, although a number of other factors need to be taken into consideration for a precise comparison. Note that the propagation delay plays a part in determining the window size required for optimum throughput. For channels with longer propagation delays window sizes need to be larger, and this could have significant implications in the performance of protocols such as TCP/IP [JACO92].

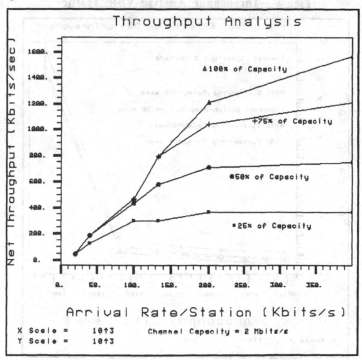

Fig. 4 Throughput Analysis-CBR Traffic

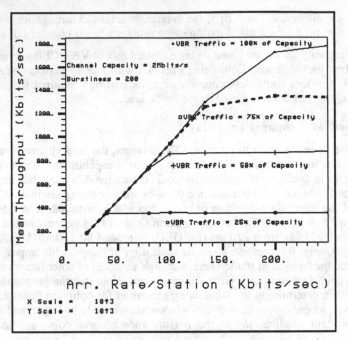

Fig. 5   Throughput Analysis-VBR Traffic

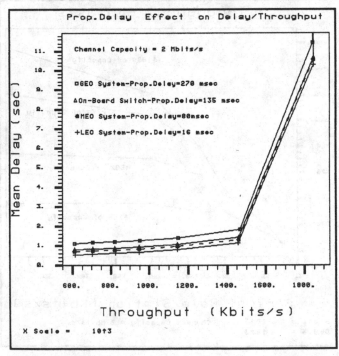

Fig. 6   Effect of Propagation Delay on Performance

4.3. TCP/IP Performance

An essential service a VSAT system offering LAN interconnection will be required to support is File Transfer, and TCP/IP is one of the most common and robust protocols for this application. To see the effect of the satellite delay and channel errors on the protocol's performance a simplified implementation of a VSAT system consisting of 10 stations and using the Reservation TDMA scheme described earlier is simulated, for both star and mesh configurations. Fig. 7 shows the throughput as a function of the channel capacity for both configurations, for the error-free channel case.

It is obvious that the window size of the TCP transport protocol is a limiting factor. With a window of 48 Kbytes, which is the maximum window the current SunOS implementation of TCP allows, the maximum achievable throughput is 1.9 MBits (which corresponds to 1.5 MBits/s net throughput) for the mesh case and 1.6 MBits/s (1.25 MBits/s net) for a star system, at the maximum channel speed of 2 MBits/s. For smaller windows we can see that increasing the channel speed does not increase the throughput and leads to low capacity utilisation. Also, for smaller windows we can see an almost linear relationship between capacity and throughput until the saturation point is reached. A star configuration (with double-hop delays) has a much lower throughput than mesh, especially as the window limitations become less significant, so there is a clear advantage if mesh architecture is used, as we would expect.

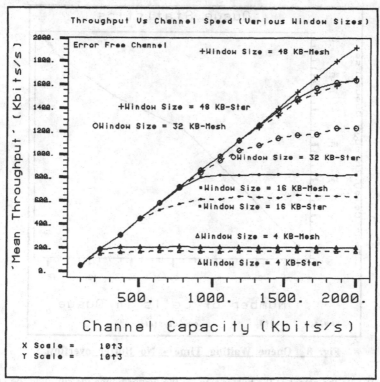

Fig. 7   Throughput Vs Capacity-Error Free Channel

272

It is important to note however that a mesh configuration would need very hard recovery algorithms for successful operation, especially in the presence of channel errors. This could have a heavy impact on the system efficiency and offset some of the advantage for mesh systems deduced from Fig. 7.

For a system with such a high propagation delay, the channel Bit Error Rate would have a significant effect on the system's throughput, especially in the case of larger window sizes. The possibility of burst errors in a satellite environment could also affect the system's performance and should be investigated in greater detail.

4.4. Effect of Buffer Size

The system discussed earlier assumed one pair of interconnected workstations, and the TCP/IP window appears to be the major bottleneck. However, if multiple workstations are interconnected via a bridge at every VSAT, and packets are buffered on a FCFS before transmission on the channel, the queueing delay imposed on the system is an additional bottleneck, which becomes increasingly more significant as the arrival rate of new messages is increased. This has been demonstrated by the analysis in [YANG92], and is also apparent in our simulation results. Fig. 8 shows the queue build up for the case of an "infinite" buffer and the delay cells experience in the queue.

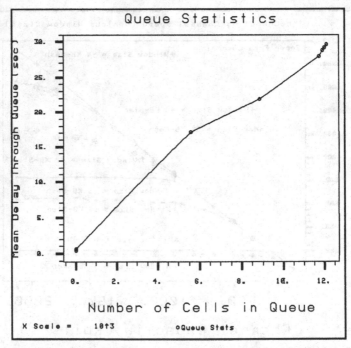

Fig. 8   Queue Waiting Time - No Buffer overflow

To investigate the effect of the buffer size on the performance we simulate the system described earlier with the addition of a fixed size buffer at every VSAT. A total channel capacity of 25 Mbits/s is assumed, and 2 Mbits/s are allocated to each VSAT while the remaining 5 Mbits/s are left for dynamic reservations. Arriving ATM cells are queued in a First In First Out (FIFO) order and transmitted via the allocated channel. If the buffer becomes full, an overflow procedure is activated, and excess cells are diverted to the shared channel, where a reservation scheme of the type described earlier operates.

Fig. 9 shows the performance for a VSAT for 4 different buffer sizes. Clearly, as the traffic load increases, by keeping the maximum buffer size small we avoid the build up of large queues and achieve much better mean delays for a particular throughput.

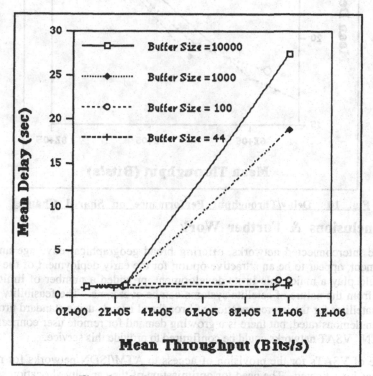

Fig. 9   Delay/Throughput per VSAT

By examining Fig.10, showing the performance of the Reservation system that uses the shared capacity, we do not observe significant differences in performance, in spite of the fact that the load is higher in the case of the smaller size buffers. This system can lead to a more efficient use of the capacity, especially if it can be adjusted to the specific requirements of individual users. If there is an available estimate of the mean total traffic expected at a station, an equivalent capacity can be allocated and an appropriate size buffer can be used. When traffic exceeds this allocation, a buffer overflow directs the extra traffic to the shared channel. In this way, the operator can guarantee a Quality of Service that a user requires, but also have the capability of providing additional capacity upon request, without any guarantees.

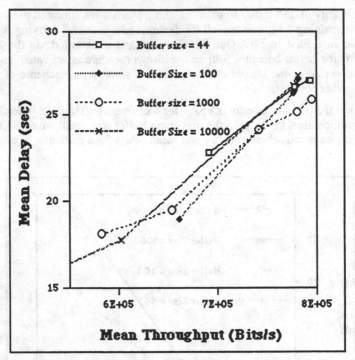

Fig. 10   Delay/Throughput Performance on Shared Channel

5. Conclusions & Further Work

Satellite interconnected networks, offering broad geographical coverage and fast deployment, appear to be an attractive option for the early deployment of the IBCN and could play a major role in its development, provided a number of limitations arising from the nature of satellite systems can be overcome. The feasibility of the use of satellites for the interconnection of broadband islands using standard protocols has been demonstrated, but there is a growing demand for remote user connection to the IBCN. VSAT networks could be configured to provide this service.

The use of VSATs for the provision of access to ATM/ISDN networks for remote users was investigated. The need for optimisation of the capacity allocation scheme was discussed using some preliminary performance results for a dynamic reservation system. Investigation on the effect of the satellite propagation delay showed that improvements in performance are possible if "next generation" satellite technology (on-board processing, LEO or MEO orbits) is used.

Some bottlenecks in the performance of protocols such as TCP/IP introduced by the satellite link were highlighted. We have seen that the link capacity, the channel error rate and the buffer size at the earth stations could have severe implications on the system's performance, and there are advantages to be gained by a more efficient sharing of the satellite resource.

Further work, aiming at a system offering full ATM connectivity, could focus on:

• Investigation of performance of more dynamic and efficient satellite access scenarios.

• The traffic characteristics across the network, including the effects of LAN and ATM traffic mixing and more complex traffic shaping mechanisms at the earth stations.

• Examination of the system's sensitivity to bursty channel errors, and, in particular the loss of synchronisation, possible undetected AAL errors and the difficulty in implementing an adequate congestion control scheme.

• Investigation of the effect of dynamic windowing and variable time-outs in the performance of File Transfer over a satellite link.

• The interface between satellite and terrestrial (fixed and mobile) networks to ensure service transparency and truly global coverage.

References

[BURR86]   Burren, J.W., "High-speed Computer Networking by Satellite: A Review of the Results from the Project UNIVERSE.", *IERE Journal*, **56**(5), May 1986, pp. 187-191.

[CELA91]   Celandroni, N., Ferro, E., "The FODA-TDMA Satellite Access Scheme: Presentation, Study of the System and Results.", *IEEE Trans. Comms.*, COM-39(12), December 1991, pp. 1823-1831.

[CHIT94]   Chitre, D.M., *et al.*, "ATM Operation via Satellite: Issues, Challenges and Resolutions", *Int. Jrnl. Sat. Comms.*, **V12**(3), May 1994, pp. 211-223.

[COAK95]   Coakley, F.P., Malinowski, P., "Fault-tolerant On-board Switches for ATM", *Proc. 10th ICDSC*, Brighton, May 1995.

[HADJ94]   Hadjitheodosiou, M.H *et al.*, "Broadband Island Interconnection via Satellite- Performance Analysis for the CATALYST Project", *Int. Jrnl. Sat. Comms.*, **V12**(3), May 1994, pp. 223-238.

[HARR91]   Harris, R.A., Koudelka, O., 1991, "Wideband Interconnection of LANs by Satellite.", *Proc. 2nd ECSC*, pp. 91-96.

[JACO92]   Jacobson, V. Braden, R., Borman, D., "TCP Extensions for High Performance", *RFC 1323*, Network Working Group, May 1992.

[ONVU94]   Onvural, R., "Asynchronous Transfer Mode Networks-Performance Issues", Artech House Ed., 1994.

[POLE94]   Polese, P., Mort, R., Combarel, L., "Satellites in UMTS and B-ISDN:Status of Activities and Perspectives", *IEE Electronics & Comms. Jrnl.*, **V6** (6), December 94, pp.297-303.

[RAYC87]   Raychaudhuri, D., Joseph, K., "Ku-Band Satellite Data Networks using VSATs-Part1:Multi-access Protocols", *Int. Jrnl. Sat. Comms.*, 1987, pp. 195-212.

[SCHO94]   Schoute, F.C., Awater, G.A., Giesbers, B.M.P., "Aloha Performance Models", *Proc. GLOBECOM'94*, San Fransisco, December 1994.

[YANG92]   Yang, O.W.W., Yao, X.X., Murthy, K.M.S., "Modeling and Performance Analysis of File Transfer in a Satellite WAN, *IEEE J-SAC*, **V10**(2), February 1992, pp. 428-436.

[ZEIN94]   Zein, T., Maral, G., Tondriaux, M., Seret, D., "A Dynamic Allocation Protocol for a Satellite Network Integrated with B-ISDN.", *Proc. 2nd ECSC*, 22 October 1991, pp. 15-20.

Acknowledgments: The greater part of the work described here was carried out while the author was with the Centre for Satellite Engineering Research, University of Surrey, U.K., and the help and support provided to him is gratefully acknowledged.

Predictive Congestion Control for Broadband Satellite Systems

Y. M. Jang[1], A. Ganz[1] and J. F. Hayes[2]

[1] Dept. of Electrical and Computer Engineering
University of Massachusetts
Amherst, MA 01003
[2] Dept. of Electrical and Computer Engineering
Concordia University,
Montreal, Quebec H3G 1M8

Abstract. In this paper, we propose a predictive and transient congestion control scheme for satellite systems that supports on-board packet
switching of multimedia traffic with predefined quality of service requirements. The congestion control scheme incorporates the unique characteristics of satellite systems, e.g. large propagation delays, no-onboard
buffer, and low computational requirements. The congestion control scheme
requires the estimation of the On-Off traffic characteristics (λ, μ) of the
traffic sources. These estimated values are used to predict the transient
cell loss probability at each downlink. In case the Quality of Service
requirements are not met the proposed congestion control scheme determines the control parameters for source traffic shaping or controls the
total number of connection in the system.
The numerical results obtained suggest that the proposed scheme is an
excellent candidate for real time burst and call level congestion prediction
and control in broadband on-board satellite networks.

Keywords: predictive congestion control, broadband satellite system, estimation
methods, Quality of Service requirements, source shaping.

1 Introduction

Geostationary Earth Orbit (GEO) Satellites with wide area coverage are expected to provide interconnections of Global Area Business Networks supporting
various applications such as point-to-point VSAT-type services, LAN interconnections, multimedia, and teleconferencing services. To accommodate these diverse services with inherent traffic fluctuations while efficiently utilizing the spectrum, packet switching capabilities should be implemented at the satellite [1,2,3].
NASA Lewis Research Center is currently investigating a geostationary communication satellite MCSPS (Multi Channel Signal Processing Satellite) which can
support packet switching capability and provide direct-to-the-user services [4].

Recall that the bandwidth in a packet switching network is allocated on demand. Therefore, congestion problems may occur when the demand for resources
on-board the satellite exceeds its capacity. Congestion can rapidly neutralize the

delay and severely limit the advantages obtained by dynamic resource sharing achieved by packet switching. To achieve both efficient utilization of the space-segment resources and acceptable user quality of service, high performance network management protocols that incorporate congestion/flow control must be provided.

The congestion problem is not unique to satellite communications and in fact has been extensively studied for terrestrial broadband ATM networks [5,6]. However, in the design of congestion control schemes for satellite systems, attention should be paid to the following unique characteristics of satellite communication:

- Satellite networks have long propagation delays between the earth-stations and the satellite (typically 125 ms), mandating predictive congestion control schemes.
- The on-board processing should be minimized (processing power and power consumption limitations), i.e., we should avoid complex, and computational-intensive procedures to be executed on-board.
- On-board storage is very expensive, i.e. congestion control schemes implemented on-board the satellite should require minimal buffering capabilities.
- All the traffic in the satellite system is routed through the satellite (which is acting like a hub). In other words, there is a central arbiter (a controller on-board the satellite) that can regulate the congestion.

The long propagation delay characteristic will cause feedback-based control to be slow and ineffective. Ramamurthy and Sengupta[7] concluded that predictive control policies perform significantly better than the statistic rate and adaptive control policies in all cases. The authors in [8] proposed the predictive approach for congestion control in satellite networks. They use statistical and neural approaches under Poisson traffic cases. There are several predictive papers for ATM and wireless networks [9,10].

The proposed predictive congestion control scheme accounts for all the above characteristics of satellite communication, and achieves both the efficient utilization of the space-segment resources while providing the required Quality of Service (QoS) (in terms of the tolerable cell loss probability) in a multimedia environment.

We estimate the state of the underlying process and predict the future cell loss probability based on estimation and measurement of the bursts in progress and the duration of each burst. This transient and predictive dynamics is used to avoid congestion while efficiently utilizing the available network capacity.

Optimal criteria are proposed on how to optimally select the parameters of individual On-Off sources or number of connections, given the transient cell loss probability and the QoS. These optimal values maximize the bandwidth utilization while maintaining the requested QoS.

In the next subsection we present the system model and in subsection 1.2, we introduce an outline of the proposed predictive congestion control algorithm.

1.1 System Model

The ground stations are interconnected through the satellite using K uplinks and K TDM downlinks (which determine the downlink time division into slots which equal the cell size). The propagation delay between the earth station and the satellite is d seconds. The uplink communication uses either TDMA, MF-TDMA, FDMA, CDMA or MF-CDMA techniques. On-board the satellite there is a switching fabric (with no buffering capabilities) capable of routing cells that successfully arrive on the uplinks to their destination downlink. Due to statistical fluctuations of the traffic and the limited bandwidth of the downlink (C bits/second), congestion may occur at the downlinks causing packets to be discarded. Therefore, it will be mandatory to implement a congestion control scheme on-board the satellite.

Although the traffic characteristics of the future on-board satellite networks are hard to predict with complete accuracy, there are a number of voice and video models reported in the literature [11,12]. Due to the statistical nature of the multimedia traffic, the modeling of incoming traffic characteristics plays an important role. As suggested in [11,12], both voice and video traffic sources can be characterized by a set of On-Off models. The QoS for each user is expressed in terms of the cell loss probability that the connection can tolerate.

Each source is therefore, modeled as an On-Off traffic model. We assume that the "On" and "Off" periods of each source are both exponentially distributed with parameters μ and λ, respectively. In this traffic model, when the process is in the "On" state, it generates cells at a constant rate of R_p bits/second. When the process is in the "Off" state, it does not generate any cells. We assume that each one of the N connections sharing a downlink has the same traffic parameters (λ, μ, R_p).

Any of these traffic parameters may be used to control the flow of the traffic. In our model, we assume that R_p is constant and λ and μ will be adjusted by our proposed congestion control algorithm. Another parameter to be shaped by the scheme is the number of connections, N. See Figure 1 for a detailed description of the system configuration.

1.2 Outline of the Congestion Control Algorithm

We next provide an outline of the predictive congestion control algorithm implemented on-board the satellite and executed per downlink:

1. *Parameter Estimation:* estimate λ and μ. For each traffic source using the downlink we collect sample data of the On-Off traffic for a number of consecutive cell samples and compute these estimates using the Maximum Likelihood (ML) method. Details are provided in Section 2.

2. *Compute (predict) the transient cell loss probability:* Based on the estimated parameters, $\hat{\lambda}$ and $\hat{\mu}$, the number of connections N, and the number of active sources at time $t = 0$, i, we compute the transient cell loss probability denoted by $P_{loss}(\hat{\lambda}, \hat{\mu}, N, i, t)$. In case $P_{loss}(\hat{\lambda}, \hat{\mu}, N, i, t) < QoS$ send congestion abatement messages. Otherwise, proceed to 3 below. Details are provided in Section 3.

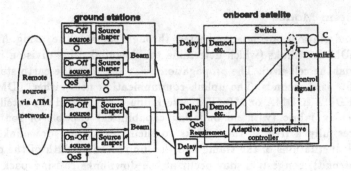

Fig. 1. Congestion prediction and control architecture for broadband satellite communication

3. *Find optimal source parameters λ^* and μ^* or number of connections N^* for given QoS requirements.* Whenever the predictive transient cell loss probability is higher than the QoS requirement, the satellite controller will compute (using results obtained in 1. and 2. above) the optimal source parameters or number of connections depending on the network environment. The satellite controller will send choke messages to the earth stations including these parameters. In case the sources are directly connected to the satellite, the satellite controller will send the source shaping parameters, λ^* and μ^*. In case the connections are established by sources that are not directly connected to the satellite, the satellite controller will disconnect $N - N^*$ connections, given there are N active connections. Details are provided in Section 4.

A flow chart of the control algorithm on-board the satellite is depicted in Figure 2. In each ground station, the following actions are taken:

1. *Choke message received:* reduce the traffic using either 1) Connection Admission Control (CAC) for disconnection of ongoing connection or rejection of the new request connections, or 2) Usage Parameter Control (UPC) to shape the source traffic using Leaky bucket scheme, window based control, or cell discarding and tagging.
2. *Congestion abatement message received:* stop the control procedures.

The congestion management at the ground station is beyond the scope of the paper. We assume that once the choke cells are received at the ground station, one of the congestion control techniques mentioned above can be executed.

2 Parameter Estimation Technique

The parameters that we estimate must be such that they can be easily evaluated and monitored by traffic measurement on-board the satellite which has limited power, bandwidth and buffering capabilities. The traffic estimation process for each connection is based on the measurement of M consecutive samples.

We next describe the Maximum Likelihood method for estimating λ and μ.

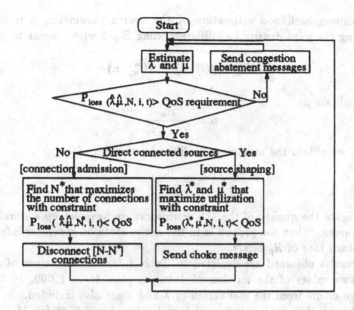

Fig. 2. Flow diagram of congestion control algorithm

According to the behavior of the traffic from each source, the sequences of active and idle periods (observed at the downlink) are exponentially distributed random variables (see Figure 3). Let $l_1^a, \cdots, l_k^a, \cdots, l_{n_a}^a$ and $l_1^i, \cdots, l_k^i, \cdots, l_{n_i}^i$ be

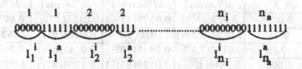

Fig. 3. Maximum Likelihood diagram

the length of successive active and idle periods, respectively. n_a and n_i are defined as the number of active and idle periods, respectively. Since we will obtain M data samples, we have:

$$\sum_{k=1}^{n_a} l_k^a + \sum_{k=1}^{n_i} l_k^i = M \tag{1}$$

Let $f(l_k^a; \mu)$ be the density function of the active period where μ is the only parameter to be estimated from a set of sample values $l_1^a, l_2^a, \cdots, l_{n_a}^a$. The likelihood function of this sample is

$$f(l_1^a; \mu) f(l_2^a; \mu) \cdots f(l_{n_a}^a; \mu) = \mu^{n_a} e^{-\mu \sum_{k=1}^{n_a} l_k^a} \tag{2}$$

The maximum likelihood estimation of the source parameter, μ is found by maximizing the joint density i.e., differentiating Eq. 2 with respect to μ:

$$\frac{\partial}{\partial \mu} f(l_1^a; \mu) f(l_2^a; \mu) \cdots f(l_{n_a}^a; \mu) = 0 \qquad (3)$$

Then we obtain $\hat{\mu}$:

$$\hat{\mu} = \frac{n_a}{\sum_{k=1}^{n_a} l_k^a} \qquad (4)$$

Similarly, we obtain the maximum likelihood estimator of λ:

$$\hat{\lambda} = \frac{n_i}{\sum_{k=1}^{n_i} l_k^i} \qquad (5)$$

To investigate the quality of the ML estimators we have set up a simulation of binary sources. When the process is in the active state, it generates information at a constant rate of R_p bits/s.

The results obtained are depicted in Table 1 for various values of λ and μ and for two values of the number of data samples, $M = 1,000, 10,000$. The percentage errors from the real values of λ and μ are also included.

We observe that for low values of λ and μ (e.g. $\lambda = 0.0085$ for $M = 10,000$) the ML estimation method exhibits large errors. However, as we increase λ and μ the percentage of error dramatically decreases. The number of samples M clearly makes a difference in the estimation quality. The more points we collect, the more accurate the estimator is.

Table 1. Estimators comparison results for M=1,000 and M=10,000

Actual parameters		M=1,000		M=10,000	
μ	λ	$\hat{\mu}$(%error)	$\hat{\lambda}$(%error)	$\hat{\mu}$(%error)	$\hat{\lambda}$(%error)
0.0132	0.0085	0.0136(3%)	0.0123(45%)	0.0155(17%)	0.0092(8%)
0.0132	0.085	0.0122(7.5%)	0.1010(1.9%)	0.0143(8%)	0.0828(2.5%)
0.0132	0.85	0.0142(7.5%)	0.7647(10%)	0.0143(8%)	0.8159(2%)
0.132	0.0085	0.1846(40%)	0.0139(63%)	0.1204(8%)	0.0097(14%)
0.132	0.085	0.1311(6%)	0.0724(15%)	0.1398(5.9%)	0.0852(0.2%)
0.132	0.85	0.1246(5.6%)	0.8720(2.5%)	0.1329(0.7%)	0.8523(0.2%)

3 Transient Cell Loss Probability

It is very important that the decisions to accept or reject calls are made in real time. That is, a simple and fast call admission control should be able to predict transient cell loss probability rather than steady state cell loss probability. We use statistical bufferless fluid flow model [13] to predict the probability that a cell loss occurs at time $t(= 2d)$, (i.e., the round-trip propagation delay) based on the traffic statistical behavior and the number of active sources at time 0. We assume N identical binary On-Off sources that share the capacity C of a downlink. Next we introduce the following definitions:

$\rho(\hat{\lambda}, \hat{\mu}, N, i, t)$ - the average arrival rate at time t given i sources are active at time 0 and there are N connections (traffic sources), assuming each traffic source is characterized by $(\hat{\lambda}, \hat{\mu})$.

$P_{loss}(\hat{\lambda}, \hat{\mu}, N, i, t)$ - the predicted transient cell loss probability at time t given i sources are active at time 0 and there are N connections (traffic sources), assuming each traffic source is characterized by $(\hat{\lambda}, \hat{\mu})$.

$OV(N, i, t)$ - excess traffic up to time t given i sources are active at time 0. and there are N connections (traffic sources), assuming each traffic source is characterized by $(\hat{\lambda}, \hat{\mu})$.

$p(\hat{\lambda}, \hat{\mu}, t)$- the probability that a source (with traffic parameters $(\hat{\lambda}, \hat{\mu})$) is active at time t given the source was idle at time 0.

$q(\hat{\lambda}, \hat{\mu}, t)$- the probability that a source (with traffic parameters $(\hat{\lambda}, \hat{\mu})$) is active at time t given the source was active at time 0

In the link overflow model, cell losses due to overflow occur if and only if the aggregated peak rate from x active sources, xR_p exceeds link capacity C. The transient cell loss probability is given by the ratio of excess traffic, $OV(N, i, t)$ and traffic load $\rho(\hat{\lambda}, \hat{\mu}, N, i, t)$ at time t (see Figure 4).

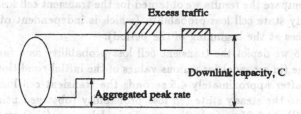

Fig. 4. Computation of the cell loss probability

$P_{loss}(\hat{\lambda}, \hat{\mu}, N, i, t)$ is given by:

$$P_{loss}(\hat{\lambda}, \hat{\mu}, N, i, t) = \frac{OV(N, i, t)}{\rho(\hat{\lambda}, \hat{\mu}, N, i, t)} \qquad (6)$$

$$\rho(\hat{\lambda}, \hat{\mu}, N, i, t) = R_p[iq(\hat{\lambda}, \hat{\mu}, t) + (N - i)p(\hat{\lambda}, \hat{\mu}, t)] \qquad (7)$$

Let $Y(t)$ denote the number of active sources at time t.

$$OV(N, i, t) = \sum_{x \in \{x | (xR_p - C) \geq 0\}}^{N} P(Y(t) = x | Y(0) = i)(xR_p - C) \qquad (8)$$

where $P(Y(t) = x | Y(0) = i)$ is given by:

$$P(Y(t) = x | Y(0) = i) = \sum_{k=0}^{x} \binom{i}{x-k}[q(\hat{\lambda}, \hat{\mu}, t)]^{x-k}[1 - q(\hat{\lambda}, \hat{\mu}, t)]^{i-x+k}$$
$$\times \binom{N-i}{k}[p(\hat{\lambda}, \hat{\mu}, t)]^{k}[1 - p(\hat{\lambda}, \hat{\mu}, t)]^{N-i-k} \qquad (9)$$

in which we assume that $\binom{i}{j} = 0$ for $i < j$.

To compute $p(\hat{\lambda}, \hat{\mu}, t)$ and $q(\hat{\lambda}, \hat{\mu}, t)$ we use the two-state Markov chain of each On-Off source. $p(\hat{\lambda}, \hat{\mu}, t)$ and $q(\hat{\lambda}, \hat{\mu}, t)$ can derived using the forward Chapman-Kolmogorov matrix differential equation [14].

$$p(\hat{\lambda}, \hat{\mu}, t) = \frac{\hat{\lambda}}{\hat{\lambda} + \hat{\mu}}(1 - e^{-(\hat{\lambda}+\hat{\mu})t}) \qquad (10)$$

$$q(\hat{\lambda}, \hat{\mu}, t) = \frac{\hat{\lambda}}{\hat{\lambda} + \hat{\mu}} + \frac{\hat{\mu}}{\hat{\lambda} + \hat{\mu}}e^{-(\hat{\lambda}+\hat{\mu})t} \qquad (11)$$

3.1 Numerical Results

For the numerical calculations of the transient cell loss probability, we consider the following system: $N = 40$ PCM coded voice sources with $R_p = 64kbps$, $\lambda = 0.5, \mu = 0.833$ multiplexed onto a downlink of capacity $C = 1.544Mbps$,

We first compare the results we obtained for the transient cell loss probability with the steady state cell loss probability (which is independent of the number of active sources at the beginning of the period).

In Figure 5 we depict the transient cell loss probability as a function of the prediction time (in seconds) for various values of the initial conditions, $Y(0)$. We observe that after approximately 4.5 seconds the transient cell loss probability will converge to the steady state cell loss probability (obtained using the result presented in [6]). At 0.25 seconds, the round trip delay in a GEO satellite system, we observe significant difference in the results obtained as a function of the different initial conditions (the number of active sources at the beginning of the estimation interval). For example, at $t = 0.25sec$ we observe that the transient cell loss probability given $Y(0) = 0$ is approximately 10^{-12} while the steady state probability is approximately 10^{-4}. We, therefore, conclude that the computation of transient cell loss probability is more accurate and can lead to significant different values of the cell loss probability and consequently of the congestion control decisions.

In Figure 6 we plot the transient cell loss probability as a function of the number of connection for different values of the ratio between the number of active sources at $t = 0$ and the number of connections $(Y(0)/N)$. We observe that for given values of (λ, μ) per connection, the transient cell loss probability increases as a function of the number of connections. This is due to the fact that the overall traffic intensity per downlink increases as the number of connections increases. As the ratio $Y(0)/N$ increases, the transient cell loss probability will increase.

These plots show the importance of considering the *transient* cell loss probability, quantity which we compare with the required QoS and consequently make our congestion control decision.

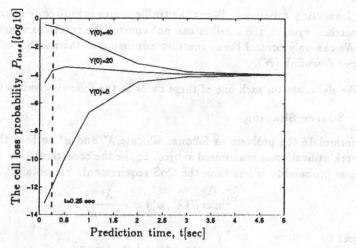

Fig. 5. Transient cell loss probability versus prediction interval for multiple values of the number of active sources at $t = 0$.

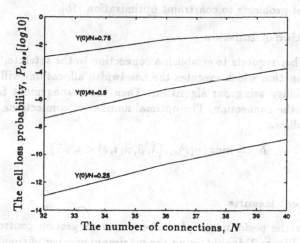

Fig. 6. Transient cell loss probability as a function of the number of connections for different values of the ratio between the number of active sources at $t = 0$ and the number of connections.

4 Optimal Control Values

In the previous sections we obtained estimators $(\hat{\lambda}, \hat{\mu})$ which we use in the computation of the transient cell loss probability $P_{loss}(\hat{\lambda}, \hat{\mu}, N, i, t)$. Once this probability exceeds the required QoS we have to take some actions, depending on the ground network structure. We distinguish between two cases:

- Source Shaping: When the traffic source is directly connected to the satellite system, the satellite controls the traffic parameters (λ, μ).

– Connection Admission: When the traffic source is indirectly connected to the satellite system, the satellite can not control the traffic parameters directly. We can only control the connection admission or the number of connections per downlink (N).

We elaborate on each one of these cases in the following subsections.

4.1 Source Shaping

We formulate the problem as follows: allocate λ^* and μ^* so that the long term network utilization is maximized subject to the the constraint that the transient cell loss probability is less than the QoS requirement. We obtain:

$$\max U(\lambda^*, \mu^*) = \frac{\lambda^*}{\lambda^* + \mu^*} \tag{12}$$

subject to

$$P_{loss}(\lambda^*, \mu^*, N, i, t) \leq QoS \tag{13}$$

To solve this optimization problem we use an extension of the damped Newton method for dual problems to constraint optimization [15].

4.2 Connection Admission

When a subscriber requests to establish a connection to the satellite, the satellite management function which executes the bandwidth allocation, will predict the cell loss probability using our algorithm. Then the management function will deny or accept the connection. The optimal number of connections, N^* will be computed as follows:

$$N^* = max\{n | P_{loss}(\hat{\lambda}, \hat{\mu}, n, i, t) < QoS\} \tag{14}$$

4.3 Numerical Results

We first obtain the performance of the proposed congestion control algorithm (as depicted in Figure 2) for obtaining the maximum number of connections, N^*. We assume the following network parameters: $t = 0.25sec$, $R_p = 64Kbps$, $C = 1.544Mbps$, $\lambda = 0.05$, $\mu = 0.833$ and $QoS = 0.01$. In Figure 7 we observe that as the initial number of active sources increases, the maximum number of connections will decrease. This is due to the fact that as the number of initial active sources increases, the cell loss probability increases. To satisfy the QoS we, therefore, have to decrease the total number of connections, N^*. We provide the behavior of the algorithm with respect to the source shaping parameters. We assume the following system parameters: $t = 0.25sec$, $R_p = 64Kbps$, $C = 1.544Mbps$, $Y(0) = 10$, and $QoS = 0.0001$. In Figure 8, we can see that as the number of connection is increased (increasing the offered traffic to the system), the optimal source shaping parameter λ^* is decreased (i.e., decrease the traffic intensity) whereas μ^* is increased (i.e., spend more time in the idle state, fact that again decreases the traffic intensity).

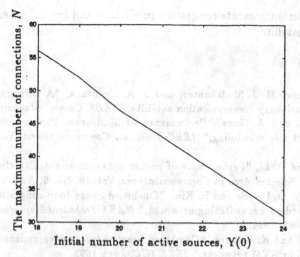

Fig. 7. The cell loss probability versus the maximum number of connections for different values of initial number of active sources

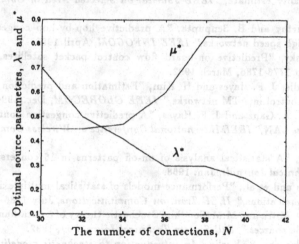

Fig. 8. Optimal source parameters versus the number of connections

5   Conclusions

In this paper we introduced a predictive congestion control procedure for future on-board satellite systems that support multimedia traffic. The proposed scheme considers the unique features of satellite communication, e.g., the congestion control scheme assumes no buffer on-board the satellite, accounts for the large propagation delays and has a relatively low computational complexity. The algorithm can be implemented in DSP chips to speed up its execution and reduce the power consumption on-board the satellite.

We have shown the importance of using a transient analysis to compute the cell loss probability as opposed to a steady state analysis. Therefore, the proposed method provides an accurate congestion prediction and control scheme executed on-board the satellite.

References

[1] W. D. Ivancic, M. J. Shalkhauser, and J. A. Quintana, "A network architecture for a Geostationary communication satellite, "*IEEE Comm. Magazine*, July 1994.

[2] I. Chlamtac and A. Ganz, "Performance of Multibeam Packet satellite systems with conflict free scheduling," *IEEE Tran. on Communications*, Vol. 36, No. 10, Oct. 1986.

[3] A. Ganz and B. Li, "Performance of packet networks in satellite clusters," *IEEE Journal on Selected Area in Communications*, Vol. 10, No. 6, Aug. 1992.

[4] P.P. Chu, W. D. Ivancic and H. Kim, "On-board closed-loop congestion control for satellite based packet switching networks," *NASA Technical Memorandum 106446*, AIAA-94-1062. 1994.

[5] W. Matragi and K. Sohraby, "Combined reactive/preventive approach for congestion control in ATM networks," *IEEE ICC*, June 1993.

[6] T. Murase and et al., "A call admission control scheme for ATM networks using a simple quality estimate," *IEEE Journal on Selected Area in Communications*, Dec. 1991.

[7] G. Ramamurthy and B. Senpupta, "A predictive hop-by-hop congestion control policy for high speed networks," *IEEE INFOCOM*, April 1993.

[8] E.A. Bobinsky, "Predictive on-board flow control packet satellites," *AIAA-92-2004-CP*, pp 1776-1786, March 1992.

[9] P. Tsingotjidis, J. F. Hayes and H. Kim, "Estimation and prediction approach to congestion control in ATM networks," *IEEE GLOBECOM*, Dec. 1994.

[10] Y. M. Jang, A. Ganz and J. F. Hayes, "A predictive congestion control for brodband wireless LAN," *IEEE International Conference on Wireless Communications*, July 1995.

[11] P.T. Brady, "A statistical analysis of on-off patterns in 16 conversations," *Bell System Technical Journals*, Jan. 1968.

[12] B. Maglaris and et al., "Performance models of statistical multiplexing in packet video communications," *IEEE Tran. on Communications*, July 1988.

[13] D. Anick, D. Mitra and M.M. Sondhi, "Stochastic theory of a data-handling system with multiple sources," *Bell System Technical Journal 61*, 1982.

[14] H. M. Taylor and S. Karlin, *An introduction to stochastic modeling*, Academic Press Inc., 1984.

[15] S. P. Han, "A globally convergent method for nonlinear programming," *Journal of Optimization Theory and Applications*, Vol. 22, July 1977.

Single-Frequency Packet Network Using
Stack Algorithm and Multiple Base Stations

N. D. Vvedenskaya * and J. P. M. G. Linnartz †

This paper evaluates a new method to combine contiguous frequency reuse with random access in wireless packet-switched networks. We consider a radio network with two base stations receiving packets transmitted by a large population of mobile users. Both stations share the same channel and therefore transmissions to one base station in one cell interfere with transmissions to the other base station in the other cell. A simulation of a two-cell system is performed, to study the interaction of retransmission traffic in two cells. To consider the performance of such system analytically, we model a one-cell system with time-varying channel properties: if only one station is receiving messages from its cell, the channel is assumed to be in "good state"; if terminals in both cells are busy the channels are assumed to be in "bad state". For conflict resolution, the stack-algorithm is used. The packet delay is calculated. Our results confirm that in wireless packet networks, it is advantageous to allow neighbouring cells to share the same channel.

Keywords : wireless networks, packet radio, multiple access, stack algorithm, receiver capture, packet delay, markov communication channels, cellular radio networks.

I Introduction

Many of the solutions for sharing communication resources among multiple users and services that have been developed for wireline networks become inefficient or require modifications if they are used for radio resource management in wireless networks. The common goal of most multiplexing, switching and multiple access schemes to is dynamically assign bandwidth during certain periods of time. If such techniques are used in radio data or multimedia networks, the radio spectrum needs to be reused spatially in an efficient (and presumably dynamic) way. However, this issue has hitherto mostly been addressed separately from allowing multiple users to share the same bandwidth–time resources within one cell. This is for instance illustrated by the fact that most existing mobile data networks use a cellular frequency reuse pattern, and within each cell a random-access scheme is used independently of the traffic characteristics in other cells.

This paper shows that if the stack-algorithm is used for random access in one cell (i.e., to handle intracell interference), it can also mitigate the effect of bursty (intercell) interference between neighbouring co-channel cells. Hence, the stack-algorithm allows substantially denser frequency reuse than currently used in radio telephony. In particular, we consider the performance of a wireless network with two base stations receiving packets

*N.D. Vvedenskaya is with Institute for Information Transmission Problems, 19 Bol'shoi Karetnyi, Moscow 101447, GSP-4, Russia, tel: +7 (095) 2995002, fax: +7 (095) 2095079, e-mail: NDV@ippi.ac.msk.su. Her work on this paper was partly sponsored by Netherlands Organization for Scientific Research NWO 713-229 and by the Russian MINNAUKA under the Project "Dostup".

†J.P.M.G.Linnartz currently is with Philips Natuurkundig Lab (Philips Research), mailstop WY 8.1, Holstlaan 4, 5656 AA, Eindhoven, the Netherlands, tel: +31 (40) 2742302, fax: +31 (40) 2744660, e-mail: linnartz@prl.philips.nl. Previously he was with the T.V.S. group at T.U. Delft.

transmitted by a large population of mobile users. The users that are in the cell area of one of the base station transmit their packets to this base station. To do that they compete for (random) access, according to the free-access stack-algorithm with feedback from the base station [1 - 3]. If both base stations share the same channel, transmissions in one cell interfere with transmissions in the other cell. The stack algorithm retransmission scheme not only avoids instability within one cell but also migitates the effect of interference between cells. We simulate the interaction of retransmissions in two interfering cells, each with its own stack algorithm. To compute the performance of a network with two base stations, we model the performance of one-station system with time-varying channel properties. Whenever only one base station is busy (or both are silent), the base stations are assumed to have a relatively reliable channel ("good state"). When both stations are busy, the stations are assumed to have an imperfect ("noisy") channel ("bad state"), due to the mutual interference. To address system performance mathematically, we approximate transitions from one state to the other as an autonomous Markov on/off process. State transitions are assumed to be independent of transmissions in the cell suffering its interference.

The performance of a two-cell system with a common channel is compared with the performance of two-cell system with different channels in each cell. We do not consider any CDMA spreading, as we have found in [4] that such modulation techniques may not be advantageous in our case of packet data transmission. To avoid interference in conventional cellular systems, two different channels would be needed, each with half the total available bandwidth. Hence, the transmission time increases by a factor two. Moreover the arrival rate expressed in (new) packets per time slot increases by factor of two. Particularly under large traffic loads, this appears to lead to a significantly larger delay than our proposal of allowing neighbouring cells to share bandwidth, thus accepting mutual interference.

II Model of Access Scheme

The uplink random access channel is considered to be time slotted and synchronized at time slot level. The slots start at $t = 1, 2, 3, \ldots$. A slot that starts at $t = n$ is called slot n. A full packet is always transmitted within one slot time. A new packet is transmitted for the first time in the first slot following its arrival (free access). The input flow of packets in a cell is modelled as a spatially and temporally uniform Poisson process with flow rate λ packets per a slot. [‡]

Any packet that captures the receiver leaves the system. Terminals are informed about packets successfully arriving at the base station through an instantaneous feedback channel. For unsuccessfully transmitted packets, i.e., for those packet that are not acknowledged, each terminal has a buffer to keep one packet for retransmission. Each packet transmitted in a slot without capturing the receiver is either retained in the user's buffer (with probability 1/2) or is retransmitted in the next slot (with probability 1/2). The main idea here is to split the backlogged terminals into different small groups after a collision. The probability of having a message collision in one of the groups rapidly vanishes after several splits. To ensure that each terminal keeps track of the group

---

[‡]More sophisticated traffic models may better reflect the burstyness of multi-media traffic, but we believe that such refinements would not fundamentally change the principal conclusions of our investigation.

to which its packet belongs, a stack counter l_n is associated with the user's buffer. The counter contents change from slot to slot according to the stack-algorithm rules, following the feedback information. Terminals transmit (only) when $l_n = 0$. Generally speaking, the stack counter increases when a conflict is reported in a slot and decreases when a slot is idle. The idea is that after a collision, all backlogged groups of packets have to wait before the current group has resolved its collision. As the current group splits, a new level is "inserted", and all existing groups increase their stack counter. Intuitively, the stack algorithm protocol is a "randomised first come - last served" protocol.

We address the particular stack algorithm that uses a ternary "idle slot/success/conflict" feedback [3]. Moreover, in case of capture, the feedback channel reports which packet captured the base station. The protocol rules of the Stack Algorithm addressed here are:

1. A packet transmitted in slot n for the first time (i.e. the packet generated in slot $n - 1$) has $l_n = 0$.

2. If $l_n = 0$ for a packet, the packet is transmitted in slot n. If $l_n > 0$, the packet is not transmitted in slot n.

3. If $l_n = 0$ for a packet and a capture is reported in slot n, for that packet, then the successful packet, leaves the system. For other transmitting packets (if any) $l_{n+1} = l_n = 0$.

4. If $l_n = 0$ for a packet and a conflict is reported in slot n, then $l_{n+1} = 1$ with probability $1/2$ and $l_{n+1} = 0$ with probability $1/2$.

5. If $l_n > 0$ for a packet and a conflict is reported in slot n, then $l_{n+1} = l_n + 1$.

6. If $l_n > 0$ for a packet and slot n is reported idle, then $l_{n+1} = l_n - 1$.

7. If $l_n > 0$ for a packet and a capture is reported in slot n, then $l_{n+1} = l_n$.

The delay of a packet is defined as the length of a time interval between the start of the packet's first transmission and the start of its successful transmission. Note, that the time a packet spends in the system includes the random waiting time till the beginning of the first transmission, the delay, and one slot of successful transmission. The mean duration T a packet spends in the system is equal to the mean delay D plus one and a half slot duration.

The value of λ_{cr} is called the throughput of the system, if there exists such λ_{cr} that for any input flow rate λ, $\lambda < \lambda_{cr}$, all packets are transmitted with finite delay with probability 1, and the mean delay D is finite, while for $\lambda > \lambda_{cr}$, delays grow beyond any finite bound. λ_{cr} depends on channel properties.

For one-cell system and for given channel probabilities, the mean delay D can be expressed in terms of the solutions of the linear algebraic equations [1 - 4].

For an ALOHA system [7 - 11] with an infinite population of user terminals without capture, it has been shown that no such $\lambda_{cr} > 0$ exists. Only for ALOHA systems with capture, the non-zero value $\lambda_{cr} = \lim \inf_{k \to \infty} \pi_k$ is found [10], with π_k the probability of capture if k signals are competing. Moreover, it has been shown that for mobile ALOHA, the capture probability in a case of an infinite number of competing terminals is non-zero only under unrealistic assumptions about the spatial distribution of terminals [5, 11]. Therefore we limit our results to the stack algorithm, which is an improved retransmission strategy of the basic ALOHA protocol. Examples of other methods to stabilize the ALOHA multiple access protocols are presented and discussed in [12 - 13].

III Simulation of Two-Cell System

In our single-frequency two-cell system, both base stations use the same stack-algorithm rules, but without mutual coordination. The cell areas overlap. The performance of this two-station system depends on the probabilities described in following paragraphs.

Let in cell $i, i = 1, 2$, slot n be idle. If in cell $j, j \neq i$, slot n is idle too, then in cell i this slot is reported idle. If in cell j slot n is not idle, and the total received power $S^{(i)}$ is below a detection threshold Q_{th}, then in cell i slot n is reported idle. Otherwise (if $S^{(i)} > Q_{th}$), the receiver in i will attempt to detect a packet, even though all available packets arrive from cell j. In cell i capture is reported with probability $p_{0s} = p_{0s}^{(i)}$ and a conflict is reported with probability $p_{0c} = p_{0c}^{(i)}$. The values of $p_{0s}^{(i)}$ and $p_{0c}^{(i)}$ depend, among other things, on the propagation distances between the base station i and the active terminals. In our simulation we use known location of terminals, which remain fixed from the first arrival of a packet until the successful reception of the packet. However, new packets arrive in terminals with randomly chosen locations.

Let one terminal transmit in cell $i, i = 1, 2$, during slot n. If in cell $j, j \neq i$, slot n is idle, then in cell i capture occurs so the packet leaves the system, with probability $p_{1s} = p_{1s}^{(i)}$, and a conflict is reported with probability $1 - p_{1s}^{(i)}$. In our simulations, we compute $p_{1s}^{(i)}$ depending on the distance between the base station and the terminal, and on noise levels. If slot n is not idle in cell j, then in cell i capture occurs (and is correctly reported over the feedback channel) with probability $p_{1s}^{(i)}$ and a conflict is reported with probability $1 - p_{1s}^{(i)}$. In this case the value of $p_{1s}^{(i)}$ is computed taking into account the distances between the base station i and all active terminals, which now are in cells i and j.

Let several, say k_i, $(k_i = 2, 3, ...)$ terminals transmit in cell i and k_j, $(k_j = 0, 1, ...)$ terminals transmit in cell j. Then, in this cell i capture is reported with probability $p_{cs}^{(i)}$ and a conflict is reported with probability $1 - p_{cs}^{(i)}$. The value of $p_{cs}^{(i)}$ depends on the distances between the base station i and the active terminals in cells i and j, and on the system noise floor.

Note, that the feedback information in cell i does not affect the behavior of terminals in cell j. In our model, transmissions within a cell either result in a capture or a conflict, but are never reported as an idle slot by the intended base station.

We adopt a generally accepted model for (outdoor) macrocellular propagation considering a Rayleigh-fading channel with "40 log d" plane-earth path loss [6]. That is, signal from the m-th $(m = 1, 2, ..., k)$ transmitting terminal with distance r_{im} is received at base station i with local-mean power $S_m^{(i)} = r_{im}^{-4}$. Because of Rayleigh fading, the instantaneous power is an exponential r.v. with mean $S_m^{(i)}$ [6].

For k, $k = 1, 2, ..$ transmitting terminals, with the distances r_{im}, $m = 1, ..., k$, the total (local mean) received power $S^{(i)}$ at base station i is $S^{(i)} = \sum_{m=1}^{k} S_m^{(i)} = \sum_{m=1}^{k} r_{im}^{-4}$. In this paper, this variable is used to determine whether the slot is idle or occupied. A refined computation may consider instantaneous, rather than local-mean powers. We assume that the received (instantaneous) power is constant during packet reception ("slow fading" [5]). For most modern mobile communication systems, including systems with GSM, IS54 or DECT parameters, this appears to be a reasonable assumption.

If several terminals (say, k, $k = 1, 2, ..$ terminals) transmit in a slot (may be in different cells), we assume that a packet captures the base station if the instantaneous signal-to-noise-plus-interference ratio exceeds a threshold z. By solving the integrals for the probability density functions of received power, i.e., considering path loss and Rayleigh fading, the capture probability for terminal 1 ('tr. 1') at distance r_{i1} from base station i can be expressed as [5]

$$p_{cs,1}^{(i)} = \Pr\{capture\ by\ tr.1\} = \exp(-zQ_n r_{i1}^4) \prod_{m=2}^{k} r_{im}^4/(zr_{i1}^4 + r_{im}^4).$$

where z is a receiver parameter, typically $z = 4$ and Q_n is the noise floor of the system. A realistic value could be on the order of a 10 dB (local-mean) signal to noise ratio at the cell boundary, where $r_{i1} = 1$, so the local-mean power is $S_m^{(i)} = 1$. So with $Q_n = 0.1$ (-10 dB), the local-mean ratio becomes $S_m^{(i)}/Q_n = 10$. For this value, packets that do not experience interference are only lost occasionally because of noise and fading. The probability that any one packet (out of k packets) captures base station i is $\Pr\{capture\ in\ cell\ i\} = \sum_{m=1}^{k} p_{cs,m}^{(i)}$, which in practice is almost identical to $\max_{m=1,...,k} p_{cs,m}^{(i)}$, i.e., only the packet from the nearest terminal is likely to capture. If only a single terminal (say, the one with index 1) transmits without interference, the probability of capturing the receiver reduces to $p_{cs,1}^{(i)} = p_{1s}^{(i)} = \exp(-zQ_n r_{i1}^4)$.

IV Approximate Markovian Model for Interference

The above (detailed) model was used for simulation, but it makes analysis tedious. To compute results for a two-cell system, we model it as a one-cell system with a two-state channel model. To reflect the idle and busy periods of sessions of transmissions in the other cell, the channel of the "victim" cell system has two states ("good" or "bad"). Markovian transitions from one-state to the other are considered: if during slot n the system is in state $i, i = 1, 2$, then it will stay in this state during slot $n + 1$ with probability q_i, and will be in different state with probability $(1 - q_i)$. The mean duration of being in state i is equal to

$$L_i = 1/(1 - q_i),\quad i = 1, 2.$$

Autonomous Markovian transitions as modelled above are only an approximation of the changes from an idle to a busy periods in the interfering cell. As we explored in our simulation, retransmissions in the two cells interfere, so busy periods interact in a very complicated manner. In the analysis, we simplify the interaction as an autonomous mechanism, however, the average values of busy and idle periods are obtained recursively, considering both intercell and intracell interference.

Let the system be in state i during a slot n. We model the effect of interference as follows: If slot n is idle, then a capture is reported in this slot with probability $\pi_{0s}^{(i)}$, and a conflict is reported with probability $\pi_{0c}^{(i)}$, the slot is reported idle with probability $(1 - \pi_{0s}^{(i)} - \pi_{0c}^{(i)})$. Here event "0s" models excessively strong signals from the other cell, to the extent that an "alian" packet captures the base station. We assume that the base does not immediately recognize that the capturing packet comes from the other cell, so it reports the capture. Event "0c" models interference power from the other cell that is strong enough to obscure the emptiness of the slot, but it does not result in any packet capturing the base station.

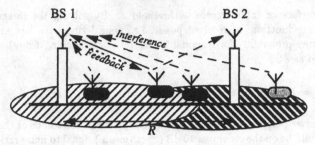

Figure 1: Multiple-access radio system with two base stations and overlapping cells

If in slot n one packet from that cell is transmitted, then a capture occurs (and is reported) with probability $\pi_{1s}^{(i)}$, and a conflict is reported with probability $(1 - \pi_{1s}^{(i)})$. Here event 1 models interference from the other cell that destroys the packet, or fading and noise that impair successful transmission.

If in slot n $k, k > 1$, packets are transmitted, then one packet captures the receiver with probability $\pi_{ks}^{(i)}$, and a conflict is reported with probability $1 - \pi_{ks}^{(i)}$. Here event "ks" models not only interference from within the cell but also interference from the other cell, and noise on the random-access channel.

We assume that in the state "1" the base station can make much better observations on transmissions in its cell, and provides almost perfect feedback channel, $\pi_{0s}^{(1)} = \pi_{0c}^{(1)} = 0$. In state "2" however, the observations are obscured due to interference from the other cell, so the base station provides imperfect feedback, $\pi_{0s}^{(2)} > 0$, $\pi_{0c}^{(2)} > 0$, $\pi_{ks}^{(2)} > 0$, and $\pi_{ks}^{(2)} < \pi_{ks}^{(1)}$. We averaged the probabilities $p_{0s}, p_{0c}, p_{1s}, p_{ks}$ over (simulated) terminal locations to obtain the values for $\pi_{0s}, \pi_{0c}, \pi_{1s}, \pi_{ks}$, respectively. Our simulation used uniformly distributed terminals, but their location is kept fixed between generation and successful (re-) transmission of a packet. In this case, the retransmission traffic becomes non-uniform, as most retransmissions are made by remote terminals (large r_{im}). Rayleigh fading was considered to be independent from transmission attempt to attempt. Note that the assumption that terminal distances are kept fixed can substantially affect the performance [11], but its effect is ignored in many other (purely analytical) investigations.

V Results and Discussion

Simulations have been performed for a one dimensional two-cell system with a common radio channel as illustrated in Figure 1: The terminals and the base stations are located on a line ("a highway model"). The base stations are in the centers of their cells.

The locations of the terminals are assumed to be uniformly distributed in the cell area, and independent. The distance between the two base station is taken to be $R, 1 < R < 2$. We assume that the terminal only considers feedback from one base station. A terminal m can be in the cell of base station i, i.e., it can rely on its feedback if the distance r_{im} is less then 1.

The results for the one-cell system with autonomous interference have been computed analytically. To calculate the performance of a one-station system, we investigate "busy

295

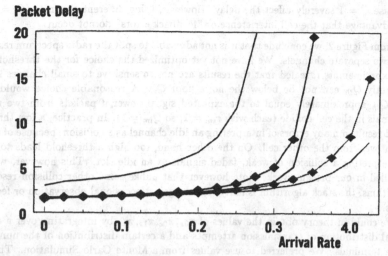

Figure 2: Delay versus normalized message arrival rate. Single frequency net: Simulation (blocks) and compuation (base station separation $R = 1.5$ and 1.9), compared with two separate channels

sessions". Roughly speaking a busy session is a time interval during which there are some packets being transmitted or waiting to be retransmitted in the stack of some terminal. A session ends as soon as all terminals have an empty stack and the channel is idle. To analyse a two-cell system with equal message arrival rates, we consider such a one-cell, two-state system. The mean length of busy session L_b is equal to the mean length L_2 of the state "2" ("bad state"). The mean duration L_1 of the system being in state "1" ("good state") is taken equal to the mean length of the slots without new packets, $L_1 = 1/(1 - e^{-\lambda})$. Therefore $q_1 = e^{-\lambda}$. For given channel properties the value of L_b can be expressed in terms of the solutions of the linear equations (similar to the computation of the delay value D); q_2 is found by iteration, to satisfy the equality $L_2 = L_b$.

The results of computations for one-cell, two-state system follow the results of simulations with 3% accuracy for λ close to .1 and with 10% accuracy for λ close to .2. Note, that our simulations accurately model the interaction between the two stack-algorithms in both cells at slot and packet level, rather than simplifying the interaction as an autonomous Markovian interference process. Apparently, for large λ this effect becomes significant, so the results from computation and simILation diverge near λ_{cr}.

The results of simulations for a one-channel two-cell system with input flow rate λ in both cells ($C = 1$,) are compared with the results for single-cell, two-state model ($C = 1$), and with the results for two-cell system with separate channels but with the same propagation and noise and capture parameters. A fair comparison requires us to accommodate the same total bandwidth to each system. As in the case with $C = 2$ reuse, each channel only has half the bandwidth, time slots need to be twice as long. Effectively, for a given arrival rate per second, this corresponds to a flow rate of 2λ new packets per slot. This increase in traffic load and also the fact that time slots are twice as long as in the case $C = 1$ severely affect the delay. However, using different channels ($C = 2$) has the advantage that the cell interference or "feedback errors" do not occur.

From Figure 2, we conclude that it is not advisable to split the radio spectrum resources into two separate channels. We have not yet optimized the choice for the threshold Q_{th}, but experimemnts revealed that the results are not so sensitive to small changes in Q_{th}. Evidently Q_{th} can not be below the noise floor Q_n. A reasonable choice would be to take Q_{th} approximately equal to the expected signal power if packets from two remote terminals in the cell collide (each with $r_{im} = 1$, so $Q_{im} = 1$). In practice, a low threshold would result in many errors of interpreting an idle channel as a collision, because of signals blown over from the other cell. On the other hand, too high a threshold leads to errors of interpreting a collision of weak, faded signals as an idle slot. This however, was not modelled in our present work. Note however that unlike some other collision resolution algorithms, the stack algorithm is fairly robust against occasional observation or feedback errors.

We could in theory obtain the values of $\pi_{0s}, \pi_{0c}, \pi_{1s}, \pi_{ks}$ by integrating over a certain spatial distribution of transmission attempts and a certain distribution of the number of active terminals. We preferred to use values from a Monte Carlo Simulation. This had not only the advantage the capture probabilities more accurately reflect that most (re) transmissions arrive from areas relatively far away from the base stations (as previously argued), but also it more accurately considers that the number of packets in a slot not exactly follows a Poisson distribution.

For the parameter values in Figure 2, i.e., for a receiver capture threshold $z = 4$, a base station separation $R = 1.9$, noise floor $Q_n = 0.1$, and idle/busy threshold of $Q_{th} = 2$, the probability that a single packet is not lost (in the absence of intercell and intracell interference) is $\pi_{1s}^{(1)} = 0.96$. So, with probability 0.04, the base station sees this transmission as a collision (it sees some radio energy, but can not detect the signal without bit errors). If multiple packets ($k = 2, 3, ..$) are transmitted in the cell, but no interference is seen from the other cell, the capture probability is found as $\pi_{kc}^{(1)} = 0.6$. In the "bad state", i.e., if the other cell causes interference, an empty slot is seen as a collision with probability $\pi_{0c}^{(2)} = 0.05$. Moreover, with probability $\pi_{01}^{(2)} = 0.01$, a packet from other cell is strong enough to capture the base station in the reference cell. Interference from other cell is rarely strong enough to destroy the transmission of a single packet in the reference cell: $\pi_{1s}^{(2)} = 0.88$, which is only marginally less than $\pi_{1s}^{(1)}$. Similarly in case of a collision in the reference cell, the capture probability is not significantly affected if intercell interference is also present: $\pi_k^{(2)} = 0.53$.

Our computations show that if the flow rate λ is sufficiently low, and the noise is sufficiently low (as is the case in the given examples), then the delay is less for the single-frequency system (with $C = 1$), both for the simulated model and for the simplified model of a one-cell, two-state system. For larger rates, it is still beneficial to share the same channel. However, the results from the simulation diverge from the computed results.

This suggestion to adopt a single-frequency concept can also be understood intuitively from considering a comparable two-cell ALOHA system, which for sake of simplicity we assume to be stable [11]. If one allows the two base station to share a common channel the delay will be small as collisions are rare. If on the other hand separate channels are used, the slot duration is doubled, which in the limiting case of very light traffic appears the only relevant effect on delay. At larger traffic load increases it furthermore becomes relevanthow heavily the channels are loaded. In the single-frequency collisions occur with probability $\exp(-2\lambda)$ because of packets from two cells. In the separaate channel system, collisions also occur with probability $\exp(-2\lambda)$, but now because the slot length is twice as long while we only see traffic from one cell. However, as capture occurs in radio networks, the former system suffers less from the collisions, as one half of all interfereing packets come from the other cell, and may be too weak to cause harmful interference.[§]

VI Conclusion

Our results show that in a narrowband two-cell system with the stack algorithm used for conflict resolution, the same radio channel can be used in both cells. At low traffic loads, its performance can be approximated by one-cell system with two-states of the system, and with autonomous Markovian transitions from one state to the other.

Our results suggest that in wireless networks with bursty traffic, it may be advantageous to allow nearby cells to use the same channels. The free-access algorithm not only efficiently controls retransmissions needed after a collision within the cell, but also makes the system robust against high levels of interference from co-channel cells. This is in contrast to conventional cellular frequency reuse used for mobile telephones, where adjacent cells need to use different frequencies.

VII Appendix

Here we present the equations used in calculations of the packet mean delay in a one-cell, two-state system. Let us begin with the rigorous definition of a session and define a session that starts at slot n. We introduce at slot n an additional packet to the system with $l_n = 1$ for this packet. Let $l_{n_1} = 0$ for the first time, $n_1 > n$. The time interval $[n, n_1)$ is called a session. A session is called a k-session if k packets are transmitted at it's first slot n. A busy session is a session with $k > 0$. Denote by $h_i(k)$ the mean length of a k-session if it starts in state i. Denote by $d_i(k)$ the mean sum of delays of all packets, that are successfully transmitted in a k-session, if this session starts in the state i. The mean packet delay D is defined as

$$D = \frac{\sum_{i=1,2}\sum_{k=0}^{\infty} p_k Q_i d_i(k)}{\lambda \sum_{i=1,2}\sum_{k=0}^{\infty} p_k Q_i h_i(k)}; \qquad p_k = \exp(-\lambda)\lambda^k/k!.$$

The values of $h_i(k)$ and $d_i(k)$ are found as solutions of linear equation systems that differ in free terms only. We present here the equation for $h_i(k)$. In the equations for $d_i(k)$ the free terms depend linearly on $h_i(k)$.

$$h_i(0) = 1 + \pi_{01} \sum_{m=0}^{\infty} p_m [q_i h_i(m) + (1 - q_i) h_j(m)] + \pi_{0c} F_{i,0,0},$$

[§]An interesting corollary is that packet-switched systems require different frequency reuse methods than used in telephony: One base station covering the entire service area gives smaller delays (by a factor of two) than a (more expensive) system with two base stations, each with only half the bandwidth available. Both systems have the same message arrival rate per time slot, but the former has shorter time slots.

$$h_i(1) = 1 + (1 - \pi_1) \sum_{m=0}^{\infty} p_m[q_i h_i(m) + (1 - q_i)h_j(m)] + \pi_1(F_{i,1,0} + F_{i,0,1})/2,$$

$$h_i(k) = 1 + \sum_{l=0}^{k} 2^{-k} C_k^l (F_{i,l,k-l} + F_{i,k-l,l}),$$

where

$$F_{i,l,k} = \sum_{m=0}^{\infty} p_m \{q_i[h_i(l+m) + s_{i,i}(l+m)h_i(k-l+m) + s_{i,j}(l+m)h_j(k-l,m)] +$$

$$+ (1 - q_i)[h_j(l+m) + s_{j,i}(l+m)h_i(l+m) + s_{j,j}(l+m)h_j(k-l+m)]\},$$

$$C_k^l = k!/l!(k-l)!, \quad i = 1, 2, \quad i \neq j, \quad k > 1.$$

Here $s_{i,j}(k)$ is the probability that a slot is in a state j if it follows a session of multiplicity k, that starts in a state $i, i, j = 1, 2$. To calculate $s_{i,j}(k)$ we first calculate the probability, that a k-session, that starts in state i is of length l. The recursion on l permits to find these probabilities. The Markov chain transition from one state to the other permits to get $s_{i,j}(k)$.

To estimate the value of D with 0.1% accuracy it is sufficient to solve a finite system of equations for systems for $h_i(k), d_i(k), k < 10 \div 12$. The value of λ_{cr} is estimated with 0.01% accuracy by the value of λ, for which $D > 500$.

VIII   References

1. N. D. Vvedenskaya and B. S. Tsybakov, "Packet delay with multi-access stack- algorithm," *Problems of Information Transmission*,2, v.16, (1984).
2. N. D. Vvedenskaya, "Multiple access stack-algorithm with imperfect feedback," *Tr. of Fifth IEEE Int.Symposium on PIMRC'94*, 4, 1126–1128 (1994).
3. N. D. Vvedenskaya, J. C. Arnbak, B. S. Tsybakov, "Improved Performance of mobile data networks using stack-algorithms and receiver capture," *March 1994 Int.Zurich Seminar on Digital Communications*, Springer-verlag, Lecture Notes in Computer Sci.,**783**, 464-475 (1994).
4. J. P. M. G. Linnartz and N. D. Vvedenskaya, "Optimizing delay and throughput in packet switched CDMA network with collision resolution using stack-algorithm," *Electronics letters, 1st Sept.* , **30**, No.18, 1470–1471.
5. J. P. M. G. Linnartz, *Narrowband Landmobile Radio Networks*, Artech House, Boston, 1993.
6. W. C. Jakes, Jr, *Microwave Mobile Communication*, New York: J. Wiley and Sons, 1978.
7. *MULTIPLE ACCESS COMMUNICATIONS. Foundations for Emerging Technologies*, Edited by N. Abramson, IEEE Press,1993.
8.A.B. Carleial and M.E. Hellman, "Bistable behavior of ALOHA-type systems", IEEE Tr. in Comm., Vol. COM-23, No. 4, Apr. 1975, pp. 401-410.
9. L. Kleinrock and S. S. Lam, "Packet switching in a multiaccess broadcast channel- performance evaluation", *IEEE Tr. on Comm.*, Vol. COM-23, No. 4, Apr. 1975, pp. 410-423.
10. S. Ghez, S. Verdu and S. C. Schwartz, "Stability properties of slotted ALOHA with multipacket reception capability", *IEEE Tr. Automatic Control*, Vol. 33, No.7, July 1988, pp. 640-649.
11. C. van der Plas and J. P. M. G. Linnartz, "Stability of mobile slotted ALOHA network with Rayleigh fading, shadowing and near-far effect", *IEEE Tr. Veh. Tech.*, Vol. 39, No. 4, November 1990, pp 359-366.
12 J.L. Massey, "Collision-resolution algorithms and random access communication in multi user communications", New York, Springer-Verlag, 1981, pp. 73-137.
13 J.I. Capetanakis, "Tree algorithms for packet broadcast channels", *IEEE Tr. I.T.*, Vol. IT-25, 1979, pp. 505- 515.

Broadband Access in RECIBA B-ISDN Experimental Platform[1]

Juan I. Solana, Jesús Mariño and Rafael Caravantes
Telefónica Investigación y Desarrollo
Emilio Vargas 6
28043 Madrid, Spain

Telf: (341) 337 4630
Fax: (341) 337 4502
e-mail: jisq@tid.es

Keywords Broadband Access, Network Terminator

Abstract

Little attention has been paid so far to broadband access Network Termination 1 equipment (NT1) due to an apparent lack of useful functions which can be performed by this type of elements. Experience gained on emerging broadband services performance have shown some problems when used through public ATM networks mainly regarding excessive peak traffic and highly bursty traffic profiles and lack of OAM facilities at the ATM layer. Most of these problems might be solved by additional functions performed by NT1 elements. This paper presents an implementation of a SDH based broadband access network emphasizing the ideas and functions which can make this type of equipment useful and details the experience gained on a number of real broadband multimedia services demonstrations.

1.- Introduction

Since 1988, Telefónica has undertaken an important activity in the broadband experiments within the RECIBA project [1]. (This acronym stands for *Broadband Communications Experimental Integrated Network*). The RECIBA project has involved four areas of Broadband Communications: Network architecture, Signalling protocols, Multimedia services development and Network Management.

The major targets of RECIBA have been: a) To provide an open and flexible platform for experimentation and benchmarking of new multimedia services, b) Availability of an experimental B-ISDN network for empirical measurements and evaluation of QoS parameters, eventually in conjunction with other European demonstrators and c) To gather a technological experience to enable Telefónica to define performance objectives in terms of network provisioning and OAM issues and actively participate in standardization bodies such as ITU-T and ETSI.

---

[1]The work described in this paper has been partially sponsored by the Spanish Goverment Program PlanBA.

This experience has helped Telefónica in the settlement of Spanish segment of the Pan European ATM pilot. Furthermore, the Spanish National Host (SNH) is based on RECIBA. The figures 1a and 1b show the network scenarios located at TI+D laboratories and at the *Universidad Politécnica de Madrid* (ETSIT).

A Spanish government initiative program on broadband communications called PlanBA has been underway since 1992 through 1995. RECIBA constitutes the backbone network of the PlanBA network. One project included in the PlanBA program was aimed at developing the technology and network elements required for broadband access networks at 622.08 Mbs. These elements and their constituting building blocks are described in detail in the following paragraphs. This project, called TR1-622, was based on the existing experience at 155.52 Mbs and finalized on June 1995. Its objectives were: a) to develop the building blocks for ATM cell transmission over SDH STM-4-4c frame structures at 622.08 Mbs according to the ITU-T I.432 recommendation, b) to develop electro-optical modules suitable for transmission both at 155.52 and 622.08 Mbs and c) to experiment with the performance monitoring functions and parameters as described in ITU-T recommendations G.826 and I.356. The result of this project is the availability of two types of elements: NT1-622 and ETM-622 to implement broadband access networks at 155.52 and 622.08 Mbs binary rates. The element ETM-622 is a line interface board compatible with the architecture and mechanical constraints of the RECIBA platform, which include physical and ATM layers functions.

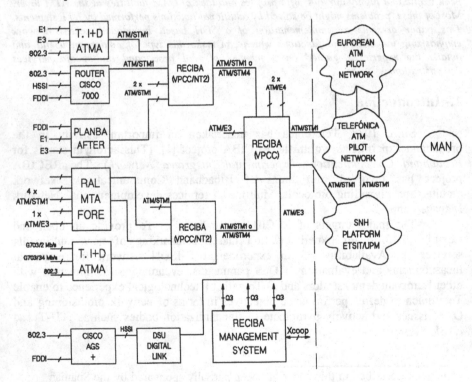

Figure 1a: Spanish National Host scenario located at TI+D

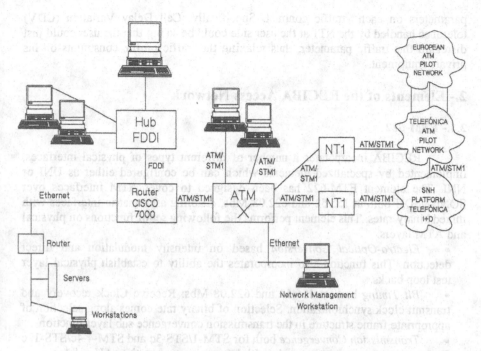

Figure 1b: Spanish National Host scenario located at UPM ETSIT

Criteria and performance objectives in a private environment (an example of a private premises scenario is shown in figure 1b) very often conflict with requirements and constraints imposed by the public network. This fact is especially acute on B-ISDN networks. For this reason it is convenient to rely on a especial purpose equipment (NT1) devoted to establish the link between the public and private networks. Typical functions performed by NT1 equipment include optical regeneration, fault management, performance monitoring and reporting at the physical layer (OAM flows F1 and F2). The experience gained on a number of broadband services (such as video-conference, video on demand, LAN interconnection, Computer Supported Cooperative Work, etc.) demonstrations held during 1994 and 1995 have shown some deficiencies in existing commercial ATM switching equipment which might be overcome by adding special functions to those typically performed by the NT1 equipment. Such functions include ATM layer OAM flows (F4 and F5) , VPI translation to avoid the use of reserved values and traffic shaping.

The following chapters describe in detail the implementation of broadband access network based on SDH at 622.08 Mbs including asymmetric interfaces 155.52 / 622.08 Mbs. It also describes the experiences on broadband services demonstrations which justify the addition of a number of functions in order to guarantee the compatibility between the public and private ATM networks. Furthermore, the traffic shaping functions may simplify the required traffic parameters on each traffic contract. Specifically, Cell Delay Variation (CDV) tolerance handled by the NT1 at the user side could be so big that the user could just disregard this traffic parameter, thus relaxing the traffic profile constraints on his private equipment.

2.- Elements of the RECIBA Access Network

2.1.- ETM-622

RECIBA incorporates a number of different types of physical interfaces, implemented by specialized elements which can be configured either as UNI or NNI. The element ETM-622 has been designed to cover ATM interfaces over SONET/SDH both at 155.52 and 622.08 Mbs, enabling asymmetric interfaces with mixed binary rates. This element performs the following set of functions on physical and ATM layers.

- *Electro-Optical Conversion* based on intensity modulation and direct detection. This function also incorporates the ability to establish physical layer test loop backs.
- *Bit Timing* both at 155.52 and 622.08 Mbs: Receive Clock recovery and transmit clock synchronization. Selection of binary rate comes along selection of appropriate frame structure in the transmission convergence sub layer function.
- *Transmission Convergence* both for STM-1/STS-3c and STM-4-4c/STS-12c frame structures: regenerator and multiplex section termination, AU pointer processing high order path termination, HEC generation, correction and verification and speed decoupling.

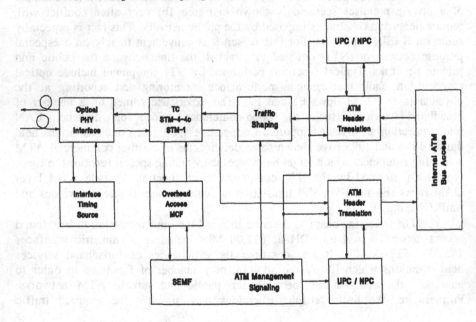

Figure 2.- ETM-622 Architecture

- *Overhead Access and Message Communication Function (MCF)*: All frame Overhead bytes (SOH and POH) are available for different applications. The present implementation offers the Message Communication Function for network management over de Data Communication Channels (DCC) channels: D1, to D3 and D4 to D12.
- *Traffic Shaping:* The ATM cell flow towards the optical interface may be modeled on a VPI or VPI/VCI basis.
- *ATM header translation*: VPI or VPI/VCI translation. This function includes the possibility to extract/insert ATM cells, which belong to reserved VPC / VCC, to/from software modules to support ATM OAM and signalling.
- *Usage parameter control / Network parameter control.* Based on the leaky-bucket algorithm.
- *Internal interface to the ATM switch.* The ATM cell flow may reach the switching fabric either through a dedicated internal bus or a statistically multiplexed bus. This multiplexed bus acts as traffic concentrator/distributor. This feature is specially suitable for the architecture of RECIBA switching elements.
- *Synchronous Equipment Management Function (SEMF) and ATM Management*: These functions are accomplished by software control modules which monitors alarm status and statistical information or performance monitoring from the TC function (OAM flows F1, F2 and F3) and from ATM Header Translation Function (OAM flows F4 and F5) according to ITU-T recommendations I.610, G.826 and I.356.
- *Signalling*: This is accomplished by software control modules by reserved VPI/VCI ATM cells which are inserted and extracted to and from the ATM flows by use of the ATM Header Translation function. The present implementation includes basic monoconection calls (Q.2931) and point to multipoint monoconection calls (Q.2971).

Figure 2 shows the architecture of the element ETM-622. The block depicted in this diagram follows the division of functions enumerated above. This architecture have been optimized by the use of five ASIC's.

- *LIBRA*: Synchronous Mux/Demux optimized for operation at 622 Mbs.
- *TRJM*: ATM mapper over SDH/SONET frame structures STS-1, STM-1 / STS-3c and STM-4-4c / STS-12c.
- *TRD*: ATM layer header translator, cell insertion and extraction.
- *PCC*: Usage / Network parameter control function chip.
- *I3*: High speed internal interface chip.

The first two have been develop specifically as part of the technology required by the project TR1-622 to cover the activity of ATM interfaces at 622.08 Mbs. The other three devices were previous results of other parts of the RECIBA project.

2.2.- NT1-622

The network terminator type 1 for broadband ISDN, NT1-622, is an optical line equipment working at STM-1 or STM-4-4c that represents the link between public and private B-ISDN networks. It monitors the quality of service (QoS) in the access network. This element is connected to the network through two interfaces: Ub connected to the public network and Tb to the subscriber network.

The functions of this element are the following. Optical conversion in both interfaces Ub and Tb, conversion of SDH frames to virtual containers (VC-4 or VC-4-4c) or ATM cells and viceversa (frame alignment, frequency decoupling and pointers processing), operation administration and maintenance (OAM flows F1 and F2).

Following the current approach of ITU-T for the NT1 equipment, the transmission path level should pass the NT1 transparently. Therefore, the SDH termination modules depicted above must interchange virtual containers. In order to increase the usefulness of the NT1 equipment, a number of functions could be added easily. In this case the SDH termination module should interchange ATM cells and the NT1 could also process F3, F4 or F5 flows. The NT1-622 is divided in different submodules:

a) *Optical transceivers*: it implements the Physical Medium sub layer. It performs the electrooptical conversion, regeneration in both interfaces Ub and Tb and receive clock recovery.

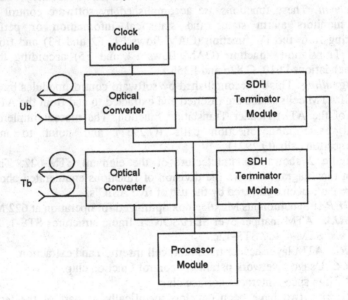

Figure 3.- NT1-622 present implementation block diagram

b) *SDH terminator module*: it implements the transmission convergence sub layer. It performs functions like frame alignment, frame adaptation and decoupling of the frequency drift between the two networks connected to Ub and Tb. This module is based on a pair of ASIC designs described below. These ASIC's can be programmed both to handle virtual containers and ATM cells.

c) *Clock module*: it generates the system clock from a reference chosen from two sources.

305

- Reception frequency obtained from the public network (Ub interface).
- Local oscillator.

The first source is selected by default. In case of failure on the Ub interface, the processor module automatically switches to the second source.

d) *Processor module*: This performs control and supervision of the others modules and processing of messages and alarms. It also collects performance monitoring information of the events occurring during the last 24 hours. The public network element, to which the NT1 is connected, has this statistical information available through a data communication channel (DCC) for management of the user access. This communication channel may also be used to transmit command to establish physical layer loops for out of service testing.

3.- Building Blocks of the RECIBA Access Network

The architecture of the elements of the access network (both NT1-622 and ETM-622) relies on three key components specifically designed and optimized for this project: a) electro-optical converter module, b) serial to parallel converter device able to work up to 622.08 MHz and c) Transmission Convergence sub layer device to cover the frame structures STS-1 at 51.84 Mbs, STS-3c / STM-1 at 155.52 Mbs and STS-12c / STM-4-4c at 622.08 Mbs for both SONET and SDH hierarchies. The components b) and c) have been implemented on ASIC's whose main characteristics are described below. The three key components can be programmed at different independent binary rates for transmission and reception, thus enabling the implementation of asymmetric broadband access interfaces.

3.1.- Electro-Optical Conversion Module (EO155/622)

The *EO155/622* transceiver [4] is an electrooptical line module that provides the most common capabilities needed to support one full-duplex STM-1 or STM-4 physical line interface requirements according to ITU-T Rec. G.957 / S-1.1 and S-4.1 with enhanced sensitivity. Both transmit and receive optical signal are carried on standard G.652 single mode fiber at 1300 nm nominal wavelength.

The module includes functions such as electrical to optical and optical to electrical conversion, clock recovery, system clock synchronization, loop-back facilities and alarms. Using two pin configuration signals, the module can support both symmetric or asymmetric modes at both bit rates 155.520 or 622.080 Mbps.

3.2.- LIBRA: *Line Interface for Broadband Applications*

This device has been designed to perform serial to parallel conversion and synchronous byte interleaving multiplexing up to 622.08 Mbs. Up to four byte streams may be multiplexed/demultiplexed into a single bit stream. It also incorporates the frame scrambling and B1 parity generation and verification of the byte interleaved signal. The LIBRA device includes a direct interface towards the TRJM device. Both devices together allow the implementation of concatenated (STM-4-4c, STM-1 loaded with AU-4) and non-concatenated (STM-4, STM-1 loaded with AU-3) frame structures.

3.3.- TRJM: ATM-SDH Mapper at VC-3, VC-4 and VC-4-4c

This device has been designed to cover the Transmission Convergence sub layer (TC) of ATM line interfaces based on SDH and SONET hierarchies from STS-1 at 51.84 Mbs to STM-4-4c at 622.08 Mbs and it is available since June 1994. The frame structure selection is independent for transmission and reception, therefore enabling the possibility of asymmetric interfaces. The functions performed by this device are:

- *Transmission frame adaptation*: The Virtual Containers (VC) are adapted to the synchronous frame structures by handling the Administrative Unit Pointers (AUP). Virtual container type (VC-3, VC-4 or VC-4-4c) is fixed according to the selected frame structure.
- *Section and Path Overhead access*: All octets included on the SOH and POH are accessible for additional external functions such as Data Communication Channels (DCC), orderwires, user data channels, etc.
- *Cell delineation, HEC header sequence generation/verification and Cell payload scrambling*: All these functions are performed according to ITU-T recommendation I.432.
- *Cell rate decoupling*: This function is accomplished by insertion and extraction of physical layer idle cells.
- *OAM functions support*: The device incorporate physical layer alarm management and statistics counter for performance monitoring on F1, F2 and F3 flows.

3.4.- Performance Parameters and Measurement Results

Both elements of the RECIBA access network (ETM-622 and NT1-622) are able to estimate the performance parameters for the Physical layers as described in ITU-T recommendation G.826 such as Errored Second Ratio, Severely Errored Second Ratio and Background Block Error Ratio. Beside these parameters, the RECIBA elements estimate raw BER and Cell Error Ratio based on the HEC verification function.

The elements: ETM-622 and NT1-622 have been both tested at 155 and 622 Mbs on the optical link between the TI+D and the UPM ETSIT buildings. This optical link is about 16 Km long. BER performance measurements were better than 10^{-12} for the 622 Mbs case and 10^{-14} for the 155 Mbs case.

4.- Experiences on real Broadband Multimedia Demonstrations

Since November 1994, Telefónica has displayed a number of multimedia services demonstrations on different sites over Spain. A multiuse Service Provider Center (SPC), located in Telefónica I+D laboratories in Madrid, supplies multimedia database accesses to remote terminals at the demo locations. [2] and [3]. Four examples of multimedia services are presented:

- Video on demand (VoD).
- Teleaudioteca: Access to music and videoclips database.

307

- Broadband multimedia Videotext with a service for hotel reservation on *Paradores Nacionales de Turismo de España.*
- Computer Supported Cooperative Work (CSCW) service.

The network architecture employed to make this remote accesses possible is shown in figure 4. At one end there is a local area network connected to the multimedia database server and one of the CSCW terminals. At the remote end there is another local area network connected to all the multimedia terminals. Both LAN's are interconnected by a broadband network including two RECIBA VP-switches (one at each extreme) linked through the public network. At both extremes, the CSCW terminals act as gateways towards the other LAN by using a commercial multi mode SDH 155 line interface board installed on the workstations. This multi mode interface is switched towards the public network by the RECIBA equipment.

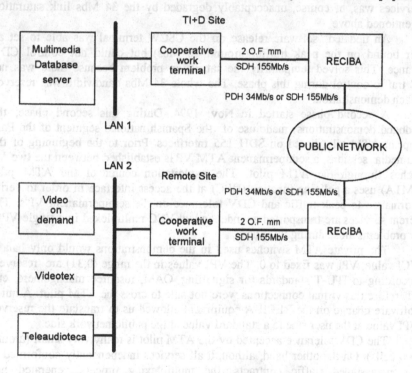

Figure 4.- Multimedia Service Demonstration scenario

The first phase of these broadband demos took place since Jan. to Nov. 1994. A classical PDH non-ATM transmission systems at 34 Mbs were used to link the broadband islands during this phase. The 34 Mbs capacity of the transmission system was the bottleneck of the network and no Usage Parameter Control (UPC) functions was set for traffic control. Several networking problems arose during this phase:

- The private ATM equipment only offered multi mode interfaces. The RECIBA equipment performs multi mode to single mode conversion plus network synchronization functions relaxing synchronization requirements at the user side.
- The average bit rate for the CSCW service was about 1.7 Mbs, but traffic measurement showed peak traffic values of 155 Mbs which saturated the 34 Mbs link. This application had not foreseen the need to limit the peak cell rate because this is not usually needed on private environments. Commercial ATM line interfaces for workstations of the first generation did not incorporate traffic shaping functions.
- The VoD and Videotext services are, in essence, of the CBR type. Measurements confirmed the well behaved conformance of the peak traffics to the preassigned traffic contracts. Performance parameters, such as CLR for this services was, of course, unacceptably degraded by the 34 Mbs link saturation mentioned above

An updated software release on the CSCW terminal was able to set an upper bound on the peak traffic (around 20 Mbs) but could not control CDV tolerance. This solved temporarily the saturation problem because there was not UPC traffic control during this phase. The whole 34 Mbs bandwidth was reserved for each demonstration.

A second phase started in Nov. 1994. During this second phase, the broadband demonstrations made use of the Spanish national segment of the Pan-European ATM pilot based on SDH 155 interfaces. Prior to the beginning of the multimedia sessions, a semipermanent ATM VP is established between the two VP switches through the ATM pilot. The exploitation center of the ATM pilot (CEMTA) uses a policing function (UPC) at the access interface in order to verify conformance to peak traffic and CDV tolerance on the semipermanent VP's. The different services are transported as independent VCC multiplexed in a single VPC. New problems arose during this phase.

- The private ATM switches used in the demonstrations would only handle VCI value, VPI was fixed to 0. The VPI values in the range (0,31) are reserved, according to ITU-T standards for signalling, OAM, resource management, etc. Therefore this virtual connections were not able to cross the ATM pilot. A quick software change on the RECIBA equipment allowed us to translate the reserved VPI value at the user side to a standard value at the public network side.
- The CDV tolerance accepted by the ATM pilot is really very small (around 500 cells). On the other hand, although all services independently conformed to the preassigned traffic contracts, the multiplexing process generated non conforming clumps of cells on the VPC. These facts yielded a very high non-conforming cell rate which were discarded by the UPC function. This problem was temporaily solved by reserving independent VPC's for each service. It is clear that this solution is only possible because the ATM service is yet in an experimental phase and the number of subscribers is still relatively small.

As a result of the work in the project TR1-622 described above, the RECIBA platform supports STM-4-4c ATM interfaces since January 1995. Unfortunately, 622'08 Mbs ATM interfaces are not yet available in ATM-Pilot. Therefore, the multimedia scenario depicted in figure 4 may only be tested with local connections at 622.08 Mbs. Within the PlanBa program, a new phase of demonstrations, including scenarios at 622 Mbs is planned for March 1996.

5.- Additional Functions for the NT1 Equipment

The second generation of broadband terminals include traffic shaping functions which are able to control the traffic parameters required in the traffic contract. However, many of these terminals do not yet handle CDV tolerance adequately. Furthermore, the commercial ATM switching equipment typically used as CPE's do not perform traffic shaping after multiplexing a number of VCC's into one VPC. Multiplexing of VCC's may generate clumps of cells which might not conform to the traffic contract at the VPC level. Therefore, aggregated traffic parameters cannot not be guaranteed on the VPC basis.

Concentrating the traffic shaping functions of a number of connections on the NT1 equipment could be a possible and elegant solution. In this way, it would be possible to get the most advantage of statistical gain on the user access and to reduce the function cost per connection. The NT1 equipment could support this function of a traffic shaping server as a natural extension of its classical functions such as optical regeneration, multi mode to single mode conversion, monitoring and reporting at the physical layer level. This enhanced NT1 equipment is aimed at optimizing bandwidth usage at the access interface. An implementation of the enhanced NT1, currently being developed which will be available in april 1996, can perform traffic shaping on 32 simultaneous virtual connections allocating up to 32 K cells for CDV tolerance. With this feature [11], the user does not need to consider CDV tolerance constraints at all on his CPE.

Present commercial small to medium capacity ATM switching equipment do not yet support OAM functions at the ATM layer (F4 and F5 flows). These functions are complex and expensive in terms of the hardware required for them. On the other hand, the experience has shown that this lack generates a lot of trouble, especially during the semipermanent VPC setup process on international connections. OAM flow F4 could also be processed by the NT1 equipment. Furthermore, from the point of view the network operator, it could be of interest that the NT1 becomes a termination point for the F4 flow in order to monitor the QoS offered to the user, on a virtual connection basis, without the need to rely on the availability of these functions on CPE's. The enhanced NT1 acts as VPC-F4 termination point and performs fault management, continuity check and performance management on up to 32 simultaneous virtual connections. It adds OAM-F4 cells to the user data flows and it is able to collect statistical performance information on selected VPC's such as Cell error Ratio (CER), Cell Loss Ratio (CLR), Cell Misinserted Ratio (CMR) according to ITU-T recommendation I.356.

6.- Conclusions

This paper describes an implementation of enhanced broadband access network elements of the type NT1 and the technology underneath for ATM interfaces based on the SDH or SONET hierarchies up to 622.08 Mbs.

The set of functions allocated in B-ISDN to the NT1 equipment by the ITU-T recommendations I.610 and G.96x does not seem very attractive to justify its use. So far little attention has been paid to the possibility to extend such functions so that the NT1 may become a useful tool to solve typical problems when connecting private environment equipment and services through the ATM public network. The experience shown in this paper justifies the addition of ATM layer functions such as OAM F4 flow processing and generation, ATM cell header translation and traffic shaping.

This enhanced NT1 equipment could be useful and attractive both to the public operator and the final user [12]. The subscriber could optimize bandwidth usage and simplify ATM layer OAM functions, reducing the connection and CPE costs. From the network operator point of view, this equipment may supervise the QoS actually offered to the customer on VP basis without relaying on the availability of complex ATM layer OAM functions, such as VP-performance management, on the CPE's.

7.- References

[1] P.J. Lizcano et al. *A Flexible Architecture Platform for B-ISDN Interfaces and Services Benchmarking*. Globecom 94, San Francisco, November 1994

[2] J Fernández-Amigo et al. *Servicios de Banda Ancha: Proyecto RECIBA*. III Jornadas Telecom I+D. Madrid 23-24 de Noviembre, 1993.

[3] J. Fernández Amigo et al. *Broadband Services: RECIBA Project, Multimedia and Broadband Services, Applications and Development Tools*. Globecom 94, San Francisco, November 1994.

[4] J. Mariño and L. Cucala. Módulo Electro-óptico 155/622 con unidad de sincronización. URSI, September 1994, Vol 1, page 203 through 208, Las Palmas, Canary Islands, Spain.

[5] ITU-T. Recommendation G.957: *Digital Networks, Digital Sections and Digital Line Systems. OPtical Interfaces for Equipment and Systems ralating to the SDH*. Melbourne 1988.

[6] ITU-T. Recommendation G.958: *Digital Networks, Digital Sections and Digital Line Systems. Digital Line Systems based on the Synchronous Digital Hierarchy for use on Optival Fiber Cables*. Geneva 1990.

[7] ITU-T. Draft Recommendation I.356: *B-ISDN ATM Layer Cell Transfer Performance*.

[8] ITU-T. Draft Recommendation I.610: *B-ISDN Operation and maintenance principles and functions*.

[9] ITU-T. Draft Recommendation G.96x: *Access Digital Section for B-ISDN*.

[10] ITU-T. Draft Recommendation G.826: *Error Performance Parameters and Objetives for International Constant Bit Rate Digital Paths at or Above the Primary Rate*.

[11] Pierre Boyer and Michel Servel. *A Spacer-Multiplexer for Public UNIs*. ISS'95. April 1995 Vol 1.

[12] Clark et al. *The AURORA Gigabit Testbed*, Computer Networks and ISDN Systems, January 1993.

LARNet, a Wavelength Division Multiplexed Network for Broadband Local Access

M. Zirngibl, C.H. Joyner, C.R. Doerr , L.W. Stulz , I.P. Kaminow
AT&T Bell Laboratories, Holmdel, NJ, 07733, USA
TEL 908 888 7153, FAX 908 888 7074, E.mail m.zirngibl@att.com

Keywords: Passive Optical Networks (PON), Wavelength Division Multiplexing (WDM), Broadband Access, Fiber-to-the-home

Abstract: We present a wavelength division multiplexed (WDM) network architecture for local access that provides broadband switched services. A central office (CO) is connected to multiple optical network units (ONU) through a shared access fiber and a passive router in a remote node (RN). A single multifrequency laser (MFL) in the CO sends data simultaneously on all wavelengths; spectral slicing and a WDM receiver is employed for upstream connectivity. The WDM layout ensures independent virtual point-to-point links up-and downstream; expensive components are shared among all ONUs; and the wavelength control problem becomes very simple. The ONU is built with inexpensive commercially available components. Aggregate rates of several Gbps downstream and several hundred Mbps upstream can be realized.

Introduction: One key question for broadband local access networks is how close to bring the fiber to the customer premises. There is a trade-off between initial deployment cost and performance. Presently, the high cost of optoelectronic components makes fiber to the home uncompetitive. On the other hand, new services, like switched digital video, require too much bandwidth to be delivered over traditional coaxial cable networks. Research into optical access networks has therefore focused on architectures that are cost-effective but provide sufficient bandwidth.

There are three physical dimensions, space, time and wavelength to link a central office (CO) to multiple optical network units (ONU), which may or may not be shared among several customers. In a space diversity system, there simply is one fiber per CO-ONU link. This solution is technically very straightforward; however, the cost associated with deploying a fiber link does not make it attractive for fiber to the home or to the curb. For this reason several proposals for local access have featured passive optical networks (PON) where a single access fiber is passively split in a remote node (RN) that connects via fiber to several ONUs.

PON systems can be roughly divided into time division multiplexed (TDM) or wavelength division multiplexed (WDM) systems. In TDM systems (1), a single optical carrier is distributed equally among all the ONUs, which electronically demultiplex the data stream and feed it to the customer. A TDM system uses relatively straightforward devices. It is, however, hard to upgrade since the same optical carrier is shared among all customers. Also, the receiver at the ONU has to process the total downstream information and, therefore, has to run at the aggregate bit rate. A TDM PON has an intrinsic N^2 power penalty due to power splitting and time-sharing, where N is the number of ONUs per access fiber. In the upstream direction, careful synchronization among the different ONUs is required in order to avoid data collisions due to different path lengths.

WDM has been proposed (2-3) as an alternative for multiple access. Different optical carrier wavelengths are now used to address the different ONUs. They are passively demultiplexed in a RN so that each ONU receives only one wavelength. A WDM system comes very close to simulating a one-fiber-per-ONU solution by providing one channel-per-ONU since the system functions as a logical star, where each ONU receives only information dedicated to it. Future network upgrades can be much more easily implemented with such a network than with TDM PONs. Other advantages include signal privacy and easier fault location. But now, device functionality and availability become major issues. For example, we require transmitters that simultaneously emit on several precisely spaced optical wavelengths.

In the past two years, significant progress has been made toward building such WDM components. In particular, multifrequency lasers (MFL) (4) and multifrequency receivers (MFR) (5) have been shown to provide WDM PON capability. These devices have the potential to be cost-effective since they are monolithically integrated on a single chip. The network architecture, discussed in this paper, relies upon these novel components and takes full advantage of their multiwavelength potential.

The network architecture: The local access router network (LARNet) is schematically displayed in Fig. 1a and b. The downstream signal with N wavelength components at the CO is generated by a MFL. A 1xN optical demultiplexer (where N denotes the number of wavelength ports) in the RN passively routes each wavelength to the appropriate ONU. In the upstream direction, we use a light emitting diode (LED) as a signal source. The RN spectrally slices (6) the LED light into different optical bands. Each wavelength band is individually detected by a WDM receiver at the CO. Coarse WDM (1.3/1.5 mm or 1.48/1.56 mm, for instance) is used to allow up-and downstream traffic to flow over the same fiber and network components. Coarse 1x2 WDM demultiplexers (not shown in Fig. 1) at the ONUs and CO separate the two data streams.

LARNet provides virtual point-to-point links for both upstream and downstream traffic which eliminates the need for TDM or subcarrier multiplexed formats and enables the network operator to upgrade the system to very high bit-rates. Although the whole architecture relies heavily upon WDM, the wavelength control problem becomes very simple. The MFL and MFR wavelength combs simply have to be matched against the router in the RN. Tuning of the two devices is achieved by controlling the device temperature, which changes only the absolute wavelength but not the relative channel spacing of the comb. A single control knob on the transmitter and receiver at the CO, thus, allows the alignment of all wavelengths. There is no wavelength control on the ONU side of the network since each LED emits essentially the same broadband light. Therefore, the ONU can be built with components that are already commercially available at low cost. The more complex MFL and MFR in the CO are shared among all ONUs.

313

The devices: The multiwavelength devices used for LARNet are all based on the integrated waveguide grating router (WGR) (7) which is schematically displayed in Fig. 2. An input port is connected through a two dimensional free space region to a waveguide grating that optically demultiplexes the signal and routes it via a second free space region to N output ports. Obviously, the router works in reverse too, combining N wavelengths into a single fiber. In our network, a 1x8 WGR with 200 GHz (1.6 nm) channel spacingi is used. A fiber insertion loss as low as 5 dB and optical crosstalk levels below -25 dB are routinely achieved by these routers which are expected to be commercially available shortly.

The MFL consists of a WGR monolithically integrated with semiconductor amplifiers on in- and output ports (Fig.3). The optical amplifiers are terminated by cleaved mirror facets that define an optical path containing both the amplifiers and the router. If the amplifiers at opposite ends of the cavity are activated by a bias current, they provide optical gain and sustain lasing at a wavelength that is determined by the path through the router. Such a laser is capable of generating a comb of precisely spaced wavelengths (Fig. 4), determined by the geometry of the router. Each wavelength can be independently modulated by the bias currents of the individual amplifiers.

The MFR looks very similar to the MFL, but the amplifiers are now replaced by detectors. An optical amplifier on the input port may be used to boost the incoming signal, which is then optically demultiplexed and each wavelength is detected individually by an array of photodetectors. The spectral characteristics of such a WDM receiver are displayed in Fig. 5 (5).

Network performance: We shall now evaluate the network performance. We assume a 1x8 split (8 ONUs per CO) and a transmission distance of 10 km. In the downstream direction, the MFL can be directly modulated at 622 Mbps (8). Fig. 6 shows an eye pattern obtained from one of the channels as received at the ONU. The MFL emits -5 dBm of power per wavelength into single mode fiber. After accounting for various losses (table 1), the ONU receives -17 dBm of power which gives a 15 dB margin if a commercial pin-FET receiver with -32 dBm of sensitivity at 622 Mbps is used. In the upstream direction, we have to contend with a large slicing loss, since most (~ 1/2N) of the LED power is filtered out by the WGR. Commercially available LEDs at 1.3 mm emit -10 dBm of power into single mode fiber but only -39 dBm will reach the CO from each ONU. The currently available MFRs have fiber insertion sensitivity of 0.5 A/W, which allows an upstream bit-rate of 52 Mbps per ONU with still sufficient system margin. The MFR could go to much higher speed if needed, as shown by the eye-pattern for one of the channels receiving a 622 Mbps signal in Fig. 7.

The aggregate bit-rates up-and downstream of the proposed network are 5 Gbps and 400 Mbps, respectively. Note, that it would be very hard to build a TDM PON with similar performance. Not only would it be difficult to satisfy the downstream power requirement because of the intrinsic N^2 power penalty, but also the ONU receivers would be much more expensive since they have to run at the full aggregate bit-rate.

Table 1: Power budget for 1x8 LARNet

	downstream	upstream
launched power	-5 dBm	-10 dBm
couplers	-2 dB	-2 dB
transmission	-2 dB	-6 dB
router insertion	-5 dB	-5 dB
splices	-3 dB	-3 dB
spectral slicing		-13 dB
received power	-17 dBm	-39 dBm

First installation and possible upgrades: Despite the fact that the CO components MFL and MFR are not commercially availabe yet, we can deploy the fiber plant for LARNet today and run a low bit-rate service over this network with low cost components (9) by using bi-directional spectral slicing. The resulting network looks very much like an ordinary TDM PON, but cheap FP lasers are replaced by LED. Experimentally, 40 Mbps rates up-and downstream have been achieved (9) with this approach. Upgrades to LARNet can then be realized by changing equipment in the CO only; the downstream LED is replaced by the MFL and the upstream receiver by the MFR. One can also envision upgrades beyond the 5 Gbps downstream and 400 Mbps upstream. The only limitation on the downstream rate is the relatively slow direct modulation rate of the MFL due to its long cavity. To make this device faster, external modulators would have to be used. If needed, the power budget in the downstream direction could be boosted by a fiber amplifier. If further capacity were needed upstream, a laser instead of the LED could be employed in the ONU. The wavelength of this laser would have to be well controlled, however. Note that all these upgrades can be done on a per-user basis.

Conclusion: We have proposed and demonstrated a novel network architecture for local access. The architecture takes advantage of the unique properties of waveguide grating routers, multifrequency lasers and multifrequency receivers. Although, LARNet is based entirely on WDM, the whole network can be spectrally aligned by tracking only one wavelength. The subscriber terminal can be built with low cost, commercially available components. Up-and downstream connections are virtual point-to-point links. Based on our experimental results, we project an aggregate downstream rate of several Gbps and aggregate upstream rate of several hundred Mbps.

References

(1)D.W. Faulkner, D.B. Payne, J.R. Stern, J.W. Ballance," Optical Networks for Local Loop Applications," J. Lightwave Technol. LT-7, 1741(1989).

(2) I.P. Kaminow, "Non-Coherent Photonic Frequency-Multiplexed Access Networks", IEEE Network Magazine, vol. 3, 4(1989).

(3) M. Zirngibl, C.H. Joyner, L.W. Stulz, C. Dragone, H.M. Presby, I.P. Kaminow, "LARNet, a Local Access Router Network", IEEE Photon. Technol. Lett. vol. 7, No. 2, pp. 215-217, 1995

(4) M. Zirngibl, C.H. Joyner, L.W. Stulz, "Demonstration of 9x200 Mbit/s wavelength division multiplexed transmitter", Electron. Lett., vol. 30, 1994, pp. 1484.

(5) M. Zirngibl, C.H. Joyner, L.W. Stulz, "A WDM receiver by monolithic integration of an optical preamplifier, a waveguide grating router and a photodiode array", Electron. Lett. vol.30, pp.581-582 (1995).

(6) S.S. Wagner, T.E. Chapuran, "Broadband High-Density WDM Transmission Using Superluminescent Diodes", Electron. Lett., Vol. 26., 696(1990),

(7) C. Dragone, C.A. Edwards, R.C. Kistler, "Integrated Optics NxN Multiplexer on Silicon", IEEE Photon. Technol. Lett., Vol.3, 896(1989).

(8)C.R. Doerr, C.H. Joyner, M. Zirngibl, L.W. Stulz, H.M. Presby, "Elimination of signal distortion and crosstal from carrier density changes in the shared semiconductor amplifier of multifrequency signal sources", submitted to IEEE Photon. Technol. Lett.

(9) P.P. Iannone, N.J. Frigo, T.E. Darcie, "WDM passive-optical-network architecture with bidirectional optical spectral slicing", OFC'95, paper TuK2.

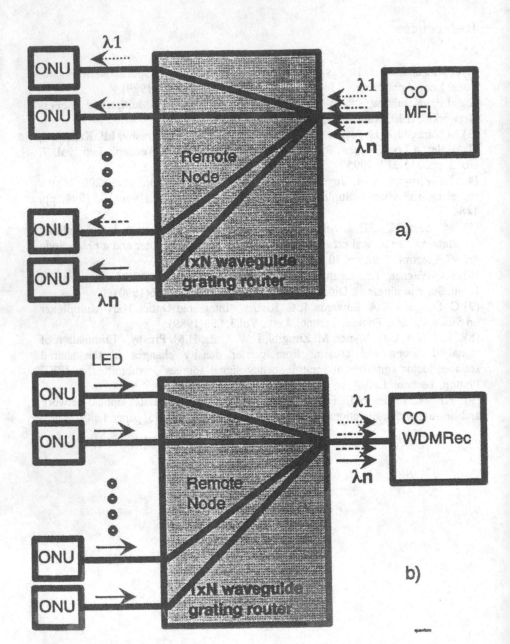

Fig. 1: Layout for the Local Access Router Network (LARNet) with N wavelength channels; downstream traffic a) and upstream traffic b). Coarse WDM couplers separate the two traffics (not shown for clarity) that run over the same fiber network.

317

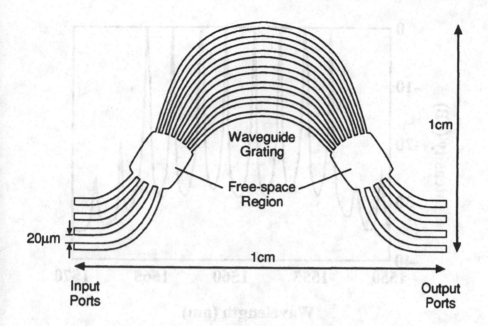

Fig. 2: Schematics of the NxN waveguide grating router (WGR).

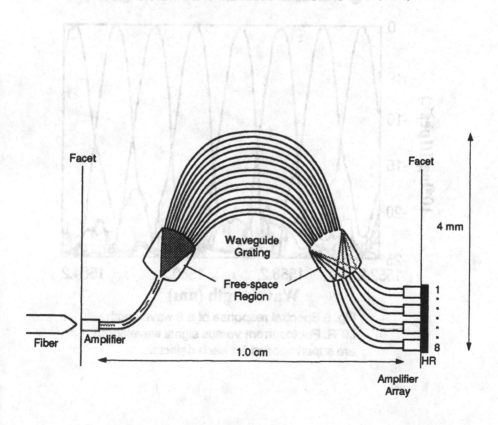

Fig. 3: Schematics of the N wavelength multifrequency lasers (MFL

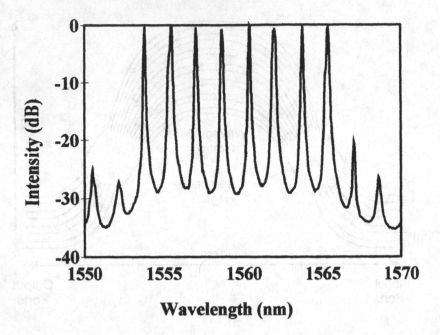

Fig. 4: Emission spectrum of a 8 wavelength MFL

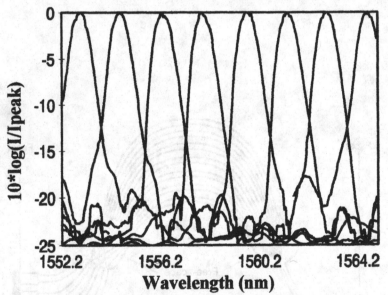

Fig. 5 Spectral response of a 8 wavelength
MFR. Photocurrent versus signal wavelength
are superimposed for each detector.

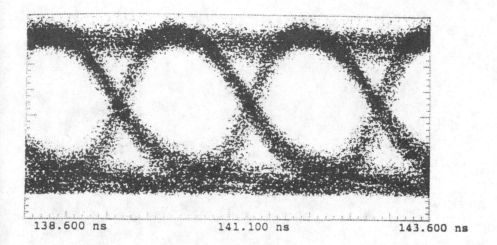

138.600 ns   141.100 ns   143.600 ns

Fig. 6: Eye diagram of received at the ONU at 622 Mbps.

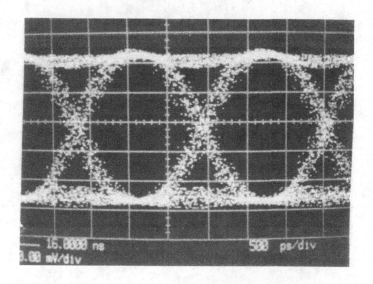

Fig. 7: Eye pattern of WDM receiver at 622 Mbps.

Design of a Large ATM Switch with Trunk Grouping *

Sushil Aryal and James S. Meditch

Department of Electrical Engineering
University of Washington
Seattle WA 98195
email: aryal@shasta.ee.washington.edu, meditch@ee.washington.edu

Abstract. We describe a large ATM switch with moderate circuit complexity which delivers packets in sequence with a small constant delay. The switch takes advantage of trunk grouping to increase throughput and has the same topology as that of a Banyan network. By using modules that look at k inputs ($k > 16$) at a time and recursively dividing the packets into two groups, the proposed switch achieves a good throughput value that is relatively independent of switch size. Cell loss on the order of 10^{-10} can be obtained using the switch. The switch has $O(n \log_2 n)$ complexity in the number of modules where n is the number of inputs. The modules of the switch have very regular interconnections, and hence are suitable for dense VLSI implementation.

Keywords: Switch, Self-routing, ATM, Trunk Groups, Packet-switching.

1 Introduction

The design of large packet switches for Broadband Integrated Services Digital Networks (B-ISDN) using the Asynchronous Tranfer Mode (ATM) is still a very challenging problem. In spite of many proposals for switch architectures, there has not been a single implementation of an ATM switch that can handle hundreds of inputs, let alone thousands of inputs. This has been the result of the fact that most of the architectures presented so far are not scalable, as their circuit complexity increases rapidly with an increase in the number of inputs. We present in this paper an architecture called the Large ATM Switch with Trunk-grouping (LAST). We show that this switch will be able to support thousands of inputs at the ATM line rate of 155.52 Mb/s.

In the next few paragraphs, we review a representative sample of the ATM switches that have been proposed. Many of the proposed switches use self-routing Banyan networks to route packets without centralized control. The Banyan network is an interconnection network consisting of 2×2 switching elements, and it routes packets by looking at the destination address bit-by-bit. A major problem

---

* This work was supported in part by the National Science Foundation under NSF Grant NCR-9103485.

with the Banyan network is that there can be conflicts in the internal links of the Banyan, even though the packets may not have conflicts in the output port addresses. This results in Banyan networks having a low throughput when the number of inputs is large. However, conflicts in the internal links of a Banyan network can be avoided if packets at the inputs are sorted and distinct. One of the earliest proposals for a wideband digital switch was the Starlite switch [1]. It uses a Batcher sorting network to sort the packets, and presents the Banyan with distinct and sorted output addresses. Duplicate packets are stored in a buffer, and are fed to the sorter together with newly arrived packets in the next time slot. The Moonshine switch [2] [3] also uses a Batcher-Banyan pair. Unlike the Starlite Switch, it uses a multiphase algorithm to determine which of the packets can make it to the outputs without internal conflicts in the Banyan, and buffers the remaining packets at the input buffers for the next attempt. The drawback of these switches is that the complexity of the Batcher sorter with n inputs is $O(n(\log_2 n)^2)$ (compared to $O(n \log_2 n)$ for a Banyan network) which increases very fast as the number of inputs increases.

Turner's switch [6] [7] takes a different approach to resolve conflicts for internal links. It uses buffers in the 2×2 switch elements of the Banyan to buffer one of the conflicting packets, and tries to transmit this buffered packet to its destination link if no new packet arrives for that link in the next time slot. Even with a single buffer per link, the hardware overhead is very high, since every 2×2 element has to store two 53 byte long ATM cells. The hardware required to store these 848 bits is about 40 times the hardware required to implement a simple 2×2 switching element. When multiple planes of buffered Banyans are used in the switch, packets may be delivered out of sequence, which is an additional undesirable characteristic.

The Hitachi switch [8] is a shared memory packet switch, in which a single memory buffer is divided into logical queues of packets destined for different outputs. This means that, in every slot, packets from each of the inputs have to be written in memory, and packets destined to each of the outputs have to be read out from the logical queues. For a switch with n inputs, $2n$ read/write operations have to occur in each slot. Hence, large switches of the shared memory type cannot be built, because of the speed limitation of the control circuitry that controls read and write operations. The Prelude switch [9] is another shared memory switch which suffers from the same speed limitation.

Another architecture that has been proposed for packet switching is the shared medium architecture. In the NEC shared bus switch [10], packets from all the inputs are multiplexed onto the shared medium, and address filters for different outputs filter the packets destined to appropriate output buffers. The speed of the shared medium limits the size of a shared medium switch that can be built. The Knockout switch [11] has an architecture similar to the shared medium switch. However, instead of having a single shared medium, there are n shared lines driven by each of the n inputs, and every output has an address filter connected to each of the n input lines. Thus, the complexity of the Knockout switch is $O(n^2)$ in the number of address filters required, and this increases dramatically as the number of inputs increases. As a result, large Knockout switches cannot be built easily.

As none of the architectures mentioned above has been found suitable for building large switches, Chiussi [12] has presented a hybrid switch architecture that can support several hundreds of inputs. The switch has three stages: the first and the last stages are shared memory switches, and the second stage consists of Banyan networks. A two-phase algorithm is used to route packets through the switch. The number of inputs to the switch is given by the product of the number of inputs of the shared memory stage and of the Banyan network. Hence, the largest switch that can be built is still limited by the largest size of a shared memory switch.

The switch architectures presented above have one or more of the following disadvantages:

1. The hardware complexity becomes unmanageable for a large number of inputs.
2. Throughput of the switch degrades rapidly with increase in switch size.
3. Packets suffer large delays through the switch.
4. Packets get delivered out of sequence.
5. Different kinds of switch modules are used, each one of which needs to be designed and tested.

In this paper, a Large ATM Switch with Trunk-grouping (LAST) is described. The LAST switch architecture has good throughput values even for switches with thousands of inputs, and has a hardware complexity that is a constant times the complexity of a Banyan network. It utilizes a single type of switch module which is suitable for dense VLSI implementation. Since there are no buffers in the switch fabric, the delay through the switch is very small. Another appealing feature of the switch is that all packets that reach the outputs are in sequence. In short, the LAST switch does not have any of the disadvantages listed above.

The organization of this paper is as follows. In Section 2, we describe a model for a self-routing switch called the Ideal Bit-by-Bit Self-routing Switch(IBSS), and compare its saturated throughput with that of a Banyan network. In Section 3, we present the architecture of the proposed LAST switch. A detailed description of the modules that make up the switch is given in the same section. We discuss some of the implementation issues of the LAST switch in Section 4. The saturated throughput and cell-loss results for the LAST switch are presented in Section 5. Finally, Section 6 summarizes the results presented in the paper.

2 A model for a self-routing switch

A model of a self-routing switch without dilation is shown in Figure 1(a). The switch is called the Ideal Bit-by-Bit Self-routing Switch (IBSS). It consists of Divide Into Two Groups Modules (DITGM) of varying sizes (the actual design of a DITGM is presented in Section 3). The first stage of the IBSS has a single DITGM that divides the incoming packets in the n inputs into two groups of

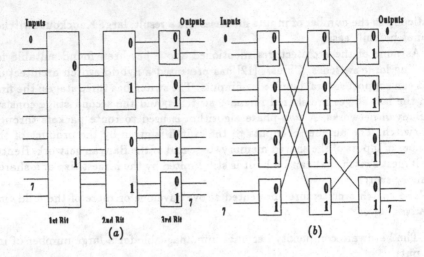

Fig. 1. Two 8 × 8 switches:(a) IBSS (b) Banyan Network

$n/2$ lines depending upon the first address bit. Similarly, the second stage has two DITGM's, each of which divide the packets in the $n/2$ lines into two groups of $n/4$ lines depending upon the second address bit and so on. This is similar to what the Banyan network in Figure 1(b) does at every stage, but by looking at only two lines at a time at every stage. Thus, in the Banyan network, a packet at the input of a 2×2 element can be routed to only one output depending upon its routing bit at every stage. However, at the ith stage of the IBSS switch, a packet may be successfully routed to any of $n/2^i$ outputs. This results in the IBSS having a better throughput than the Banyan network. The IBSS switch is similar to the Multinet switch [13] at the top level block diagram, but the modes of operation are totally different. Also, there are no buffers in the modules of the IBSS switch while the Multinet modules use FIFO (First-In-First-Out) buffers.

2.1 Calculation of cell-loss in a DITGM under Bernoulli traffic

In this subsection, we calculate the cell-loss in a DITGM under the assumption that the traffic has a Bernoulli distribution, that is, the traffic is uncorrelated. Let n be the number of inputs of a DITGM, and let there be $n/2$ outputs labelled 0 (group 0) and $n/2$ outputs labelled 1 (group 1). We define p, p_0 and p_1 as

p: the probability that there is a packet at an input

p_0: the probability that there is a packet in an input destined for group 0

p_1: the probability that there is a packet in an input destined for group 1

Since there are two groups of outputs with each group having $n/2$ outputs, packets are lost when

1. there are more than $n/2$ packets destined to group 0, or
2. there are more than $n/2$ packets destined to group 1.

Since both of these events are mutually exclusive, the average number of packets lost in a slot is given by

$$\sum_{i=n/2+1}^{n} (i - n/2)\binom{n}{i}p_0^i(1 - p_0)^{n-i} + \sum_{i=n/2+1}^{n} (i - n/2)\binom{n}{i}p_1^i(1 - p_1)^{n-i}$$

The average number of packets input to the DITGM in a slot is given by $n(p_0 + p_1)$. Then P_{loss}, the ratio of the average number of lost packets to the average number of packets input to a DITGM, is given by

$$P_{loss}(n, p_0, p_1) = \frac{\sum_{i=n/2+1}^{n} (i-n/2)\binom{n}{i}p_0^i(1-p_0)^{n-i} + \sum_{i=n/2+1}^{n}(i-n/2)\binom{n}{i}p_1^i(1-p_1)^{n-i}}{n(p_0+p_1)}$$

$$(1)$$

For the case of uniform traffic when packets are destined to either output group with equal probability, $p_0=p_1=p/2$,

$$P_{loss}(n, p) = \frac{\sum_{i=n/2+1}^{n}(i - n/2)\binom{n}{i}(p/2)^i(1 - p/2)^{n-i}}{np/2}$$

$$(2)$$

Let p_i be the probability that there is a packet at an input of stage i of the switch. Let n_i be the number of inputs in a module at stage i. Then, the probability that there is a packet at an input of stage $i + 1$ is given by,

$$p_{i+1} = (1 - P_{loss}(n_i, p_i))p_i \qquad (3)$$

Equations (2) and (3) can be used iteratively to calculate the probability of having a packet at different stages of the switch, given the probability of having a packet at the inputs. The saturated throughput is defined as the probability of having a packet at an output of the switch, given that the probability of having a packet at the inputs is 1.

2.2 Saturated Throughput Comparison of IBSS and Banyan Network

Using Equations (2) and (3), we can calculate the saturated throughputs of the IBSS and the Banyan network. Figure 2.2(a) shows a comparison between the saturated throughputs of the Banyan network and the IBSS switch for different numbers of inputs. Note that the x-axis in Figure 2.2(a) is the logarithm to the base 2 of the number of inputs. The graph shows that the IBSS switch achieves an asymptotic saturated throughput of about 56% (which is 7.2% less than 63%- the saturated throughput of an ideal, internally non-blocking, unbuffered switch), while the throughput of the Banyan network decreases as the number of inputs increases. The saturated throughput of an ideal, non-blocking, unbuffered switch which looks at all the routing bits before routing the packets to an output address is also shown in Figure 2.2(a) for comparison.

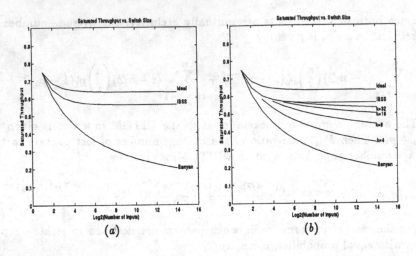

Fig. 2. Throughput comparisons (a) IBSS (b) modified IBSS

3 The LAST Switch

It is clear from Figure 2.2(a) that the throughput of the IBSS switch is good,
and deteriorates only gradually as the number of inputs increases. However, as
the number of inputs increases, it is also true that we have to make larger and
larger DITGM's, particularly in the first few stages. This may be difficult if not
impossible, as the DITGM's are not scalable. Also, the size of a DITGM varies
from stage to stage. This leads us to the questions, " Can similar throughput values be obtained by replacing the large DITGM's by moderately sized DITGM's
of size k? What will happen if modules of the same size are used throughout the
switch?" We again use Equations (2) and (3) from subsection 2.1 to answer the
first question. The throughput of a modified IBSS, obtained by replacing the
DITGM's of size greater than k in the IBSS by multiple DITGM's of size k, is
given in Figure 2.2(b) for various values of k. As long as $k > 16$, the saturated
throughput is very close to that of the IBSS. We consider 128 to be an upper
limit on the size of a DITGM for two reasons: 1) the throughput of a modified
IBSS with 128 input DITGMs is very close to that of a IBSS 2) the DITGMs
become very complex when $k > 128$.

The second question can be answered by eliminating the DITGM's of size
less than k in the modified switch. The result of this is that a switch that routes
packets to a group of size $k/2$ is obtained, and this is the LAST switch. A
128×128 LAST switch is shown in Figure 3(a) with $k = 32$. It uses 32-input
DITGM's, and packets on the 32 inputs of a DITGM are divided into two groups
of 16 each at each stage. Packets are routed to 8 groups of 16 each by the switch.
The topology is the same as that of an 8×8 Banyan switch, but its switching
elements are of a larger size than the 2×2 elements used in the Banyan network.
As in the Banyan network, routing is done by looking at the destination group

address bit-by-bit. The LAST switch has no buffering in its switch elements. Also, there is no dilation used in the LAST switch because every module has as many outputs as there are inputs.

It is true that the LAST switch cannot be used by itself to route to a specific output. However, if routing to a specific output is desired rather than to a trunk group, then two IBSS's(or two shared memory switches) with $k/2$ inputs can be added after each DITGM in the last stage of the LAST switch. We will then be using DITGM's of different sizes in the last few stages, but the largest of them will possibly be of size 64 (which is half the size of the largest DITGM to be used in the LAST switch), which is much smaller than the size of the DITGM in the first stage of an IBSS. The saturated throughput of such a switch, a LAST switch followed by IBSS modules, is the same as given in Figure 2.2(b) for various values of k. The curves show that the asymptotic saturated throughput of the IBSS switch is almost equalled by a switch obtained by replacing the earlier DITGM's of the IBSS by LAST modules of size 32 or larger.

The IBSS switch and the Banyan network represent two extremes in the design of a self-routing network without redundant paths that routes packets on a bit-by-bit basis. The LAST switch lies in between the two designs, in the sense that it looks at a subset of inputs of size k where k is much larger than two. The design problem is that of choosing k so that the modules are of manageable size, but at the same time the loss in the modules is small. Also, by choosing k to be much larger than two, the throughput advantage of a large trunk group size is obtained.

Next we consider the design of a module that divides n inputs into two groups of $n/2$ each. Figure 3(b) shows the block diagram of such a module where n is 8. The module consists of three blocks: the switching block, the 0 group resolver and the 1 group resolver. The switching block consists of four 2×4 switches. Each 2 input switch sub-element has 4 outputs: 1 (high priority 1), 1d (low priority 1), 0 (high priority 0) and 0d (low priority 0). Each of the outputs has a line associated with it that indicates whether there is a packet at that output. The 2×4 switching element routes one packet to the output 0 and the other to the output 1 if the packets at its inputs differ in the destination address bits. Otherwise, one packet is routed to the low priority output, and the other to the high priority output corresponding to their address bits. For example, if both packets have 1 as their destination bit, then one packet is routed to output 1 , and the other packet is routed to output 1d. We should note that the notion of high priority means that a packet has already been assured an output of the DITGM, and low priority means that the packet has to look for a output in the resolver.

The 0 and 1 group resolvers have the same structure. The diagram of an 8 input resolver is shown in Figure 3(c). The resolver consists of circuit blocks of a single type called the Pass/Cross block and has 4 high priority inputs (w, x, y, z) and 4 low priority inputs (wd, xd, yd, zd) together with lines indicating whether a packet is present on those inputs. In the resolver, high priority packets proceed directly to the output of the resolver horizontally, while low priority

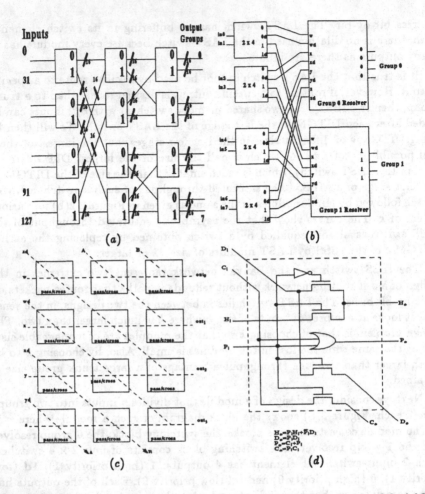

Fig. 3. (a) A 128 × 128 LAST switch using 32 × 32 modules (b) A module of the LAST switch with 8 inputs (c) A Group Resolver in a module with 8 inputs (d) A Pass/Cross circuit

packets move diagonally and try to turn horizontally to an output which has no packet on it. They attempt this operation at every output. A low priority packet that succeeds in making a turn towards an output and is able to proceed horizontally becomes a high priority packet from then on, because it is now assured a DITGM output. Thus, it is not possible for two low priority packets to contend for an output at any Pass/Cross block. Any low priority packet that cannot proceed to an output is lost. Since all packets with a destination bit of 0 (1) go to the group 0 (1) resolver, it is guaranteed that up to 4 ($k/2$ in a module of size k) packets are routed correctly by each resolver.

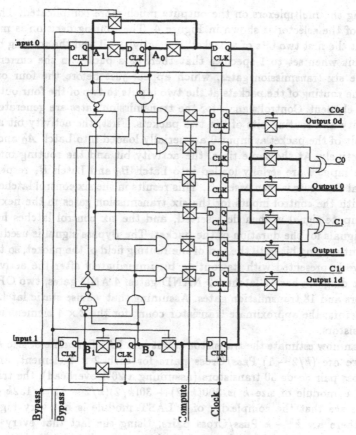

Fig. 4. Logic Diagram of a 2 × 4 element

4 Implementation Considerations

A logic implementation of the Pass/Cross circuit is shown in Figure 3(d). There
are four inputs: the horizontal input H_i, the diagonal input D_i, the horizontal
input packet-present indicator P_i and the diagonal input packet-present indicator
C_i. The corresponding outputs are H_o, D_o, P_o and C_o. The horizontal output
H_o has a packet if H_i or D_i has a packet on it (either P_i or C_i is true) and
this is indicated by P_o being true. If there is a packet at the horizontal input
H_i as indicated by P_i being true, then H_o is connected to H_i, and the packet
at diagonal input D_i is connected to the diagonal output D_o. Only if P_i is not
true is H_0 connected to D_i. The implementation shown in Figure 3(d) requires
12 transistors.

Implementation of the 2 × 4 switch element is similar to the implementation
of the 2 × 2 Banyan element in [4]. Instead of two outputs of a Banyan element
being selected from the two inputs, two out of the four outputs are selected
from the two inputs in the 2 × 4 switch element. This makes the decision logic

controlling the multiplexers on the outputs much more complicated. The logic diagram of the selector is shown in Figure 4. The routing decision is made by looking at the first two bits of a packet: the activity bit and the routing bit (the activity bit when set to 1 specifies that there is a packet in the current time slot). The six transmission gates, which appear just before the four outputs, control the routing of the packets at the two inputs to two of the four outputs of the 2×4 element. Control signals for the transmission gates are generated from the activity and routing bits of the two packets. First, the activity bit and the routing bit of the packet at input 0 are serially loaded into Latch A_0 and Latch A_1 respectively. At the same time, the activity bit and the routing bit of the packet at input 1 are serially loaded into Latch B_0 and Latch B_1 respectively. The signal Compute is then asserted. This results in the six control latches being loaded with the control inputs for the six transmission gates in the next cycle. The Compute signal is then de-asserted, and the six control latches hold the control signals for the duration of the packet. The Bypass signal is used to trap the current routing bit until the end of the routing field of the packet, so that the next stage is presented with its routing bit immediately after the activity bit. The logic diagram has 12 latches, 2 NAND gates, 4 AND gates, two OR gates, 4 invertors and 18 transmission gates. Assuming that we use static latches with 16 transistors, the approximate transistor count for the 2×4 element is about 300 transistors.

We can now estimate the transistor count in a LAST module of size k. Since that there are $(k/2 - 1)$ Pass/Cross pairs for each 2×4 element, and each Pass/Cross pair needs 30 transistors (assuming a 20% overhead), the transistor count of a module of size k is $300(k/2) + 30(k/2)(k/2 - 1) = 132k + 15k^2$. Thus, we see that the complexity of a LAST module is not very high, even though there are $k^2 - k$ Pass/Cross pairs. Using the fact that every module of the LAST switch replaces $k/2$, 2×2 elements with 114 transistors of the Banyan Network (the Banyan element in [4] uses 114 transistors), the ratio of the transistor count of the LAST switch compared to the Banyan network is given by $300/114 + (k/2 - 1)30/114 = 0.132k + 2.368$ where k is the module size. For $k = 32$, 64 and 128, this ratio is about 6.6, 10.8 and 19.2, and is independent of the number of inputs of the switch.

5 Results

For any given number of inputs, the throughput of the LAST switch is a function of the module size. When the size of a module is chosen, the trunk group size is half that of the module size. Hence, for different switch sizes, the saturated throughput can be studied in terms of the group size. Figure 5(a) shows the saturated throughput of the LAST switch for different switch and group sizes. Equations (2) and (3) from Subsection 2.1 were used to calculate the throughput values for switch sizes of 1024, 2048, 4096 and 8192. The group sizes were chosen to be 8, 16, 32, 64 and 128. It is seen that, for a given group size, the throughput variation is less than 4% among the switch sizes considered. This means that the throughput degrades very slowly as the switch size is increased. It can also be noted from Figure 5 (a) that the group size needs to be 32 in a switch with 8,192 inputs in order to achieve a saturated throughput of 85%. The throughput of an ideal, non-blocking, unbuffered switch with trunk grouping is also shown in Figure 5(a) for comparison. The results for cell-loss, when DITGM's of differ-

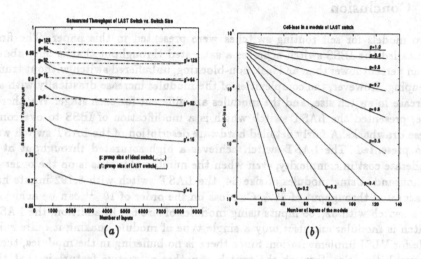

Fig. 5. (a) Throughput for different group sizes (b) Cell-loss of different module sizes

ent sizes are used, are given in Figure 5(b). The different curves correspond to different values of input load p, the probability of having a packet at the inputs of a module. To build a 32,768 input switch which can handle a load of $p = 0.9$ using 64 input modules, two switch planes each supporting a load of $p = 0.45$ are needed. Since the cell-loss probability of a module of size 64 is 3.5×10^{-8} for $p = 0.45$, each of the planes would have 10 stages, and the overall cell-loss per plane in the LAST switch would be approximately $10 \times 3.5 \times 10^{-8} = 3.5 \times 10^{-7}$. Such a switch will have trunk groups of size 32. If a lower cell-loss is desired, then three switch planes can be used, in which case the load per plane would be 0.3. With a load of 0.3, the cell-loss per module of size 64 is 1.1×10^{-12}. The cell-loss per plane is then $10 \times 1.1 \times 10^{-12} = 1.1 \times 10^{-11}$. On the other hand, if we insist on using only two planes, we can use modules of size 128. At $p = 0.45$, the cell-loss per 128 input module is 1.3×10^{-13} giving a cell-loss of $9 \times 1.2^{-13} = 1.2 \times 10^{-12}$ per switch plane. The disadvantages of using modules of size 128 are: 1) the group size of the switch would now be 64 instead of 32, and 2) the transistor count of a switch with two planes consisting of 128 input modules is greater than that of a switch with three planes consisting of 64 input modules. The design is summarized in Table 1.

Table 1. Design of a 32,768 input LAST switch, load $p = 0.9$

Module size	Number of stages	Number of planes	Load per plane	Cell-loss per plane
64	10	2	0.45	3.5×10^{-7}
64	10	3	0.3	1.1×10^{-11}
128	9	2	0.45	1.2×10^{-12}

6 Conclusion

Two models for self routing switches were presented in this paper. The first one called the IBSS switch achieves a saturated throughput that is only about seven percent lower than an ideal, non-blocking, unbuffered switch without trunk grouping. However, the complexities of the modules increase drastically with an increase in switch size, and the modules are different for each stage. We, therefore, presented the LAST switch which is a modification of IBSS to overcome these drawbacks. A fairly detailed hardware description of the LAST switch was also presented. The LAST switch achieves a high saturated throughput at a moderate cost in complexity, even when the number of inputs is on the order of a thousand. Using modules of size 64, the LAST switch with 8,192 inputs has a saturated throughput of 85%. Cell-loss on the order of 10^{-11} can be achieved for a switch with 32,768 inputs using modules of size 64. In addition, the LAST switch is modular and uses only a single type of module, making it quite suitable for VLSI implementation. Since there is no buffering in the modules, there is very little delay through the switch. Another attractive feature is that the switch delivers all packets to the outputs in sequence. In conclusion, it is our strong belief that the architecture of the LAST switch is suitable for building large switches with thousands of inputs.

References

1. A. Huang and S. Knauer, "Starlite: a wideband digital switch," in *Proc. GLOBE-COM'84*, Atlanta, GA, Nov. 1984, pp. 121-125.
2. J. Hui, "A broadband packet switch for multi-rate services," in *Proc. ICC'87*, Seattle, WA, Jun. 1987, pp. 782-788.
3. J. Hui and E. Arthurs, "A broadband packet switch for integrated transport," *IEEE J. Select. Areas Commun.*, vol. 5, no 8, pp. 1264-1273, Oct. 1987.
4. W. S. Marcus, "A CMOS Batcher and Banyan chip set for B-ISDN packet switching," *IEEE J. Solid State Circuits*, vol. 25, no. 6, pp. 1426-1432, Dec. 1990.
5. J. H. Oneill, B. D. Ackland, S. Rao and M. Hatamian, "A 200 Mhz CMOS broadband switching chip," *IEEE J. Solid State Circuits*, vol. 28, no. 3, pp. 269-275, Mar. 1993.
6. J. Turner, "Design of a broadcast packet network," in *Proc. INFOCOM'86*, Miami, FL, Apr. 1986, pp. 667-675.
7. J. Turner, "Design of a broadcast packet network," *IEEE Trans. Commun.*, vol. 36, no. 6, pp. 734-743, Jun. 1988.

8. H. Kuwahara, N. Endo, M. Ogino and T. Kozaki, "Shared buffer memory switch for an ATM exchange," in *Proc. ICC'89*, Boston, MA, Jun. 1989, pp. 118-122.

9. M. Devault, J. Couchennec and M. Servel, "The Prelude ATD experiment: assessments and future prospects," IEEE *J. Select. Areas Commun.*, vol. 6, no. 9, pp. 1528-1537, Dec. 1988.

10. H. Suzuki, *et al*, "Output-buffer switch architecture for asynchronous transfer mode," in *Proc. ICC'89*, Boston, MA, Jun. 1989, pp. 99-103.

11. Y. S. Yeh, M. Hluchyj and A. Acampora, "The Knockout switch: a simple, modular architecture for packet switching," *IEEE J. Select. Areas Commun.*, vol. 5, no. 8, pp. 1274-1283, Oct. 1987.

12. F. M. Chiussi and F. A. Tobagi, "A hybrid shared-memory/space division architecture for large fast packet switches," in *Proc. SUPERCOMM/ICC '92*, Chicago, IL, Jun. 1992, pp. 905-11.

13. H. S. Kim, "Multinet switch: Multistage ATM switch architecture with partially shared buffers," in *Proc. INFOCOM'93*, San Francisco, CA, Mar. 1993, pp. 473-480.

End-to-End Performance Evaluation of Datagram Acceptance Control in DQDB-ATM-DQDB CL Network

Rugang Vogt and Ulrich Killat

Department of Digital Communication Systems
Technical University of Hamburg-Harburg
D–21071 Hamburg, Germany

Abstract. In this paper we extend a buffer overflow control scheme called datagram acceptance control (DAC) to shared buffers, then evaluate its end-to-end performance and compare it with another two methods. To manage CL traffic in an interconnected CL overlaid network operated in on-the-fly mode, one can use loss mechanisms by discarding cells at a congested buffer. However, since the size of a typical datagram is larger than the payload size of a cell, cell discarding without taking into account the integrity of a datagram can produce corrupted datagrams. As a corrupted datagram is dropped at its final destination, the network resources used on its route are wasted. Moreover, it is observed in our studies that when the network is overloaded even with a simple buffer overflow control scheme the waste still can be a large portion. Thus, cell discarding at congested buffers has to be treated with the consideration of concentrating discarded cells upon as few datagrams as possible to achieve reducing the waste but not at the expense of the overall datagram loss ratio. It is shown by our simulation results that DAC, by managing a buffer together with its associated outgoing bandwidth, can achieve the stated objectives end-to-end. It is also observed in our studies that increased buffer size can only improve the datagram loss ratio for load levels up to 100%. Once overload occurs an *eightfold/sixteenfold* increased buffer size, which is 10 times of the mean frame length, has virtually no impact on the datagram loss ratio. Thus, losses due to overload cannot be avoided by large buffer capacity.

Keywords: Buffer management for B-ISDN, Congestion control for B-ISDN, Performance evaluation, DQDB/ATM internetworking

1 Introduction

In a public ATM network connectionless server (CLS), due to its simplicity in controlling the impact of bursty LAN/MAN data traffic and its flexibility in scaling up the capacity, will likely be used at first to provide CL service for interconnecting LANs/MANs (see an example in Fig. 1). An interworking unit (IWU), between a DQDB MAN and the ATM network, can simply forward inter-MAN datagrams to its local CLS. CLSs are responsible for routing of datagrams. To reduce connection setup delays, semipermanent VP/VCs can be used. It is

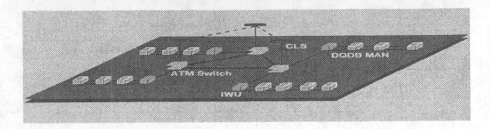

Fig. 1. An Example of DQDB-ATM-DQDB CL Network

supposed that ATM Adaptation Layer (AAL) 3/4 is employed to support the datagram service (though it has an overhead of four bytes compared with AAL5), because it provides not only the possibility of multiplexing at adaptation layer but also the reliable identification of a routing relevant cell.

Due to the payload size of a cell (a DQDB slot as well), a datagram may have to been carried in up to 210 of them. After passing through DQDB MAC or ATM switching facilities, datagrams are interleaved. In such an environment, buffer overflow control becomes critical especially for IWUs and CLSs operated in the so-called on-the-fly mode (without reassembling), because discarded cells at congested internal buffers may belong to different datagrams. If only one cell of a datagram is lost, all the other cells of the datagram become useless and start wasting the buffers and bandwidth on their way after. Therefore, uncontrolled buffer overflow may corrupt a large number of datagrams at heavy load and overload, which not only reduces the datagram throughput of a network dramatically but may also waste a large portion of network resources by the useless cells.

Thus, central objectives of an overflow control scheme are to avoid discarding cells of different datagrams, and to reduce the number of useless cells transmitted in the network. But, the waste reduction should not come at the expense of the overall datagram loss ratio. Furthermore, a control scheme should not trade in an overcontrolled performance at light or normal load for a better one at heavy load and overload.

Four buffer overflow control schemes have been proposed. They are: basic scheme (BAS); discard (DIS) algorithm [1]; pushout buffer [2]; and datagram acceptance control (DAC) by the authors in [3]. [1] and [4] have studied the performance of DIS and compared it with BAS. [2] has studied the performance of the pushout buffer. [5] has compared BAS, pushout buffer and DAC but without any on-line estimation of cell/slot arrival rate. The authors in [6] have studied DAC with an on-line estimation of cell/slot arrival rate but for a dedicated buffer, and have compared it with BAS and DIS. However, all studies have been conducted only at a single hop. Moreover, none of them, except [4], have looked into the efficiency of the schemes in reducing the wasted resources in terms of useless cells transmitted. This paper, however, will show: the end-to-end simulation studies of datagram delay and datagram loss of three schemes (BAS,

337

DAC, and DIS); their waste reduction efficiency at every hop as well as on the end-to-end basis. In addition, DAC is extended here for a shared buffer.

This paper does not consider the pushout buffer, because it may be rather too complex to implement the pushout buffer to operate at a high speed [5]. Another scheme (a modified bit rate reservation) discussed in [7] is not considered either, because this scheme will not provide high multiplexing efficiency when interleaved datagrams are transmitted at a high bit rate, say 150 Mbps, as it is a possible case in a DQDB MAN.

This paper is organized as follows: In section 2 an overview of BAS, DIS and DAC is given, with an extension of DAC for a shared buffer; Section 3 describes end-to-end simulation models; Results of the simulation studies are presented and discussed in section 4; And finally, the paper concludes at section 5.

2 Buffer Overflow Control Schemes

Note again that one of the objectives of the overflow control schemes is to concentrate discarded cells on a few datagrams in an interleaving environment. Therefore, they differ from each other in how the concentration is done.

BAS is a purely reactive scheme. Whenever a cell of a datagram is discarded at a congested buffer, this dropping is marked in its routing information data at the hop so that its subsequent cells are automatically discarded at this hop when they arrive.

DIS adds a threshold method on top of BAS. To control the cell discarding, DIS checks the buffer's actual level of occupancy whenever the first cell (Beginning of Message, BOM) of a datagram arrives. If this level is below a given threshold, the arrived BOM is served. If the threshold is crossed, the datagram size is checked against the actually available space. If the space is not sufficient, the BOM is discarded and its following cells as well.

DAC adds a fast dynamic buffer reservation on top of BAS. A detailed state machine description can be found in [6]. DAC is invoked whenever a BOM cell arrives. It, then, derives an estimate of buffer demand. If overflow is anticipated, the BOM and its following cells are discarded; otherwise the datagram, ie. all its cells, is said to be accepted. The estimation is based on:

- The current level of buffer filling;
- The number of cells accepted but not yet served plus the size of the datagram just arrived;
- An estimated arrival rate of cells accepted and under consideration.

As the estimation algorithm developed in [6] supports only a dedicated buffer, we now extend it to a shared buffer. Suppose that a link with a buffer B_{max} supports M peak rate allocated VPs. Let $b_i(t)$ denote the current buffer filling level by the ith VP's traffic, $p_i(t)$ denote the number of cells already accepted by the ith VP but not yet served, and $\lambda_i(t)$ denote an on-line estimated accepted cell arrival rate of the ith VP. Then, when a BOM arrives at time t and is routed to, say, the kth VP, the total buffer demand, denoted by $\mathcal{B}$, is

$$\mathcal{B} = max\{0, (\lceil p_k(t) + l - \mu_k \mathcal{T} \rceil + \sum_{i=1;\, k+1}^{k-1;\, M} \lceil min\{p_i(t), \lambda_i(t)\mathcal{T}\} - \mu_i \mathcal{T} \rceil)\} \quad (1)$$

where μ_i is the peak cell rate of the ith VP, l is the size of the just arrived datagram and the $\mathcal{T}$ is

$$\mathcal{T} = \frac{p_k(t) - b_k(t) + l - 1}{\lambda_k(t)} \quad (2)$$

The datagram is accepted if and only if $B_{max} - \mathcal{B} \geq 0$. As it has been derived in [6], $\lambda_i(t) = r_{a_i}(t) + (1 - \frac{r_{a_i}(t)}{r_{p_i}(t)})$, where $r_{a_i}(t)$ is the measured average accepted cell arrival rate in the interval of $t - t_0$, and $r_{p_i}(t)$ is the measured peak cell rate in the same interval. t_0 is the point when a BOM arrives at a VP whose all accepted cells have arrived, ie. it is at $p_i(t_0) - b_i(t_0) = 0$ state.

3 Simulation Models

The end-to-end simulation model used in our simulation studies is depicted in Fig. 2. Since the purpose of our study is to compare the three overflow control schemes, thus IWU can be located freely as long as its location stays the same during the comparative study. By locating the sending IWU (SIWU) at the end of bus A in the sending MAN (SMAN) and the receiving IWU (RIWU) at the

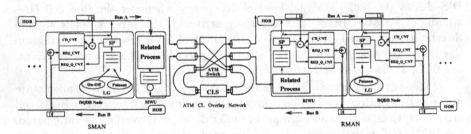

Fig. 2. End-to-end Simulation Model

head of bus A in the receiving MAN (RMAN), we need only to simulate the information transmission on bus A.

There is a MAC layer at every DQDB node, which consists of a segmentation process, a local segment queue and three counters (countdown counter (CD-CNT), request counter (REQ-CNT) and request queued counter (REQ-Q-CNT)) for one priority level operation. At the local segment queue there is no space limitation. Besides, at each node of SMAN there are two independent traffic generators. One is for generating intra-MAN traffic, which is modelled by batch Poisson. The other is for generating inter-MAN traffic, which is modelled by on-off source. Intra-MAN frame length is equal to the batch size, while inter-MAN frame length is equivalent to the duration of on-period. The batch size as well as the on and the off period are all geometrically distributed. A frame may be larger than a datagram. If that occurs, a couple of datagrams are made out of that frame, and they are passed over one after another. Intra-MAN traffic generated at each node is proportionate to the number of nodes at its downstream direction, while inter-MAN traffic is generated equally likely at every node of SMAN. At every node of RMAN there is only the batch Poisson source, since inter-MAN traffic on RMAN comes from the ATM network. For detailed formulas please refer to [6].

SIWU is modelled with a cell processing and overflow control block, and an output buffer associated with a server. The buffer is space limited and is served with the peak rate of its associated outgoing VP/VC. Therefore, a cell spacing or rate control function is implemented. RIWU is just as a DQDB node except that the space at the local segment queue is limited, and the load generator is replaced by the cell processing and overflow control block.

To avoid linking our studies to any particular ATM switch architecture and call admission control (CAC) algorithm, the ATM switching node is assumed to be equipped with a perfect output queuing switch (no loss). And CL VPs are allocated with their peak rates.

Interfering traffic arrived at the ATM switching node is generated by a number of independent on-off sources. All the on-off sources transmit cells at the VP's peak rate. Each source is assigned a destination address which is drawn uniformly between CLS's outgoing VPs.

Fig. 3 depicts the model for CLS. Its space limited buffer is shared by two outgoing VPs, and each VP's logical buffer is served at most by the VP's peak cell rate. In case of simultaneous link service requests, the link will pick up one

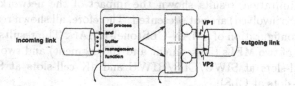

Fig. 3. Model for CLS

of the VPs randomly.

To observe the end-to-end performance, a pair of nodes must be chosen, one on SMAN and the other on RMAN. Then, the inter-MAN datagrams generated at the chosen SMAN node, called datagrams under test (DUT), are all addressed to the chosen RMAN node and are numbered so that the end-to-end datagram loss can be measured at the chosen RMAN node. In order to measure the end-to-end datagram delay, every EOM/SSM segment of DUT carries a time-stamp which is the time when its BOM is transmitted onto the bus at the chosen SMAN node. Other SMAN nodes' inter-MAN traffic is uniformly distributed between the outgoing VPs of CLS.

One remark still has to be made for our choice of simple on-off and batch

Poisson sources. As an overflow control scheme is located at hops, traffic arrived at a hop is already multiplexed either by DQDB MAC or ATM switching facilities. That traffic preserves few source characteristic. Thus, in our opinion, simple source models are good enough for the purpose.

4 Simulation Results

In our simulation runs, 150 Mbps is used for both ATM links and DQDB buses. On each MAN, due to the simulation run time (days) needed, only 8 nodes plus IWU are connected. The distance between neighbour nodes is equivalent to one slot time long, and IWU is right beside its neighbour. Propagation delay between SIWU and RIWU is not considered because for our studies this delay contribution is the same for all datagrams as well as for all schemes. Furthermore, all VP/VCs' peak rates are fixed at 1/4 of the link rate, ie. 37.5 Mbps. The aggregated DQDB bus load is fixed at 80% of the bus rate for both SMAN and RMAN, while inter-MAN traffic load varies. Such a setting is chosen because we are interested in how our DAC performs under loads which are often talked as an example (80% load on MAN and load towards ATM being $\leq$ 25% of MAN's). Based on this, we think that to let a CL VP/VC take a peak rate of 25% of the link rate is the best way to study how the three control schemes behave under such a conventional bandwidth allocation at DQDB/ATM IWU and at CLS. Note that the aggregated load level of the interfering traffic entered at the ATM switch is the same as the inter-MAN traffic load level on the SMAN relative to the VP/VC's peak rate.

The results presented here show the impact of the buffer size and of the mean frame length (MFL) on performance of the three control schemes. Due to the page limitation, results shown the impact of the network size (ie. the number of CLSs involved) are not presented. Therefore, all shown results are from the network configuration of SMAN-CLS(one)-RMAN. The results are from the combinations of two MFLs (50-segment and 100-segment) and two sets of buffer sizes (500 cell-slots at SIWU/CLS/RIWU and 4K cell-slots at SIWU/RIWU while 8K cell-slots at CLS).

4.1 End-to-End Performance

Fig. 4 shows the results of the end-to-end datagram loss ratio (DLR) as a function of the offered load. The 95% confidence intervals are shown by the bars. A datagram is said to be lost if one of its cells is discarded. The results are based on 10 end-to-end observations, and each is defined by the complete arrival of 1,000 uncorrupted DUT. Obviously, DAC in all cases yields the lowest end-to-end DLR. In contrast, DIS (with a threshold of 85% the buffer size) results in the worst end-to-end DLR for the case of 500-buffer and 50-segment MFL (noted in the Fig. as 500B & 50seg), and almost the same end-to-end DLRs as those of BAS for the other three cases. One of the reasons for this, besides the threshold level, is that no additional check is done after the threshold is crossed. However,

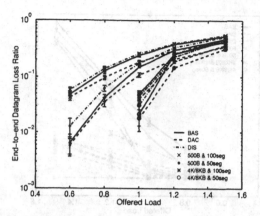

Fig. 4. End-to-end DLR as a function of offered load with 95% confidence intervals

there is still more important information shown in Fig. 4. One is the sensitivity of the DLR to the change in the MFL. In general, the lower the load level is, the more sensitive the DLR to the change is. But, BAS and DIS are sensitive only when the buffer is small, while DAC is sensitive at both buffer sizes with a different degree–the smaller the buffer is, the more sensitive DAC is. The other information which is also the most important one shown in Fig. 4 is the impact of the buffer size on the DLR. The divergence of the DLRs shown at the load levels up to 1.0 and the convergence shown at the load levels of 1.2 and 1.5 clearly evidences that, for all the three control schemes, an *eightfold* increased buffer size at SIWU/RIWU and a *sixteenfold* increase at CLS have improved the DLR only for load levels up to 1.0, but have almost no impact on the DLR once overload occurs. Thus, it shows that the conventional believe of using large buffer to improve the DLR at overload is wrong.

With only the performance of the DLR, one may come to the conclusion that after all the improvement of DAC is only moderate. But, a better DLR is only one part of the central objectives (stated in section 1). The other part, the efficiency of waste reduction is shown in Fig. 5. In general, BAS produces the worst waste, e.g., at overload it leads to a large portion of wasted network resources by cells of corrupted datagrams. With DAC and DIS this phenomenon is significantly reduced. Such as at overload level 1.5 BAS, for all the four combinations of buffer size and MLF, has a waste around 40%, while DAC has kept its waste at around 10%, which is a 50% improvement in meaningful usage of network resources. As it is shown, for BAS, the longer the MFL and the smaller the buffer are, the higher the waste is. However, both the buffer size and the MFL have more impact on the waste of BAS at normal to heavy load regions than at the overload region. For DAC, its waste reduction is much less sensitive to the changes in the two parameters than what BAS has shown. The most sensitive of all is DIS. At the 500-buffer, the waste reduction of DIS, even with its much higher DLR, is worse than that of DAC. And the doubled MFL has triggered off a more than doubled waste, except at the load level of 0.6. But at the 4K/8K-buffer, DIS

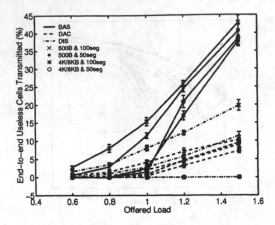

Fig. 5. Waste as a function of offered load with 95% confidence intervals

with almost the same DLR as BAS and somewhat higher DLR than DAC has kept the waste at *zero* for both MFLs. This means that if DIS is used at a large buffer, an additional check after the threshold is crossed or a higher threshold may bring its end-to-end DLR somewhat down though at the expense of the zero waste.

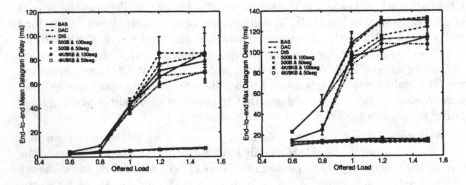

Fig. 6. Mean datagram delay with 95% confidence intervals

Fig. 7. Max datagram delay with 95% confidence intervals

Not only the end-to-end DLR and the end-to-end efficiency of waste reduction have been investigated, the end-to-end datagram delay has also been studied. Fig. 6 and Fig. 7 show the measured mean and maximum end-to-end datagram delays, respectively. For the mean datagram delay, as can be seen, for all the control schemes significant difference is caused only by different buffer size. Moreover, the large buffer has introduced negligible increases at load levels of 0.6 and 0.8, but more than tenfold increase at load levels above 1.0. As for the maximum delay, it shows the same behaviour as the mean delay at load levels above 1.0, while at load levels of 0.6 and 0.8 the doubled MFL together with the 4K/8K buffers has produced the largest maximum delay. Moreover, the difference between the maximum delay and the mean delay is larger at normal load than at overload. For example, at the load level of 0.6 the maximum delay differs from the mean delay by a factor of more than four, while at the 1.5 load level they differ by only a factor of around two. The difference between the maximum and the mean is due to the datagram length, DQDB MAC and the buffer size at hops.

4.2 Performance at Hops

Not only the end-to-end performance, but also the performance at every hop have been observed. Observations at hops are made at every second on the total traffic passed by. Note that no buffer overflow has been observed at RIWU. Fig. 8 shows the *goodput* [4] versus the *observed* load on outgoing VP/VC. The goodput is the *effective* cell throughput which excludes cells of corrupted datagrams. Goodput is a much more meaningful measurement than the uncorrupted

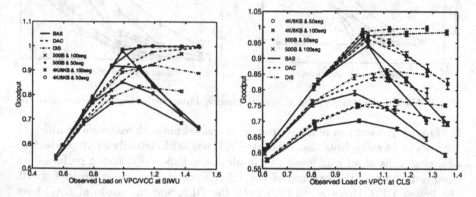

Fig. 8. Goodput at hops v. observed load with 95% confidence intervals

datagram throughput since traffic offered to VP/VC is measured in cells per cell-time unit, not datagrams per cell-time unit. Besides, datagram size is an upper bounded random variable, thus cannot be used to convert the datagram throughput to the goodput.

As expected, in overload region the goodput of BAS, even with large buffers, is severely degraded at both hops. As for DAC and DIS, though their goodput varies from hop to hop, their improvement on the goodput is quite impressive. As a general rule, the buffer size has a significant influence on the goodput, for example, the larger the buffer is, the better the goodput becomes. Whereas the size of the MFL has an effect only when small buffer is in use, for example, with 500-buffer, the longer the MFL is, the worse the goodput gets. As shown at SIWU, DAC in all the four cases yields the best goodput which increases steadily to the 100% of the outgoing VP/VC's capacity, while DIS performs worse with the 500-buffer than DAC does since its goodput starts dropping at load levels above 1.2. But with the 4K-buffer DIS performs as good as DAC. However, at CLS, DAC is no longer the best performer, rather DIS is. The goodput of DAC at load levels above 0.8 is below what DIS has produced, and it has dropped in the overload region especially with the 8K-buffer. Although Fig. 8 is the most straightforward way to show the different performance of the schemes as well as the impacts of the buffer size and the MFL, it can hardly be used to find out whether the dropping at the goodput is due to excessive useless cells transmitted or excessive blocking of entire datagrams, because the difference between the goodput and the observed load includes both the useless cells and the discarded cells. Fig. 9 and Fig. 10 can serve the purpose of separating the two.

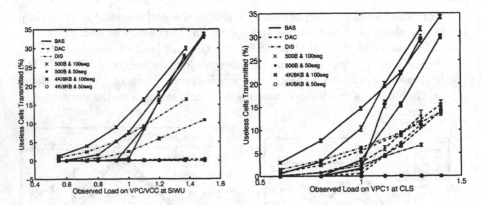

Fig. 9. Useless cells transmitted at hops v. observed load with 95% confidence intervals

From Fig. 9 one can read the amount of wasted network resources by different schemes right away. Note that DAC at SIWU has kept virtually a null production of useless cells at all load levels and in all cases. Especially, such a performance does not come at the expense of the DLR since it has, as shown in Fig. 10, the lowest DLR. However, at CLS both the DLR and the waste of DAC have worsened. It can only keep the waste below the 16% mark. This indicates that the formulae (1) and (2) developed in section 2 for the extended DAC underestimate the buffer demand. Because of the shared link at CLS, a non-empty logical buffer is no longer served by a constant rate, ie., its VP's peak cell rate, as it is the case at SIWU. Thus, (1) can provide only a lower bound of the buffer demanded, which results in accepting more datagrams than what actually could be dealt with. Therefore, modifications for the formulae are needed. As for DIS, its waste, depicted in Fig. 9, has shown the same behaviour at both hops. With the large buffer it has kept the waste at null, but with the small buffer it can only keep its waste below the 17% mark. In contrast, its DLR, depicted in Fig. 10, has shown a bit different behaviour at the two hops. At SIWU its DLR is almost the same as the DLR of BAS, but at CLS it is somewhat better in the overload region. As for BAS, with no exceptions, it has produced the worst portion of waste at both SIWU and CLS. At the load levels of 1.4 and 1.5, it produces a more than 30% waste of network resources.

Again, it is clearly to see in Fig. 10 the divergence of the DLRs shown at the load levels up to 1.0 and the convergence shown at the load levels above

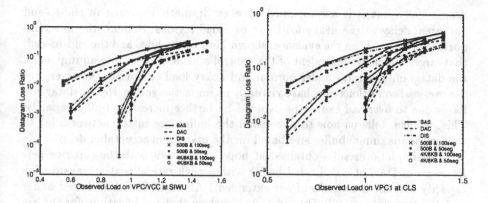

Fig. 10. DLR measured at hops v. observed load with 95% confidence intervals

1.0, which have been observed already at the end-to-end DLRs. This means that datagram losses due to overload cannot be avoided by large buffer capacity! However, as it is shown in Fig. 8, large buffer brings a significant improvement in the goodput (the effective network utilization). Therefore, the maximum buffer size should be chosen only based on the maximum acceptable delay. A similar opinion has also been expressed in [8]. It should be noted at this place that the measured performance at CLS's VP2 is the same as what has been measured at its VP1.

5 Conclusion

In this paper, we have considered performance evaluation and comparison of three buffer overflow control schemes (BAS, DAC and DIS) for interconnected CL networks. An extension of our DAC (a fast dynamic buffer reservation scheme) to a shared buffer has been presented. Although the emphasis has been given to the end-to-end performance comparison, how the three schemes behave at hops has also been observed and compared.

The end-to-end performance has been evaluated in an interconnected three-hop CL network which consists of two DQDB MANs and an ATM overlaid CL network. Our simulation results from normal load to overload regions, from 50 segments mean frame length to 100 segments, and from 500 cell-slots buffer at each hop to a 4K-buffer at SIWU/RIWU and a 8K-buffer at CLS have revealed that DAC is the only one fulfilled the central objectives identified in section 1. The objectives are to reduce the number of useless cells transmitted in the network but not at the expense of the overall datagram loss ratio, and not to trade in an overcontrolled performance at normal load for a better one at heavy load and overload.

Furthermore, it is observed that the doubled mean frame length has an impact on the performance of the control schemes only when the equipped buffer is small, ie., as 5 times of the mean frame length. Moreover, our results show that by increasing buffer capacity the effective network utilization is improved dramatically, but it is accompanied by even dramatic increases in end-to-end datagram delay at the heavy load and overload regions. Besides, the most important of all has been the evidence, shown both at the hops and the end-to-end, that larger buffers, independent of the control schemes, have quite an impact on the datagram loss ratio under normal and heavy load conditions, however, under overload conditions they have virtually no impact anymore. Hence, datagram losses due to overload cannot be avoided by further increasing buffer capacity. This, in turn, tells us how to dimension the buffer size in the network, ie., to choose the maximum buffer size based on the maximum acceptable delay.

The simulation results obtained at hops have shown both the superior performance of DAC at a dedicated buffer with a peak rate allocated transmission capacity and the drawbacks of our extended DAC in CLS with a shared buffer and a shared link as well. These drawbacks show that modifications for the extended DAC are needed. Thus, further work is under way to obtain a better estimation of buffer demand for the extended DAC. And future research in the area of buffer overflow control in CL network will be focussed on a hopwise feedback control mechanism which will let CL traffic be served in ATM networks on "available bit rate" basis.

References

1. G.Boiocchi, P.Crocetti, L.Fratta, M.Gerla, and M.A.Marsiglia, "ATM Connectionless Server: Performance Evaluation," in Modelling and Performance Evaluation of ATM Technology (C-15), 1993 IFIP, pp. 185-195.
2. S.Aalto, I.Norros, J.Virtamo, K.Kvols, and S.Manthorpe, "Performance Aspects of Streaming and Message Modes of Interworking," in Proc. of Interworking'92, Bern, Switzerland, Nov. 1992.
3. R.Vogt, U.Killat, J.Ottensmeyer, and M.Rümekasten, "A Concept for Interconnecting DQDB MANs through ATM-based B-ISDN and Related Issues with respect to Simulation," in Proc. of Interworking'92, Bern, Switzerland, Nov. 1992.
4. A.Romanow and S.Floyd, "Dynamics of TCP Traffic over ATM Networks," in IEEE J. Sel. Areas Commun., Vol. 13, No. 4, May 1995, pp. 633-641.
5. S.Manthorpe, "Buffering and Packet Loss in the DQDB to ATM Interworking Unit," in Proc. of Integrated Broadband Communication Networks and Services, Copenhagen, Apr. 1993, pp. 31.2.1 - 31.2.12.
6. R.Vogt and U.Killat, "Performance Evaluation of Datagram Acceptance Control in DQDB to ATM Interworking Unit," in Proc. of ISS'95, vol. 1, Berlin, Germany, Apr. 1995, pp. 82 - 86.
7. U.Briem, "Performance Comparison of Resource Sharing Schemes in a Connectionless Server on Top of ATM," in Proc. of the 14th International Teletraffic Congress, Antibes Juan-les-Pins, France, June 1994, pp. 1109-1120.
8. H.J.Fowler and W.E.Leland, "Local Area Network Traffic Characteristics, with Implications for Broadband Network Congestion Management," in IEEE J. Sel. Areas Commun., Vol. 9, No. 7, Sept. 1991, pp. 1139-1149.

Comparison of Explicit Rate and Explicit Forward Congestion Indication Flow Control Schemes for ABR Service

Aleksandar Kolarov and G. Ramamurthy

C&C Research Laboratories, NEC USA Inc.,
4 Independence Way, Princeton, NJ 08540, USA

Abstract. An adaptive end-to-end rate based congestion control scheme to support a class of best effort service known as Available Bit Rate Service (ABR) is being proposed by the ATM Forum. In this paper we investigate two variants of this control scheme: the Explicit Forward Congestion Indication (EFCI) scheme and the Explicit Rate (ER) scheme. We show that for EFCI based switches the performance of virtual channels traversing large number of hops in WANs can be substantially improved by giving priority to network transit traffic over traffic entering the network. The EFCI scheme with priority exhibits a robust behavior, and ensures fair share of the bandwidth for all VCs, regardless of the number of hops they traverse. The ER scheme however is very stable, even under extreme loading conditions, and ensures fair sharing of resources.

1 Introduction

High speed networks based on the Asynchronous Transport Mode (ATM) are at the fore front because of their ability to support multi rate switching and transmission in an efficient manner. Since an ATM based switch can simultaneously switch traffic with real time constraints such as voice or video along with data traffic, it provides an ideal platform for supporting multi-media based applications. Thus, there is tremendous interest in the development of ATM-based LANs that will provide high bandwidth connectivity between work-stations and servers, and ATM Hubs and WANs that will interconnect geographically separated LANs.

Most data applications are based on the concept of connectionless service, and cannot describe their traffic characteristics adequately. Further, data traffic is generally very bursty. These two characteristics result in traffic patterns where large bursts of data can arrive without prior notification. For comparable performance with current technologies we require that the bursts encounter very low blocking and/or loss, and the need to retransmit is minimal. These requirements lead to serious concerns in ATM switches from a performance point of view. To ensure the required quality of service, the network should reserve resources either on a statistical basis or on a deterministic basis (i.e., without sharing of resources). This approach requires that the offered traffic be adequately characterized and monitored.

ATM-based switches have been generally characterized by small internal buffers, ranging from a few hundred cells to a few thousand cells. Further, to provide a fast and lean transport, all link level controls have been eliminated, and one has to resort to end-to-end based controls [1] . If there are traffic hot spots (i.e., traffic from several bursty sources are directed to a single output port), statistical multiplexing of ATM cells with no mechanism to control the source rate will lead to severe cell loss. This loss will be aggravated if the input port speeds are comparable to the output port speeds and the burst sizes are large. Such cell loss will lead to even larger frame loss compared to other technologies that have a media access control. Since ATM interleaves cells from different connections, when cell loss occurs, it can lead to loss of all the frames that are currently being transmitted. Frequent retransmissions due to cell loss increases the effective load on the system, resulting in an end-to-end throughput that is several times less than that of a shared medium network [2].

The ATM Forum has defined a new service class for data applications called the Available Bit Rate Service (ABR) [3]. Users of this service dynamically share the available bandwidth. While this service does not provide any strict guarantees, it attempts to minimize the cell loss at the expense of delay. The dynamic sharing of bandwidth between competing users has to be achieved via appropriate set of distributed controls. Two end-to-end rate based feedback congestion control mechanisms have been proposed. In the first scheme, a single bit is used to indicate if a switch in a virtual channel's (VC) path is congested or not. In the Explicit Forward Congestion Indication (EFCI) scheme, switches use the ECI bit in the header of ATM data cells to notify the destination if they are congested. The destination filters this information and signals the source through a special control cell. This binary scheme is similar to the DEC bit scheme for packet-switched networks [4]. In the second scheme, instead of a single bit feedback, the switches explicitly specify the maximum rate each VC is allowed to use. The switches will compute the rate for each VC based on the state of its queues, the available capacity for ABR service and the number of active sources.

While the thrust for ABR service comes from LAN providers, such services should also be capable of being operated across wide area networks. This makes it necessary that controls chosen to support ABR service not only function well in a LAN environment where the propagation delays are small, but they must also work over larger distances where the feedback latencies are large, as in the case when LANs are interconnected by WANs. However, the effectiveness of any feedback control scheme is limited by the latency of the feedback loop. Hence, such end-to-end controls are likely to be less effective as the propagation delay and the bandwidth of the network increase. In particular, when there are virtual channels that traverse several hops, extreme unfairness can result. Virtual channels whose feedback delays are smaller and thus have more up to date information can have an unfair advantage over virtual channels that have larger feedback delays. In a wide network environment, the latency of the feedback loop coupled with the fact that the amount of buffers at each node can be less than the bandwidth delay product, can lead to significant degradation in the network performance. In fact, the network throughput can collapse at high offered loads if one take into account the retransmission traffic. In this paper we compare the performances of the EFCI and ER based end-to-end congestion control schemes in a WAN environment. We investigate the performance under both steady state and transient conditions, including the presence of high priority Variable Bit Rate (VBR) cross traffic.

2 Reference Models

In order to compare the performance of different competing control schemes, we define a reference network model and a reference traffic model.

2.1 Reference Network Model

Figure 1 shows one of the reference network models, adopted by the ATM Forum [5, 6], that consists of a multi node network, with three local switches (1, 3 and 4), and one transit switch (2). Terminals generating traffic are attached to the input ports

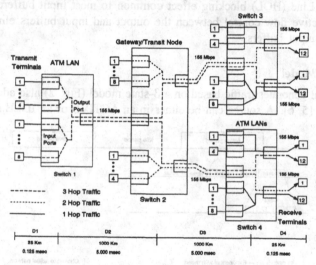

Fig. 1. Reference network model

of all four switches, while terminals receiving traffic are attached to the output ports of switches 3 and 4 only. The terminals are connected to the switches through links running at 155 Mbits/sec. The switches are interconnected via links running at 155 Mbits/sec. Switches receive traffic from terminals that are attached to their input ports. Switches 1, 3 and 4 also receive traffic from upstream switches. The routing of traffic is such that traffic originating at switch 1 traverses 3 hops, traffic originating at switch 2 traverses 2 hops and traffic originating at switches 3 and 4 traverse 1 hop. Further, all n $(n = 1, 2, 3)$ hop traffic compete with m $(m = 1, 2, 3, \ m \neq n)$ hop traffic. Each transmit terminal has one virtual channel that is terminated on a unique receive terminal. Virtual channels constituting the n $(n = 1, 2, 3)$ hop stream are grouped into two groups of 8 and 4 virtual channels respectively. The distance between the terminals and their respective access switch is $D_1 = D_4 = 25$ km. The distance D_2 between switch 1 (representing a LAN) and switch 2 (representing a transit switch), and the distance D_3 between switch 2 and the terminating switches 3 and 4, is 1000 km.

This reference model captures:

a. The interference between traffic traveling $1, 2$ and 3 hops.
b. The effect of large propagation delay on the effectiveness of the feedback control.
c. Fairness between $1, 2$ and 3 hop traffic.

2.2 Switch Architecture

Each switch is modeled as a generic input/output buffered switch with 1000 cell buffers at each input port. Each output port has 128 cell buffers. Internally, the switches use a Random In Random Out (RIRO) queue management to move cells from the input buffers to the appropriate output buffers. This RIRO scheduling eliminates the Head of Line (HOL) blocking effect common to most input buffered switches. Further, selective flow control between the output and input buffers eliminates cell loss at the output buffers.

2.3 Source Model

Each terminal generates traffic based on a 3-state model (Fig. 2) also adopted by the ATM Forum [5, 6]. A source can be either in an ACTIVE or an IDLE state. In the

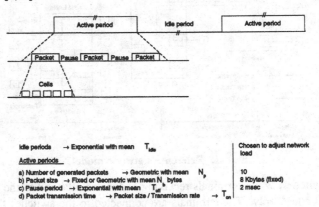

Fig. 2. Source traffic model

IDLE state no traffic is generated. While in an ACTIVE state, the source generates a series of packets or bursts which are interspersed by short pauses. Each packet can either be fixed (8 or 64 Kbytes) or can be a truncated exponential. Each pause period is drawn from a negative exponential distribution. The number of packets generated during an ACTIVE period is geometrically distributed. The IDLE periods can have any general distribution. In the numerical examples of Sect. 4 where we examine the steady-state performance of the congestion control mechanisms, the pause periods between packets have an average value of 2 msec and the packets have a fixed length of 8 Kbytes. The length of the idle period is determined by the offered load on the link. To investigate the transient response of the control mechanisms we assume the sources have an infinite backlog which assures a continuous traffic flow during an observation time interval.

We use a simple retransmission protocol in case of cell losses, which occur during periods of congestion. A packet is presumed to be lost even if a single cell is lost. Packets that are received by the receive terminal with missing cells are retransmitted by the transmit terminal until successful delivery. Packet retransmission is scheduled with a back of delay equal to twice the current (estimated) round trip delay. The useful throughput represents the actual throughput of packets (in Mbits/sec) that are eventually delivered to the destination without cell loss (after retransmission if necessary).

2.4 Basic Source and Destination Operations [3, 7]

The rate at which an ABR source is allowed to schedule cells for transmission is denoted by Allowed Cell Rate (ACR). The ACR is initially set to the Initial Cell Rate (ICR) and is always bounded between the Minimum Cell Rate (MCR) and the Peak Cell Rate (PCR). Transmission is initiated by the sending of a Resource Management (RM) cell followed by data cells. The source will continue to send RM cells after every $(N_{RM} - 1)$ data cells are transmitted. The source rate is controlled by returning RM cells, which contain information about the state of the network. The source places the rate at which it is currently transmitting (the ACR) in the CCR field of the RM cell, and the rate at which it wishes to transmit cells (usually the PCR) in the ER field. The RM cell travels forward through the network, providing the switches in its path with information about the state of the source. The switches use this information to determine the allocation of bandwidth among ABR connections. The switches can also decide to reduce the value of the Explicit Rate (ER) indicated by the ER field, or set the Congestion Indication (CI) bit to 1. Switches only supporting the EFCI mechanism will ignore the content of the RM cell. When the RM cell arrives at the destination, the destination changes the direction bit in the RM cell and returns the RM cell to the source. If the destination is congested and cannot support the rate in the ER field, the destination can reduce the value of ER to whatever rate it can support. When returning an RM cell, if the destination had observed that the last data cell received had its EFCI bit set, then it should set the RM cell's CI bit to indicate congestion. As the RM cell travels back to the source through the network, each switch may examine the cell and determine if it can support the rate ER for this VC. If ER is too high, the switch can reduce it to a rate that it can support. No switch can increase the value of ER in the RM cell since prior switch congestion information would be lost.

When the RM cell arrives back at the source, the source should modify its rate, ACR, based on the information carried back by the RM cell. If the congestion indication bit is not set (i.e., CI = 0), then the source is allowed to increase its rate ACR by an amount $N_{RM} * AIR$, where AIR is additive increase in rate. ACR can be increased up to the ER value contained in the last received RM cell, but

352

never exceeding PCR. If the congestion indication bit is set (CI = 1), then the source should decrease its rate by at least $ACR * N_{RM}/RDF$ where RDF is Rate Decrease Factor (an exponential decrease). ACR is further decreased to the returned ER value, although never below the MCR.

When a source starts transmitting after being idle, if the elapsed time since the last RM cell was sent is greater than $T_{rm} = 100$ msec, and during this interval if the source did not send more than $M_{rm} = 2$ cells, then the next cell to be sent out is an RM cell. Before sending an RM cell, if the time T that has elapsed since the last RM cell was sent is greater than $T_{of} * N_{rm}$ (Time Out Factor $T_{of} = 2$) cell intervals (of 1/ACR), and if ACR is greater than ICR, then the ACR should be reduced to at most the rate obtained as follows: $\frac{1}{ACR_{new}} = \frac{1}{ACR_{old}} + \frac{T}{RDF}$. If the new rate is smaller than ICR, then it should be set to ICR. The last constraint aims at protecting the network from the impact of sources that, having gone idle at a high ACR, do not claim large bandwidth as soon as they become active, and lead to possible congestion.

3 Switch Behavior

We compare two different types of switches. The first type of switches are EFCI-based switches which set the EFCI bit in data cells to indicate congestion. The second type of switches are Explicit Rate-based switches which modify the ER field of the RM cell to indicate the rate at which a VC may transmit. In this paper we assume that the ER setting is performed on backward RM cells. We now describe in detail a control algorithm for each type of switch.

3.1 Switch Control with EFCI Marking

The EFCI signal is generated at a switch on a per VC basis by setting the EFCI bit in the cell header of data cells when the output port is congested. The output port is marked congested if the total number of ABR cells waiting at all the input ports and are destined to be routed to this output port exceeds a high water mark H_2 (500 cells). We will refer to these waiting ABR cells as the global queue fill with respect to a given output port. The output port remains in the congested state until the global queue fill falls below the low water mark L_2 (300 cells). This hysteresis ensures that the oscillations are minimized. When a cell arrives at an input port, the EFCI bit is set if the output port to which the cell must be routed is congested. The cell is then routed towards its destination.

In [2, 8] we show that with end-to-end based controls, VCs traversing two or more hops can see a throughput collapse at high loads. To overcome the unfair advantage that VCs traversing small number of hops have over VCs traversing larger number of hops, we proposed an access priority mechanism where transit traffic from upstream switches are given priority of service over new traffic entering the network at each switch. With this priority control mechanism, the EFCI scheme not only performs well over large distances, but traffic travelling longer number of hops receive nearly the same throughput as traffic travelling fewer number of hops [8].

3.2 Switch Control with ER Marking

The primary consideration in the design of a congestion controller for ER based switches is to ensure good dynamic characteristics, high utilization and fairness in resource allocations amongst competing VCs.[9]. We call our explicit rate controller a Predictive Explicit Rate Controller (PERC), since we use a predictive control law for the congestion controller design. The main idea is to compute a rate which will bring the global queue fill of a given output port to a desired threshold level x^0 in a fixed number of update intervals denoted by D.

The level of congestion at the output port is estimated by monitoring the difference between the global queue fill x and a queue set point x^0. Based on this difference, the congestion controller associated with each output port periodically calculates an explicit rate that is common to all VCs using the given output port. The control algorithm updates the explicit rate once every T msec. In the examples presented in this paper, T is equal to the time required to transmit 400 cells at rate 155 Mbits/sec, which is approximately 1 msec. In the numerical example, the queue set point x^0 is set equal to 90 cells. We choose D to be 10 which corresponds to an average round trip time of 10 msec. We assume that time is slotted, where each slot is T msec long. The rate controller equation is given by

$$R(n) = \left[\widehat{C} - \frac{(x(n) - x^0)}{D} \right] \frac{1}{\widehat{N}} \qquad (1)$$

where $R(n)$ is the explicit rate for each VC, computed at the n-th time slot ($n = 0, 1 \ldots$), $\widehat{C}$ is the estimated link capacity available for ABR service, $x(n)$ is the global queue occupancy at the n-th time slot, x^0 is the desired level of queue occupancy, D is a constant described above, and $\widehat{N}$ is the estimated number of active ABR VC's using the given port. The available capacity for ABR service can be computed as the difference between the total port capacity and the bandwidth reserved for other services, such as CBR and VBR. The estimate $\widehat{N}$ can be obtained by monitoring the arrivals of RM cells in the forward direction, on a time interval that is longer than the basic update interval T (e.g. $4T$). At each switch the ER field in the backward RM cell is modified as $ER = min\{R(n), ER^*\}$ where ER^* is the value of ER in the received RM cell at the switch.

4 Numerical Results

To investigate the performance over larger distances as in WANs, we set the inter switch distances D_2 and D_3 to 1000 km (5 msec propagation delay). First, we compare the EFCI schemes with and without priority for transit traffic. The source parameters are chosen as follows: $PCR = 365$ cells/msec (155 Mbits/sec), $ICR = 25$ cells/msec (10 Mbits/sec), $MCR = 4$ cells/msec (1.5 Mbits/sec), $N_{rm} = 32$, $RDF = 1024$, and $AIR = 0.15$ cells/msec (0.06 Mbits/sec). This choice is in line with the ATM Forum recommendation. Figure 3 shows the plots of 1, 2, and 3-hop source useful throughput (at the packet level) in the case of EFCI scheme with and without priority for transit traffic, as a function of end-to-end delay. The numbers in parenthesis indicate the overload factor on each link. For example an overload factor of 1.4 means that the average offered load on a link is 1.4 times its maximum capacity. As the offered load increases, the packet level throughput decreases and the end-to-end delay increases.

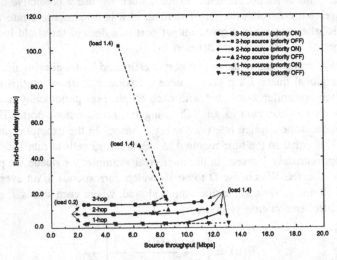

Fig. 3. End-to-end packet delay vs. per hop source throughput under EFCI

From Fig. 3 we see that without priority for transit traffic the 2 and 3-hop VCs experience a throughput collapse under overload. However, the 1-hop source throughput is high with low delay even under overload. Thus, VCs traversing two or more hops have significant performance degradation. This performance degradation results from the fact that a VC traversing a larger number of hops is more likely to loose a cell due to blocking at some intermediate switch, compared to VCs traversing fewer hops. In particular, the 3-hop source throughput collapses when the loading factor of the link is 0.8. Note that the end-to-end delay for 3-hop VCs at a low link load is close to the propagation delay of 10 msec. At a high load of 1.4 times the link capacity, the end-to-end delay increases to 100 msec (a 10 fold increase) while the useful throughput reduces by 33 % from its peak throughput. The 1-hop VCs incur the least degradation. While the 1-hop throughput monotonically increases, the end-to-end delay even under overload only shows a small increase. On the other hand, with priority for transit traffic, the 1, 2 and 3-hop traffic throughputs are nearly equal at all offered loads. For example, at an offered load of 0.8 times the link capacity, the throughput of all VCs are around 8 Mbits/sec. Note that the end-to-end delays are different since the propagation delays are different. Even with an overload factor of 1.4 , the 2 and 3 hop VCs do not show any throughput collapse, and the end-to-end delay shows only a marginal increase. In the rest of the text, we will only consider the EFCI scheme with priority for transit traffic, and compare it with the PERC scheme.

In order to allow a more aggressive source behavior in networks with ER based switches, we change the two source parameters: $RDF = 512$ and $AIR = 0.70$

cells/msec (0.3 Mbits/sec). Figure 4 shows the plots of 1, 2, and 3-hop source useful throughput (at the packet level) in case of the EFCI scheme with priority for transit traffic and the PERC scheme, as a function of end-to-end delay. Note that in the PERC scheme, transit traffic do not have priority over traffic entering the network.

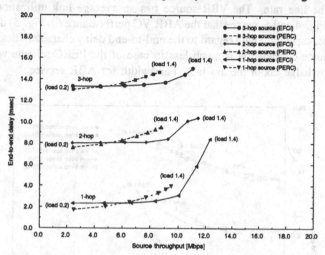

Fig. 4. End-to-end packet delay vs. per hop source throughput under EFCI and PERC

We observe that at higher loads, all VCs under the EFCI scheme have a higher throughput than under the PERC scheme. On the other hand, the end-to-end delay characteristics are worse in case of the EFCI scheme. It is important to emphasize the PERC scheme shows more fairness in bandwidth allocation amongst the competing VCs than the EFCI scheme. In case of PERC, the 1, 2 and 3-hop traffic throughput are nearly identical at all offered loads. The reason why VCs have a lower throughput under the PERC scheme, is due to the fact that the Allowed Source Rate (ACR) is obtained as a minimum function of explicit rates obtained from all switches on the VC route. Since at any given time some switch is likely to be in congestion, the source rate is always constrained by the bottleneck rate on its path. The PERC scheme is closer to a pure rate control and does not fully utilize the store and forward feature of packet switched networks. In attempt to minimize cell loss by keeping the queues small, at the expense of overall utilization. The EFCI scheme on the other hand uses the buffers better and can achieve higher utilization at the expense of cell loss. One must note that we are comparing the EFCI scheme with priority with the PERC scheme without priority.

The examples considered so far assume that all the link capacity was available for ABR service. In reality, there will be CBR and VBR traffic that have higher priority than ABR traffic. To reflect this condition, at each output port of switches 3 and 4 we replace one 1-hop ABR source by a VBR source. We model the VBR traffic as a first order autoregressive process [10]. We assume that the VBR traffic arrival rate changes every 33 msec, and the rate $R_{vbr}(n)$ in the n-th frame is related to the rate $R_{vbr}(n+1)$ in the $(n+1)$-th frame through $R_{vbr}(n+1) = \alpha R_{vbr}(n) + y(n)$, where

$y(n)$ is normally distributed. In our example, we choose the VBR mean arrival rate $\bar{R}_{vbr} = 60$ cells/msec (25 Mbits/sec), the peak rate $R_{vbr}^{peak} = 365$ cells/msec (155 Mbits/sec), the correlation coefficient $\alpha = 0.9$, and the squared coefficient of variation in the arrival rate $C^2 = 1$. With these parameters the VBR source peak rate can be as high as the line rate. The VBR source has an average link utilization of 16 %. Comparing Fig. 4 and 5, we see that the ABR VC performance (in Fig. 5) degrades for both schemes, especially with regard to the end-to-end delay characteristics. However, the performance degradation is much less in case of the PERC scheme which better uses the knowledge about the available bandwidth for ABR service.

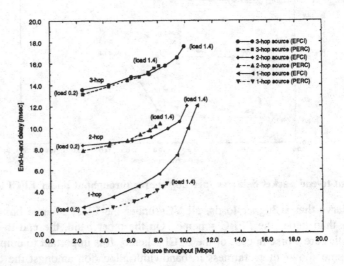

Fig. 5. Steady-state performances in the presence of high priority VBR traffic

The last two examples show the steady-state performances of the two schemes when driven by ON-OFF traffic sources. The transient analysis of network response is analyzed in the next example. We assume that all sources have infinite backlog. Four 3-hop sources start transmission at time $t = 1$ msec. Eight 2-hop sources start transmission at time $t = 300$ msec. Finally, four 1-hop sources start transmission at time $t = 600$ msec. It is important to note that 3-hop and 2-hop sources share the same path through an output port of switch 2, which becomes congested when 2-hop sources become active. Figure 6 shows the transient response of a 3-hop source rate whereas Fig. 7 shows the transient responses of the queue at input port of switch 2 for access traffic (i.e. User Network Interface (UNI)), and the queue at input port of switch 2 for transit traffic (i.e. Network Network Interface (NNI)), respectively. It can be seen that during transient periods, the PERC scheme shows a stable operation with a fast rise time in source rate and very short period of buffer overshoot (small cell loss occurs in the NNI port). The source rate is fairly constant, and after the short transient period the queue fills are close to zero and do not fluctuate anymore. On the contrary, the EFCI scheme has an oscillatory behavior. The source rate fluctuates between PCR and MCR. The queue fill also shows a similar oscillatory behavior, with loss periods lasting for up to 50 msec. A higher cell loss occurs in the UNI ports of switch 2. The EFCI based scheme is a reactive control that allows congestion to set in first and then reacts to it by reducing the load. Each congestion period is then followed by an underload period which causes the control to increase the load again. The large propagation delays directly contribute to the amplitude of oscillations in source rate and queue fill. The PERC scheme on the other hand is conservative and always acts in a preventive mode. While the conservative approach leads to a small reduction in utilization, it is very stable in its operation, with very low loss.

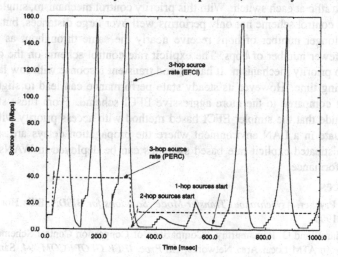

Fig. 6. Source transient response

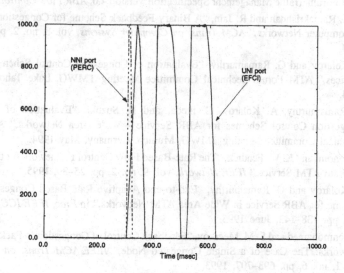

Fig. 7. Buffer transient response

5 Conclusion

We investigate the performance of two end-to-end rate based flow control mechanisms to support ABR service being proposed by the ATM Forum. In the case of the EFCI scheme, we observe that VCs traversing two or more hops can see a throughput collapse at high loads. The unfair advantage that VCs traversing small number of hops have over VCs traversing larger number of hops can be overcome by priority mechanism where transit traffic from upstream switches are given priority of service over local traffic at each switch. With this priority control mechanism, single bit EFCI congestion control scheme not only performs well over large distances, but the traffic travelling longer number of hops receive nearly the same throughput as the traffic travelling fewer number of hops. The explicit rate control scheme on the other hand requires no priority mechanism. It has a good transient response with low latency and quick settling time. However its steady state performance can lead to slightly lower throughput compared to the more aggressive EFCI scheme. From these studies one may conclude that the simpler EFCI based method with access priority will be more than adequate in a LAN environment where the propagation delays are small. The more sophisticated explicit rate based switches can be deployed in WANs to obtain stable performance.

References

[1] M. D. Prycker, *Asynchronous Transfer Mode: Solutions for BISDN*. Ellis Horwood, New York, 1993.

[2] A. Kolarov and G. Ramamurthy, "Comparison of Congestion Control Schemes for ABR Service in ATM Local Area Networks," in *Proc. IEEE GLOBECOM '94*, San Francisco, CA, pp. 913–918, Dec. 1994.

[3] ATM Forum Traffic Management Specification Version 4.0, *ABR Flow Control*, Mar. 1995.

[4] K. K. Ramakrishnan and R. Jain, "A Binary Feedback Scheme for Congestion Avoidance in Computer Networks," *ACM Trans. on Computer Systems*, vol. 8, no. 2, pp. 158–181, 1990.

[5] A. Kolarov and G. Ramamurthy, "Evaluation of Congestion Control Schemes for ABR Services," ATM Forum Technical Committee Meeting TMWG, Lake Tahoe, CA, Jan. 1994.

[6] G. Ramamurthy, A. Kolarov, C. Ikeda, and H. Suzuki, "Evaluation of Rate Based Congestion Control Schemes for ABR Service - Wide Area Networks," ATM Forum Technical Committee Meeting TMWG, Munich, Germany, May 1994.

[7] F. Bonomi and K. W. Fendick, "The Rate-Based Flow Control Framework for the Available Bit Rate ATM Service," *IEEE Network*, vol. 9, no. 2, pp. 25–39, 1995.

[8] A. Kolarov and G. Ramamurthy, "End-to-end Adaptive Rate Based Congestion Control Scheme for ABR Service in Wide Area ATM Networks," in *Proc. IEEE ICC '95*, Seattle, WA, pp. 138–143, June 1995.

[9] L. Benmohamed and S. M. Meerkov, "Feedback Control of Congestion in Packet Switching Networks: The Case of a Single Congested Node," *IEEE/ACM Trans. on Networking*, vol. 1, no. 6, pp. 693–707, 1993.

[10] G. Ramamurthy and B. Sengupta, "A Predictive Congestion Control Policy for High-Speed Wide Area Networks," in *Proc. IEEE INFOCOM '93*, San Francisco, CA, pp. 1033–1041, May 1993.

Author Index

Lecture Notes in Computer Science

For information about Vols. 1–970

please contact your bookseller or Springer-Verlag

Vol. 1007: A. Bosselaers, B. Preneel (Eds.), Integrity Primitives for Secure Information Systems. VII, 239 pages. 1995.

Vol. 1008: B. Preneel (Ed.), Fast Software Encryption. Proceedings, 1994. VIII, 367 pages. 1995.

Vol. 1009: M. Broy, S. Jähnichen (Eds.), KORSO: Methods, Languages, and Tools for the Construction of Correct Software. X, 449 pages. 1995. Vol.

Vol. 1010: M. Veloso, A. Aamodt (Eds.), Case-Based Reasoning Research and Development. Proceedings, 1995. X, 576 pages. 1995. (Subseries LNAI).

Vol. 1011: T. Furuhashi (Ed.), Advances in Fuzzy Logic, Neural Networks and Genetic Algorithms. Proceedings, 1994. (Subseries LNAI).

Vol. 1012: M. Bartošek, J. Staudek, J. Wiedermann (Eds.), SOFSEM '95: Theory and Practice of Informatics. Proceedings, 1995. XI, 499 pages. 1995.

Vol. 1013: T.W. Ling, A.O. Mendelzon, L. Vieille (Eds.), Deductive and Object-Oriented Databases. Proceedings, 1995. XIV, 557 pages. 1995.

Vol. 1014: A.P. del Pobil, M.A. Serna, Spatial Representation and Motion Planning. XII, 242 pages. 1995.

Vol. 1015: B. Blumenthal, J. Gornostaev, C. Unger (Eds.), Human-Computer Interaction. Proceedings, 1995. VIII, 203 pages. 1995.

VOL. 1016: R. Cipolla, Active Visual Inference of Surface Shape. XII, 194 pages. 1995.

Vol. 1017: M. Nagl (Ed.), Graph-Theoretic Concepts in Computer Science. Proceedings, 1995. XI, 406 pages. 1995.

Vol. 1018: T.D.C. Little, R. Gusella (Eds.), Network and Operating Systems Support for Digital Audio and Video. Proceedings, 1995. XI, 357 pages. 1995.

Vol. 1019: E. Brinksma, W.R. Cleaveland, K.G. Larsen, T. Margaria, B. Steffen (Eds.), Tools and Algorithms for the Construction and Analysis of Systems. Selected Papers, 1995. VII, 291 pages. 1995.

Vol. 1020: I.D. Watson (Ed.), Progress in Case-Based Reasoning. Proceedings, 1995. VIII, 209 pages. 1995. (Subseries LNAI).

Vol. 1021: M.P. Papazoglou (Ed.), OOER '95: Object-Oriented and Entity-Relationship Modeling. Proceedings, 1995. XVII, 451 pages. 1995.

Vol. 1022: P.H. Hartel, R. Plasmeijer (Eds.), Functional Programming Languages in Education. Proceedings, 1995. X, 309 pages. 1995.

Vol. 1023: K. Kanchanasut, J.-J. Lévy (Eds.), Algorithms, Concurrency and Knowlwdge. Proceedings, 1995. X, 410 pages. 1995.

Vol. 1024: R.T. Chin, H.H.S. Ip, A.C. Naiman, T.-C. Pong (Eds.), Image Analysis Applications and Computer Graphics. Proceedings, 1995. XVI, 533 pages. 1995.

Vol. 1025: C. Boyd (Ed.), Cryptography and Coding. Proceedings, 1995. IX, 291 pages. 1995.

Vol. 1026: P.S. Thiagarajan (Ed.), Foundations of Software Technology and Theoretical Computer Science. Proceedings, 1995. XII, 515 pages. 1995.

Vol. 1027: F.J. Brandenburg (Ed.), Graph Drawing. Proceedings, 1995. XII, 526 pages. 1996.

Vol. 1028: N.R. Adam, Y. Yesha (Eds.), Electronic Commerce. X, 155 pages. 1996.

Vol. 1029: E. Dawson, J. Golić (Eds.), Cryptography: Policy and Algorithms. Proceedings, 1995. XI, 327 pages. 1996.

Vol. 1030: F. Pichler, R. Moreno-Díaz, R. Albrecht (Eds.), Computer Aided Systems Theory - EUROCAST '95. Proceedings, 1995. XII, 539 pages. 1996.

Vol.1031: M. Toussaint (Ed.), Ada in Europe. Proceedings, 1995. XI, 455 pages. 1996.

Vol. 1032: P. Godefroid, Partial-Order Methods for the Verification of Concurrent Systems. IV, 143 pages. 1996.

Vol. 1033: C.-H. Huang, P. Sadayappan, U. Banerjee, D. Gelernter, A. Nicolau, D. Padua (Eds.), Languages and Compilers for Parallel Computing. Proceedings, 1995. XIII, 597 pages. 1996.

Vol. 1034: G. Kuper, M. Wallace (Eds.), Constraint Databases and Applications. Proceedings, 1995. VII, 185 pages. 1996.

Vol. 1035: S.Z. Li, D.P. Mital, E.K. Teoh, H. Wang (Eds.), Recent Developments in Computer Vision. Proceedings, 1995. XI, 604 pages. 1996.

Vol. 1036: G. Adorni, M. Zock (Eds.), Trends in Natural Language Generation - An Artificial Intelligence Perspective. Proceedings, 1993. IX, 382 pages. 1996. (Subseries LNAI).

Vol. 1037: M. Wooldridge, J.P. Müller, M. Tambe (Eds.), Intelligent Agents II. Proceedings, 1995. XVI, 437 pages. 1996. (Subseries LNAI).

Vol. 1038: W: Van de Velde, J.W. Perram (Eds.), Agents Breaking Away. Proceedings, 1996. XIV, 232 pages. 1996. (Subseries LNAI).

Vol. 1039: D. Gollmann (Ed.), Fast Software Encryption. Proceedings, 1996. X, 219 pages. 1996.

Vol. 1040: S. Wermter, E. Riloff, G. Scheler (Eds.), Connectionist, Statistical, and Symbolic Approaches to Learning for Natural Language Processing. Proceedings, 1995. IX, 468 pages. 1996. (Subseries LNAI).

Vol. 1041: J. Dongarra, K. Madsen, J. Waśniewski (Eds.), Applied Parallel Computing. Proceedings, 1995. XII, 562 pages. 1996.

Vol. 1042: G. Weiß, S. Sen (Eds.), Adaption and Learning in Multi-Agent Systems. Proceedings, 1995. X, 238 pages. 1996. (Subseries LNAI).

Vol. 1043: F. Moller, G. Birtwistle (Eds.), Logics for Concurrency. XI, 266 pages. 1996.

Vol. 1044: B. Plattner (Ed.), Broadband Communications. Proceedings, 1996. XIV, 359 pages. 1996.

Vol. 1045: B. Butscher, E. Moeller, H. Pusch (Eds.), Interactive Distributed Multimedia Systems and Services. Proceedings, 1996. XI, 333 pages. 1996.

Vol. 1046: C. Puech, R. Reischuk (Eds.), STACS 96. Proceedings, 1996. XI, 690 pages. 1996.

Vol. 1047: E. Hajnicz, Time Structures. IX, 244 pages. 1996. (Subseries LNAI).

Vol. 1048: M. Proietti (Ed.), Logic Program Syynthesis and Transformation. Proceedings, 1995. X, 267 pages. 1996.